WORKSHOPS IN COMPUTING
Series edited by C. J. van Rijsbergen

Also in this series

**Building Interactive Systems:
Architectures and Tools**
Philip Gray and Roger Took (Eds.)

Functional Programming, Glasgow 1991
Proceedings of the 1991 Glasgow Workshop on
Functional Programming, Portree, Isle of Skye,
12–14 August 1991
Rogardt Heldal, Carsten Kehler Holst and
Philip Wadler (Eds.)

Object Orientation in Z
Susan Stepney, Rosalind Barden and
David Cooper (Eds.)

Code Generation – Concepts, Tools, Techniques
Proceedings of the International Workshop on Code
Generation, Dagstuhl, Germany, 20–24 May 1991
Robert Giegerich and Susan L. Graham (Eds.)

Z User Workshop, York 1991, Proceedings of the
Sixth Annual Z User Meeting, York,
16–17 December 1991
J.E. Nicholls (Ed.)

Formal Aspects of Measurement
Proceedings of the BCS-FACS Workshop on
Formal Aspects of Measurement, South Bank
University, London, 5 May 1991
Tim Denvir, Ros Herman and R.W. Whitty (Eds.)

AI and Cognitive Science '91
University College, Cork, 19–20 September 1991
Humphrey Sorensen (Ed.)

5th Refinement Workshop, Proceedings of the 5th
Refinement Workshop, organised by BCS-FACS,
London, 8–10 January 1992
Cliff B. Jones, Roger C. Shaw and
Tim Denvir (Eds.)

**Algebraic Methodology and Software
Technology (AMAST'91)**
Proceedings of the Second International Conference
on Algebraic Methodology and Software
Technology, Iowa City, USA, 22–25 May 1991
M. Nivat, C. Rattray, T. Rus and G. Scollo (Eds.)

ALPUK92, Proceedings of the 4th UK
Conference on Logic Programming,
London, 30 March – 1 April 1992
Krysia Broda (Ed.)

Logic Program Synthesis and Transformation
Proceedings of LOPSTR 92, International
Workshop on Logic Program Synthesis and
Transformation, University of Manchester,
2–3 July 1992
Kung-Kiu Lau and Tim Clement (Eds.)

NAPAW 92, Proceedings of the First North
American Process Algebra Workshop, Stony Brook,
New York, USA, 28 August 1992
S. Purushothaman and Amy Zwarico (Eds.)

First International Workshop on Larch
Proceedings of the First International Workshop on
Larch, Dedham, Massachusetts, USA,
13–15 July 1992
Ursula Martin and Jeannette M. Wing (Eds.)

Persistent Object Systems
Proceedings of the Fifth International Workshop on
Persistent Object Systems, San Miniato (Pisa),
Italy, 1–4 September 1992
Antonio Albano and Ron Morrison (Eds.)

**Formal Methods in Databases and Software
Engineering,** Proceedings of the Workshop on
Formal Methods in Databases and Software
Engineering, Montreal, Canada, 15–16 May 1992
V.S. Alagar, Laks V.S. Lakshmanan and
F. Sadri (Eds.)

Modelling Database Dynamics
Selected Papers from the Fourth International
Workshop on Foundations of Models and
Languages for Data and Objects, Volkse, Germany,
19–22 October 1992
Udo W. Lipeck and Bernhard Thalheim (Eds.)

14th Information Retrieval Colloquium
Proceedings of the BCS 14th Information
Retrieval Colloquium, University of Lancaster,
13–14 April 1992
Tony McEnery and Chris Paice (Eds.)

Functional Programming, Glasgow 1992
Proceedings of the 1992 Glasgow Workshop on
Functional Programming, Ayr, Scotland,
6–8 July 1992
John Launchbury and Patrick Sansom (Eds.)

Z User Workshop, London 1992
Proceedings of the Seventh Annual Z User
Meeting, London, 14–15 December 1992
J.P. Bowen and J.E. Nicholls (Eds.)

Interfaces to Database Systems (IDS92)
Proceedings of the First International Workshop
on Interfaces to Database Systems,
Glasgow, 1–3 July 1992
Richard Cooper (Ed.)

continued on back page...

Kevin Ryan and Richard F.E. Sutcliffe (Eds.)

AI and Cognitive Science '92

University of Limerick,
10–11 September 1992

Published in collaboration with the
British Computer Society

Springer-Verlag
London Berlin Heidelberg New York
Paris Tokyo Hong Kong
Barcelona Budapest

Kevin Ryan, BA, BAI, PhD, MBCS, C.Eng

Richard F.E. Sutcliffe, BSc, PhD

Department of Computer Science and Information Systems
University of Limerick
National Technological Park
Plassey, Limerick, Ireland

ISBN 3-540-19799-0 Springer-Verlag Berlin Heidelberg New York
ISBN 0-387-19799-0 Springer-Verlag New York Berlin Heidelberg

Library of Congress Cataloging-in-Publication Data
AI and cognitive science '92 : University of Limerick, 10-11 September 1992 /
Kevin Ryan and Richard F.E. Sutcliffe, eds.
    p. cm. – (Workshops in computing)
"Published in collaboration with the British Computer Society."
Includes bibliographical references and index.
ISBN 3-540-19799-0 : $79.00   – ISBN 0-387-19799-0 : $79.00
    1. Artificial intelligence–Congresses. 2. Cognitive science–
Congresses. I. Ryan , Kevin, 1949– . II. Sutcliffe, Richard F.E., 1960–
III. British Computer Society. IV. Series.
Q334.A444 1993   93-26018
006.3–dc20   CIP

Apart from any fair dealing for the purposes of research or private study, or
criticism or review, as permitted under the Copyright, Designs and Patents Act
1988, this publication may only be reproduced, stored or transmitted, in any form,
or by any means, with the prior permission in writing of the publishers, or in the
case of reprographic reproduction in accordance with the terms of licences issued
by the Copyright Licensing Agency. Enquiries concerning reproduction outside
those terms should be sent to the publishers.

©British Computer Society 1993
Printed in Great Britain

The use of registered names, trademarks etc. in this publication does not imply,
even in the absence of a specific statement, that such names are exempt from the
relevant laws and regulations and therefore free for general use.

The publisher makes no representation, express or implied, with regard to the
accuracy of the information contained in this book and cannot accept any legal
responsibility or liability for any errors or omissions that may be made.

Typesetting: Camera ready by contributors
Printed by Antony Rowe Ltd., Chippenham, Wiltshire
34/3830-543210 Printed on acid-free paper

Preface

This book contains the edited versions of papers presented at the Fifth Irish Conference on Artificial Intelligence and Cognitive Science, held at the University of Limerick on 10–11 September 1992.

These conferences bring together Irish and overseas researchers and practitioners from many areas of artificial intelligence and cognitive science. The aim is to provide a forum where researchers can present their current work, and where industrial and commercial users can relate this research to their own practical experience and needs.

This year the conference was divided into six sessions: Knowledge Representation, Cognitive Foundations, Natural Language 1 and 2, Learning and Expert Systems, and Novel Aspects of AI and CS. Because of the high number of paper submissions for the conference the organisers decided to run a poster session in addition to the plenary sessions. Posters are represented in this volume as four page abstracts, while each plenary session paper appears in full.

The keynote speaker for AICS'92 was Professor Erik Sandewall from Linköping University, Sweden. Professor Sandewall, a world authority on artificial intelligence, chose the frame problem as his topic.

While many of the participants at the conference were from companies and universities in Ireland, there were also delegates from Australia, Canada, France, Germany, Italy, Sweden, UK and USA.

We are especially grateful to our sponsors: Digital Equipment Ireland, Hitachi, IBM Ireland, K&M Technologies and the Shannon Free Airport Development Company.

January 1993   Kevin Ryan
                                                                                              Richard F.E. Sutcliffe

Contents

Keynote Address

The Frame Problem – Past, Present, and Future
E. Sandewall 3

Session 1: Knowledge Representation

A Model-Based Theory of Conceptual Combination
F. Costello and M.T. Keane 7

An Hierarchical Structure for Modelling Real-Time Domains Using Object-Oriented Techniques
H. Dai, J.G. Hughes and D.A. Bell 16

Argumentation in Weak Theory Domains
K. Freeman and A.M. Farley 31

Coordination and Dynamic Organization in the ALM System
F. Abbruzzese, E. Minicozzi, G. Nardiello and G. Piazzullo 44

Session 2: Cognitive Foundations

The Nature and Development of Reasoning Strategies
R.M.J. Byrne and S.J. Handley 59

Semantic Transparency and Cognitive Modelling of Complex Performance
J.G. Wallace and K. Bluff 71

Others as Resources: Cognitive Ingredients for Agent Architecture
M. Miceli, A. Cesta and R. Conte 84

Session 3: Natural Language 1

Word Recognition as a Parsing Problem
P. McFetridge 101

Using Actions to Generate Sentences
P. Matthews 111

Terminology and Extra-Linguistic Knowledge
J. Pearson 121

Session 4: Learning and Expert Systems

A Knowledge-Based Autonomous Vehicle System for
Emergency Management Support
J.F. Gilmore 141

Learning to Play Connect 4: A Study in Attribute Definition
for ID3
B.W. Baird and R.J. Hickey 157

Machine Learning Under Felicity Conditions: Exploiting
Pedagogical Behavior
J.D. Lewis and B.A. MacDonald 166

A Blackboard Based, Recursive, Case-Based Reasoning
System for Software Development
B. Smyth and P. Cunningham 179

Session 5: Natural Language 2

The Construction and Use of Scope Neutral Discourse Entities
N. Creaney 197

Constructing Distributed Semantic Lexical Representations
Using a Machine Readable Dictionary
R.F.E. Sutcliffe 210

Some Transfer Problems in Translating Between English and Irish
A. Way 224

Session 6: Novel Aspects of AI/CS

Why Go for Virtual Reality if You Can Have Virtual Magic?
A Study of Different Approaches to Manoeuvring an Object
on Screen
R. Cowie and F. Walsh 237

A Context-Free Approach to the Semantic Discrimination
of Image Edges
E. Catanzariti 251

Computer-Based Iconic Communication
C. Beardon 263

Toward Non-Algorithmic AI
S. Bringsjord 277

Poster Session

A Unified Approach to Inheritance with Conflicts
R. Al-Asady 291

From Propositional Deciders to First Order Deciders:
A Structured Approach to the Decision Problem
A. Armando, E. Giunchiglia and P. Traverso 295

Modelling Auditory Streaming: Pitch Proximity
K.L. Baker, S.M. Williams, P. Green and R.I. Nicolson 299

A Simulated Neural Network for Object Recognition in
Real-Time
B. Blöchl and L. Tsinas 303

Constraints and First-Order Logic
J. Bowen and D. Bahler 307

Programming Planners with Flexible Architectures
A. Cimatti, P. Traverso and L. Spalazzi 311

On Seeing that Particular Lines Could Depict a Vertex
Containing Particular Angles
G. Probert and R. Cowie 315

Terminological Representation and Reasoning in TERM-KRAM
D. D'Aloisi and G. Maga 319

Integrating Planning and Knowledge Revision:
Preliminary Report
A.F. Dragoni 323

Learning Control of an Inverted Pendulum Using a Genetic
Algorithm
M. Eaton 327

Aspectual Prototypes
A. Galton 330

Formal Logic for Expert Systems
H. McCabe 334

Exceptions in Multiple Inheritance Systems
C. Oussalah, M. Magnan and L. Torrès 338

Sublanguage: Characteristics and Selection Guidelines for M.T.
E. Quinlan and S. O'Brien 342

On the Interface Between Depth-From-Motion and
Block-Worlds Systems: Some Psychophysical Data
A.H. Reinhardt-Rutland 346

A Connectionist Approach to the Inference of Regular
Grammars from Positive Presentations Only
D. O Sullivan and K. Ryan 350

Author Index 355

Keynote Address

THE FRAME PROBLEM – PAST, PRESENT, AND FUTURE

Erik Sandewall

Department of Computer and Information Science
Linköping University
58183 Linköping, Sweden
E-mail: `ejs@ida.liu.se`

Common-sense reasoning about time and action is a significant and enigmatic research topic in artificial intelligence. A number of approaches have been proposed in recent years, many of them using preference-based non-monotonic logics. Like for other aspects of common-sense reasoning, the strength and the weakness of each approach has been analyzed by a combination of its philosophical plausibility, and its success or failure for a number of representative examples of common-sense reasoning where the intended conclusions are considered to be obvious.

In a forthcoming book[San93a] I argue that this methodology is not sufficient, and that for each of the existing logical approaches to the frame problem there are a number of fairly simple examples where it fails to give the intended set of conclusions. Research in this area will only be useful if we have results regarding the *range of correct applicability* for each logic that is being considered.

I also propose a systematic methodology based on the notion of *underlying semantics* which is used to specify in a formal way what is the intended set of models for a given scenario description. In this way it is possible to formally characterize classes of discrete dynamical systems, and to establish precisely what logic is able to correctly characterize what class of dynamical systems.

Presentations of this approach can be found in references[San93a, San93b, San93c].

References

[San93a] Erik Sandewall. *Features and Fluents*. To appear, 1993.

[San93b] Erik Sandewall. The range of applicability of nonmonotonic logics for the inertia problem. In *International Joint Conference on Artificial Intelligence*, 1993.

[San93c] Erik Sandewall. Systematic assessment of temporal reasoning methods for use in autonomous agents. In Z. Ras, editor, *Proceedings of ISMIS 1993*. Springer Verlag, 1993.

Session 1: Knowledge Representation

A Model-Based Theory of Conceptual Combination

Fintan Costello and Mark T. Keane
University of Dublin,Trinity College
Dublin, Ireland[1]

Abstract

Traditionally, the meaning of a combined concept like "wooden spoon" has been treated *compositionally*; that is, the meaning of the combined concept is seen as being solely determined by the meaning of its constituents. However, in the last decade, research has shown that other knowledge -- background, world knowledge -- also plays a role in comprehending combined concepts. However, few models specify how this knowledge comes into play. The present paper proposes a theory of conceptual combination which demonstrates how world knowledge is accessed.

1. Introduction

Conceptual combination can be broadly defined as "the process occuring when a compound phrase is understood". Most often, the compound phrases in question are of the form *adjective-noun* (e.g., red apple, fake gun), *noun-noun* (e.g., "pet fish", "horse race") and *adverb-adjective-noun* (e.g., "very red fruit", "slightly large bird"). More rarely *noun-verb* phrases (e.g., "birds eating insects"), or even sentences, such as "The old man's glasses were filled with sherry" are used (Conrad & Rips, 1986). The central problem in concept combination is to account for how the distinct "simple" concepts in a compound phrase are combined to comprehend the phrase. The traditional view is that such an account will arise from considering the contribution made by each of the simple concepts to the phrase. However, research has shown there is also a contribution from another source, from background world knowledge (see Medin & Shoben, 1988). This paper presents a theory of conceptual combination which demonstrates how such world knowledge can be brought into play.

1.1 Representing Concepts for Combination

The way in which concepts are represented is clearly a central part of any theory of conceptual combination. Traditionally, concepts have been represented as sets of singly necessary and jointly sufficient attributes (Smith & Medin, 1981; Medin, 1989). This viewpoint predicts that categorisation simple and clearcut : that an instance is a member of a category if and only if it possesses the defining features of that category. Theories of concept combination based on these sorts of representations propose that a combined

---

[1]E-mail: fcostllo@cs.tcd.ie, mkeane@cs.tcd.ie Phone:+353-1-772941 Fax:+353-1-772204

concept is the union of the defining attributes of each of the constituent concepts (see Costello & Keane, 1992).

This classical theory of concepts is appealing but wrong. Research has shown that membership of a category is not clear-cut but is fuzzy (see Rosch & Mervis, 1975; Eysenck & Keane, 1990, for a review). Category membership is also graded, with some instances having higher degrees of membership than others. For example, the instance "robin" is a more typical member of the category *bird* than the instance "ostrich". Prototype theories have been proposed to account for these observations. One version of prototype theory argues that concepts are not represented by defining attributes but rather by characteristic attributes, which vary in the degree to which they define a category. In this model, an instance's degree of membership of a category is given by its *membership score*, which is usually the sum of the definingness quotients of the attributes that instance possesses.

1.2 The Concept-Specialisation Model of Combination

Several models of concept combination make use of the proposals of prototype theory. The most broadly applicable and explanatory of these is the Concept Specialization model (Cohen and Murphy, 1984). Rather than viewing concept combination as the union of the attribute sets of two simple concepts, this model proposes that combination involves a process of specialization of the attributes of one concept (the last or *head* word of the phrase) by another (the first word or *modifier*). The essential insight in this model is that concept combinations are formed by finding a *mediating relation* between two concepts. For example, in the combination "morning flight" the mediating relation is that the flight *occurs* in the morning, in "morning coat" the relation is that the coat is typically *worn* in the morning and in "red apple" it is a *colour* relation that mediates between the two concepts.

Cohen and Murphy use frame-based knowledge representation systems from artificial intelligence (see Finnin, 1980). Concepts are implemented as structured lists of *slot* and *slot-filler* pairs. The slots represent the possible mediating relations the concept can have, and the slot-fillers indicate the concept's typical values for those mediating relations. For example, the concept *apple* could have the slot COLOUR. This slot could be filled by the value GREEN, indicating that the prototypical colour of an apple is green. In the Concept-Specialisation model, each slot of a prototypical concept is associated with a list of default-value concepts, each of which has a certain degree of typicality. Thus for the concept *apple*, the most typical filler of the COLOUR slot might be GREEN, the next most typical RED and so on. These degrees of typicality correspond to the degrees of definingness of attributes in prototype theory.

1.3 World knowledge is involved in concept specialisation

Concept combination is achieved in this model when a dominant concept is changed by placing a modifying concept into the mediating relation of that concept (Murphy, 1988). For example, in the combined concept "house boat" the mediating role of the concept "boat" might be FUNCTION, with the value "house" inserted in it. The process whereby the correct mediating relation between the modifier and the head concept is found is called "role fitting" (Finnin, 1980). Role fitting can be carried out by a simple exhaustive matching process of values to possible slots, although Cohen and Murphy suggest that it may also be guided by background world knowledge. This world knowledge would be consulted to decide which slot is the most appropriate one to modify or specialise. The choice of slot depends on at least two factors; (a) which slot is the most relevant in the context and (b) of which slot would the modifier be a typical filler. For example, in forming the combined concept "apartment dog", world knowledge might be used to determine that it is more likely to refer to dogs that *inhabit* apartments rather than dogs that *look like* apartments or that *bite* apartments. One reason Cohen and Murphy reject the idea of an exhaustive search through slots because the number of possible mediating relations in a concept may be unlimited. Kay & Zimmer (1976) discuss the ambiguity of phrases such as "finger cup", which have a large number of possible interpretations. For example, a *finger cup* could be "a cup shaped like a finger", "a cup holding a finger of whisky", "a cup used for washing the fingers", "a cup held between the fingers" and so on. The possible mediating relations are not limited to readily recognisable properties of concepts. They may be extended to any possible scenario. We would like to argue, along with others, that world knowledge is needed to generate these possible mediating relations.

World knowledge may also be used in another way in concept specialisation. It could be necessary to elaborate and "clean up" a combined concept's representation to make it more coherent and complete (Murphy, 1988). Consider the *apartment dog* concept; world knowledge may also be used to infer that an apartment dog is likely to be smaller, quieter and cleaner than say a hunting dog or a farm dog (see Murphy, 1988). Medin and Shoben (1988) give examples where this cleaning up process involves changing values in slots other that the one specialised. In the concept "wooden spoon", the length of the spoon changes from that of a typical spoon as well as the material it is made of. Murphy (1988) describes instances of totally new properties being added to concepts by elaboration (e.g., "overturned chairs" are typically on tables).

The Concept-Specialisation model acknowledges that world knowledge plays an important role in both the construction and elaboration of combined concepts, yet the structure and representation of this knowledge is not specified and methods whereby it is accessed and used are not explained. Indeed Fodor (1983) argues that no such explanations are impossible because general access to world knowledge is a "central process", which can never be explained scientifically. Murphy (1988) rejects this stance by proposing that while conceptual combination requires broad reference to world knowledge, constraints and generalisations can be made about the way it is used. This is evident because people can construct novel compound phrases (e.g. "black apple", "seven-legged chair"), which listeners can interpret correctly. He suggests that models of concept combination could be developed which could exploit these constraints and generalisations.

1.4 World knowledge is involved in categorisation

Conceptual combination is not the only area in which the use of and access to general knowledge has become an issue. A number of researchers have come to the conclusion that an appeal to world knowledge is necessary in theories of simple, non-combined concepts also. There are a number of reasons for this (see Medin 1989). First, prototype theories treat concepts as context independent. But Roth and Shoben (1983) have shown typicality judgements for categories can vary as a function of context. Second, prototype theories assume that classification is driven by a summing of (weighted) features. This implies that learning a category involves getting the attribute weights right so that the correct classifications of members and nonmembers occur. For some types of categories (called "linearably seperable categories"), this learning process should be easier and quicker than for others. However, these is no evidence that linear seperability has any impact on the speed at which categories are learned (Medin 1989).

Another problem for prototype theories is the growing realisation that the formation of categories does not depend on general perceptual similarity. That is, things are not classified together because they look the same (see Lakoff, 1986, 1987; McCauley, 1987; Neisser, 1987). Rather our construction of categories is shaped by our expectations and by what we see as relevant. This realisation has led to an approach to concepts and categorisation based on the use of world knowledge in the form of *models* and *theories*. By this account, concepts are seen not as independent category descriptions, but as entities filling roles in models of the world. Rather than being static abstractions of prototype theory, concepts are shaped by the roles that they play and as those roles change, the concepts change with them.

This move away from classification by general perceptual similarity towards classification by explanatory theories solves many of the problems discussed above. The models which define a concept can change according to context; linear seperability is unimportant in learning models as opposed to forming characteristic prototype descriptions and categorisations which are not based on perceptual similarity are easily explained. An important advantage of model-based theories of concepts from the point of view of conceptual combination is that they allow constrained access to and use of world knowledge in classification and concept construction; contrary to Fodor's pessimistic view of central processes, an attempt can be made to explain and model such access.

2. A Model-Based Theory of Combination

Our aim is to integrate the best proposals from the concept-specialisation model with a model-based view of concepts. If we get the synthesis right then the product should be a theory of concept combination that is fundamentally based on world knowledge, rather then one which has world knowledge as a ill-specified adjunct. We will present this theory in four parts. First, we discuss critically a model-based theory's tenets on the internal structure of concepts, using Lackoff's (1986,1987) Idealised Cognitive Model (ICM) theory. Second, we talk about inter-concept structure and the organisation of models. For theories of conceptual combination, as opposed to theories of simple concepts, the relationships and connections between concepts are all important. As a starting point, we use Kedar-Cabelli's (1987) description of how concepts can be shaped by the roles they play in interactions with other concepts. Third, we present our proposals on a model-based theory of combination. Finally, we outline the merits and limitations of this approach to conceptual combination.

2.1 A Model-Based Theory of Concepts

One model-based theory of concepts is Lakoff's ICM theory (Lakoff 1986,1987). This theory is intended to show how prototype effects (such as graded category membership) could arise from the use of world knowledge organised and structured in the form of cognitive models. Each element in such a mental model can correspond to a concept. There are a number of different types of models to which Lakoff refers (propositional, image-schematic, metaphoric, metonymic and symbolic) but for simplicity we will just deal with propositional models.

According to Lakoff, concepts do not consist of sets of characteristic attributes each with a different degree of definingness. Rather a concept is given strict, classical definition with respect to an Idealised Cognitive Model. A concept can play a *role* in an ICM if and only if it possesses the defining attributes of that role. Each role has just a few defining attributes, it is a limited definition of only one aspect of a concept. Lakoff describes a number of ways in which concepts with a prototype-like internal structure could emerge from such classically-defined roles. For instance, a given concept can be defined in a number of different ways by a number of different ICMs. This cluster of ICMs gives the concept the appearance of having a prototype structure. For example, the concept "mother" could be defined in one ICM as "the woman who gave birth to a child", while another ICM could define it as "the woman who cares for the child". Yet another model could define the concept as "the wife of the child's father" or "the woman who donated half of the child's genes". The overlapping between these ICMs gives rise to graded concept membership for the concept "mother". A woman who has conceived a child, given birth to it and cared for it is a more typical "mother" than a woman who has given birth to a child, but not brought it up (an adoptive mother), or who has not donated genes, but has given birth (a surrogate mother). This is because the latter two cases only satisfy the definition of one of the "mother" ICMs while the first satisfies three.

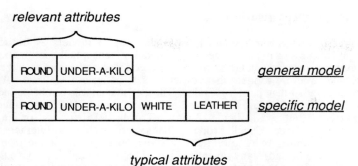

Figure 1. Two overlapped models of the concept *football*

Several problems arise when we attempt to extend Lakoff's ICM theory of concepts with regard to conceptual combination. Concepts in ICM theory do not have typical but rather has only attributes which are *relevant*. In his example of the genetic model for the concept 'mother' for example, the only attribute used is "gives genes to child", because that is the only thing relevant to the model. However, concepts may have attributes which are not relevant to any model, and yet are typical of the concept (see Smith and Medin 1981). Consider the concept *football*. The attribute SPHERICAL is defining for the concept *football*' because it is relevant to the concept's role - a football must roll, therefore it should be round. Another typical attribute of the concept *football* is the attribute COLOURED-WHITE. However, that attribute is totally irrelevant to the concept's role. The attribute COLOURED-WHITE is part of the concept *football* because it is an attribute that instances of the concept are *likely* to have. Since Lakoff's ICMs do not represent such typical yet irrelevant attributes, it cannot account for their definingness.

One of the main problems for theories of conceptual combination is determining what attributes are typical of a combination. Thus, a model-based account of concept combination must allow typical yet irrelevant attributes to be a part of its models. The ICM theory cannot accommodate attributes of this type, because it assumes that all attributes of each model in an ICM cluster are equally relevant. I therefore propose an alternative way in which classically defined models can be "overlapped" to produce prototype effects, which will allow models with typical yet irrelevant attributes.

This proposal is based on the idea of models with varying degrees of specificity. A highly-general model defines concepts using only a few attributes - those which are relevant. Such a model of the concept *football* could define it as ROUND (so it will roll) and WEIGHING-UNDER-A-KILO (in order to be kickable). A more specific model could define *football* as ROUND, WEIGHING-UNDER-A-KILO, COLORED-WHITE, and MADE-OF-LEATHER. These models could be overlapped to create a concept with graded membership. This type of overlap happens in a way different from Lakoff's cluster overlap; it is overlapping by *inclusion* rather than *juxtaposition*. The general 'football' definition can *include* the more specific definition, since the more specific one possesses all the defining attributes of the general concept. The most relevant attributes of the concept *football* are those used in the general definition, the typical attributes are those in the more specific model's definition only (see Figure 1).

In fact, given a number of specific concepts included in one general concept, attributes can be given degrees of typicality, the attribute which occurs in most of the specific models is more typical than that which occurs in few of them. If thereare three different specific models of the concept *football* included in a more general model, if two give the colour as BLACK-AND-WHITE and the other as WHITE, then BLACK-AND-WHITE is a *more* typical colour that WHITE, although WHITE is typical too.

The importance of this type of model structure for conceptual combination is that it can allow a process of concept combination based on these models access to information about the typical attributes of concepts rather than just their relevant attributes. By indicating how such models can be used to derive prototype effects for concepts, I am justified in basing a theory of conceptual combination on them.

2.2 Inter-concept Organisation

In my initial discussion cognitive models I was interested in models as limited definitions of single concepts or roles. To develop a model of conceptual combination we need to refer to the connections and relations between concepts rather than just their internal structure. We require models that describe sets of strictly defined roles involved in interactions rather than just single roles. The idea that an role can be defined by the interaction in which it takes part is presented in Kedar-Cabelli's (1987) account of purpose-based concepts (but see Miller & Johnson-Laird, 1976, for the original development of this idea). Kedar-Cabelli proposes that to participate in an interaction, an entity must fulfil a number of requirements. For example, to participate in the interaction "contains (Cup,Liquid)", any potential filler of the role "cup" *must* have the attributes IS-SOLID and IS-CONCAVE. The interaction "contains(X,liquid)" can be said to give a limited definition of the role "Cup" in the same way that Lakoffs ICMs give limited definitions of roles.

Any interaction contains a number of roles. Consider the interaction "drinks(Agent,Liquid)". The limited definition of "Liquid" in this model could be FLOWS and NON-POISIONOUS. These are the attributes an instance *must* possess if it is to fill that role. However, an interaction may involve other interactions. "Drinks(Agent,Liquid)" could contain the sequence of interactions

> contains(Cup, Liquid),
> picks-up(Agent,Cup),
> has(Agent,Mouth),
> pours(Agent,Cup,Liquid,Mouth),
> swallows(Agent,Liquid).

At this more detailed level, other roles involved in the interaction become apparent. These roles are defined by the *set* of interactions in which they take part. Take the role "cup" for example. To fill this role an instance must be able to contain liquid (possess the attributes IS-SOLID and IS-CONCAVE), be picked up (HAS-HANDLE), and have liquid poured out of it (HAS-OPEN-TOP). Of course, *these* interactions may themselves contain others, introducing new concepts; "contains(Cup,Liquid)" could expand to

> Has-part(Cup,Bowl),
> Is-concave(Bowl),
> Made-of(Cup,Material),
> Solid(Material).

adding still more roles ("Material", "Bowl"), and more detail to the model. Models of this type could have the structure shown in fig. 2.

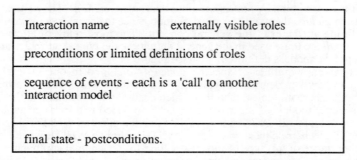

Figure 2. Structure of an interaction model.

As we saw above, concepts can be constructed and shaped by the different definitions they have in different models. The roles that concepts play in these models are strictly defined with respect to the attributes and properties relevant to the model, but concepts can play roles in many models at once. The overlap of definitions gives a concept's attributes varying degrees of relevance and typicality.Concepts can be given graded membership in exactly the same way by models of interactions. The roles a concept plays in different interactions can be 'overlapped' in the same way ICMs are overlapped, to derive the degrees of definingness of the concepts various attributes.

The problem we are faced with is how to find the set of models from which a combined concept can be derived. We require a set of roles associated in some way with the compound phrase in question, and which is identified by some mechanism using the words that make up the compound phrase. The overlap of this set of models will define the combined concept in the same way a simple concept is defined.

2.3 Model-Based Concept Specialization

The theory of concept combination we want to put forward is based on the concept specialization model. The problem is how to adapt this type of theory to a model-based view of concepts. We saw that the core process in concept-specialization models is the selection of a slot in the head concept to be filled by the modifier. These slots represent possible mediating relations of the head concept; a filling of one slot by the modifier indicates that the modifier stands in a certain relation to the head concept. In a model-based theory, the slots of a concept correspond to roles which are related to the concept in some way. The possible mediating relations of the concept are represented by the models in which the concept takes part. Suppose we are trying to interpret a combined concept such as *coffee cup*. If we find a model in which the dominant concept plays a role (the "drinks(Agent,Liquid)" model above), and fill a role in that model with the modifier concept (i.e. the "Liquid" role is filled by the concept *coffee*), the modifier concept stands in a certain mediating relation to the dominant concept. This seems almost identical to the slot-fillng process of the concept specialization theory. However, it does have an advantage. Take the combination "wooden spoon" for example. In the specialization model this would be interpreted as the specialisation of the MADE-OF slot on the concept *spoon*. However, suppose we have a number of interaction models containing the concept *spoon*. One describes the use of a spoon to stir soup, or stew or something hot. The requirements for the concept "spoon" in this model are that it be long (so it won't fall into the pot) and wooden (so that it won't conduct heat). One of the roles in this model is the material of which the spoon is made - by fitting the modifer "wood" to that role, we create a plausible interpretation of the concept *wooden spoon*. Because this interpretation is made within a model, we get extra information about the combined concept *wooden spoon;* that it has a long handle, that it is used to stir things, that is is used in the kitchen and so on.

2.4 Objections to a Model-Based Theory of Combination

There are two objections to this idea of a combination referring to an already existing concept. The first proposes that a combined phrase refers directly to an existing concept, so the argument goes, then all possible combined phrases must also refer directly to existing concepts. Understanding compound phrases must be no different from understanding simple words, the meaning is just directly associated. But since the number of compound phrases is unlimited, we do not have space in our minds to store the concept associated with each phrase. Therefore, compounds cannot refer to existing concepts, they must be constructed. This argument does not apply here. The phrase "wooden spoon" does not refer directly to its already existing concept. Rather, it is used to find that concept. Other phrases could have been uses to find the same concept: "stirring spoon" or "kitchen spoon". Of course the degree of fit between the phrase and its referent would be different.

The other objection arises from the fact that people can understand combined phrases for which they have no existing referent concept. The phrase "blue dog", for example, cannot refer to anything existing in a model, and so cannot be understood in an approach of this type. However, this phrase is easily interpreted using models. Assume a simple model of the colour blue as a property of objects in general. The model contains two roles; the object and the colour, connected by the mediating relation VISIBLE-ON-SURFACE. The concept *dog* can be easily fitted into the role of *object*, thus providing a model which interprets the phrase as a dog with an exterior-coloured blue.

What is required then to interpret a combined concept is to find a model containing one role which can be referred to by the head of the concept and another which can be referred to by the modifier. This is a problem of role-fitting; given a model containing a role which has a good fit with the head category, find a role in that model in which the modifier category can fit (or vice versa). Therefore, the critical part of the process is that of judging the fit or typicality of a given concept for a given role. There are two ways in which a concept and a role can fit together. The concept can fit into the role (i.e. above "dog" fitting into the role *object*), or the role can fit into the concept (i.e. non heat-conducting material being classified as wood). The concept-into-role type of fit occurs in the construction of new concept interpretations, while the role-into-concept type occurs when finding existing concepts. For a concept to fit into a role, it must possess *all* of the defining attributes of that role to some degree. If the concept does not possess any one of the role's necessary and sufficient attributes, the concept cannot fit the role. If it possesses the defining attributes of the role with a high degree of characteristicness, it has a good fit with the role. In fitting a role into a concept on the other hand, the process is different. The role need not possess all the characteristic features of the concept, only some. The degree of goodness of the fit is the sum of characteristicness of the attributes that the role possesses. The sequence of actions in interpreting a compound phrase in a model-based theory of combination would thus be as follows.

1. Get all models containing roles which define the head concept (the *head-concept* models).

2. Find another role in one of those models which can fit into the modifier concept. If such a role is found, the combined concept can be taken as referring to an *existing* concept in that model (as in the case of *wooden spoon* above).

3. If such a role related to the dominant concept cannot be found, retrieve all the models containing roles which define the modifier (the *modifier-concept* models).

4. Find a role in one of these models into which the head concept can fit. If such a role can be found, the combined concept can be taken as referring to a newly constructed concept (as in *blue dog*).

3. Conclusions

This paper sketches a developing theory of conceptual combination. It shows how this theory is built on a model-based approach to concepts, and how it developed as an extension of the concept specialization model of conceptual combination. There are a number of advantages to this model. It describes a constrained way in which the process of conceptual combination can refer to and use world knowledge. This world knowledge is used to determine the typical attributes of a combined concept. The paper also describes models of the world that contain these typical attributes. Finally, it shows how novel combinations can be interpreted in a model-based theory.

The theory presented here is limited in a number of ways however. Its problems lie in the area of *role-fitting*. The theory does not give a detailed account of how the models which define the modifier and head concepts are found (the *modifier-* and *head-concept* models referred to above). This is because such an account is really a question of how concepts are *constructed* from such models. No detailed account of such concept construction, which describes how both the relevant and the typical attributes of concepts emerge, has been developed. The development of a detailed theory of concepts would provide a sound basis on which future theories of concept combination could rest.

4. References

Cohen, B., & Murphy, G. L. (1984). Models of concepts. *Cognitive Science*, **8**, 27-58.
Conrad, F. G., & Rips, L. J. (1986). Conceptual combination and the given/new distinction. *Journal of Memory and Language*, **25**, 255-278.
Costello, F. & Keane, M.T. (1992). Conceptual combination: A theoretical review. *Irish Journal of Psychology,* **13**(2), 125-140.
Eysenck, M. W. & Keane. M. T. (1990). *Cognitive psychology a student's handbook.* Hove, England: LEA.
Finin, T. (1980). The semantic interpretation of nominal conpounds. In *Proceedings of the First Annual National Conference on Artifical Intelligence.* 310-312.
Fodor, J. A. (1983). *The modularity of mind.* Cambridge, MA: MIT press.
Kay, P. & Zimmer, K. (1976). On the semantics of compounds and genitives in English. In *Sixth California Linguistics Association Proceedings,* San Diego.
Kedar-Cabelli, S. T. (1987). Formulating Concepts According to Purpose. In *Proceedings of the AAAI,* 477-481.
Lakoff, G. (1986). *Women, Fire and Dangerous Things.* Chicago: University of Chicago Press.
Lakoff, G. (1987). Cognitive models and prototype theory. In U. Neisser (Ed.) *Concepts and Conceptual Development.* Cambridge: Cambridge University Press.
McCauley, R. N. (1987). The role of theories in a theory of concepts. In U. Neisser (Ed.), *Concepts and Conceptual Development.* Cambridge: Cambridge University Press.
Medin, D. L. (1989). Concepts and Conceptual structure. *American Psychologist,* **44** (12), 1469-1481.
Medin, D. L., & Shoben, E. J. (1988). Context and structure in conceptual combination. *Cognitive Psycology*, **20**, 158-190.
Miller, G. A. & Johnson-Laird, P. N. (1976). *Language and perception.* Cambridge, MA: Harvard University Press.
Murphy, G. (1988). Comprehending complex concepts. *Cognitive Science*, **12**, 529-562.
Neisser, U. (1987) *Concepts and Conceptual Development.* Cambridge: Cambridge University Press.
Rosch, E. & Mervis, C. D. (1975). Family resemblance studies in the internal structure of categories. *Cognitive Psychology,* **7**, 573-605.
Roth E. M., & Shoben E. J.,(1983). The effect of context on the structure of categories. *Cognitive Psychology,* **15**, 346-378.
Smith, E. E., & Medin, D. L. (1981). *Categories and Concepts.* Cambridge, MA: Harvard University Press.

An Hierarchical Structure for Modelling Real-Time Domains using Object-Oriented Techniques

Haihong Dai[*], John G. Hughes and David A. Bell

Dept. of Information Systems, University of Ulster at Jordanstown
Newtownabbey, N. Ireland

Abstract

Real-time applications are always one of the most important area in artificial intelligence (AI). Many real-time knowledge-based systems have been developed as results of past research. In those systems, however, emphases were mainly put on aspects such as knowledge representations and fast execution speed. Although these aspects are very important, another key issue that of *guaranteed response time* is often overlooked. In this paper, we discuss some essential techniques for developing real-time systems. Issues of real-time AI problem-solving are addressed. An hierarchical structure for modelling real-time application domains is then presented in which the response time can be guaranteed. We also show that object-oriented methods can be used to implement such a model effectively and efficiently. It is believed that the hierarchical model provides a framework for the future development of a real-time AI environment.

1. Introduction

Real-time applications have always been one of the most important and challenging topic in AI research. As the real-time problem-solving is evidently different from that in other traditional application domains such as medical diagnosis, it presents a number of special requirements. Classical AI problem-solving methods face difficulties in coping with real-time situations (Dai, Anderson, and Monds, 1989).

First, the usefulness and correctness of a real-time system depends not only on the logical results of computation, but also on the time at which the results are physically produced. In most cases, deadlines or other explicit timing constraints are attached to tasks, and systems must be able to generate some results in strictly limited time periods. In other words, response time must be guaranteed. Second, unlike other traditional AI applications, real-time domains are not usually well structured. A large amount of information may be

---

[*]*Tel.: +44 (0) 232 365131 Ext.3030; Fax: +44 (0) 232 362803*
Email: CBJC23@UK.AC.ULSTER.UJVAX

produced by a variety of sources with different characteristics. Inferencing processes tend to be complicated, and therefore sophisticated inference strategies are needed in order to deal with such situations. Finally, information used by a real-time system typically are input from electronic sensors. This presents additional difficulties to the system. In order to be able to cope with such difficult situations, many kinds of knowledge should be incorporated in the system and the processing of tasks should be accomplished effectively and efficiently at many different levels.

Much research effort has been put into the area of real-time AI applications (Dai, Anderson, and Monds, 1990; Laffey, Cox, and Schmidt, 1988) and the results have been fruitful. A number of special requirements of real-time AI systems have been identified as follows: 1) *Continuous operation*; 2) *Interfacing to external environment*; 3) *Dealing with uncertain or missing data*; 4) *Ensuring high performance*; 5) *Supporting time-based reasoning*. Successful methods of meeting these requirements can be found in both the software and hardware categories. However, one key aspect of the requirements has often been overlooked in the past - *guaranteeing response time*. The meaning of guaranteed response time is often confused with what is usually called high performance in this content, but they in fact are different. The purpose of high performance is to decrease the *average* response time of all tasks in the system, while the guaranteed response time is constantly placed on all *individual* tasks. The high system performance is just a necessary condition for the latter but not sufficient. Although advances in computer hardware have greatly increased a computing system's execution speed, the requirement of guaranteed response time will not be automatically satisfied. Many other factors need to be considered, such as development of corresponding software.

Recently, much research has been focused on the issue of guaranteed system performance, especially on developing operating systems for managing and scheduling real-time tasks (Krishna and Lee, 1991). For the AI field, however, less effort has been put on developing tools or environments for real-time applications in which response time must be guaranteed. Current methods generally lack the capability to model a real-time domain efficiently and do not guarantee the response time.

In this paper, we will firstly present some techniques used in developing real-time systems. Then an hierarchical model for real-time applications will be proposed. We will show how to use object-oriented programming (OOP) methods to model a real-time domain and to implement strategies for a real-time AI system to guarantee the response time. An example will be given and some future work will be indicated at the end of this paper.

2. Techniques for developing real-time systems

In real-time applications, systems may usually be divided into two categories called *hard-real-time systems* and *soft-real-time systems* according to different timing requirements. This classification is important for designing and developing software for systems with different real-time requirements (e.g., Faulk and Parnas, 1988).

In a hard-real-time system, there is a so called *0/1 constraint* on timing requirements. It means that if a computation cannot be finished before a certain deadline expires, it must be abandoned completely. Then the system must produce an error message to indicate the failure of accomplishing the task. In contrast, in a soft-real-time system, there may not be a 0/1 constraint. Although a task cannot be completed in time, partial results obtained so far may be useful to some extent. Thus, a computation is not necessarily abandoned completely even though it is not finished in time.

In developing real-time operating systems and programming languages, two methods have been proposed for coping with the timing requirements (Kenny and Lin, 1990, 1991; Liu, Lin, and Shih, 1991). They are *imprecise computations*. and *performance polymorphism*. These two methods can provide functional and logical correctness, and guarantee the response time. In fact, they produce an appropriate trade-off between the result quality and the time needed for computation.

2.1 Imprecise computations

As mentioned above, partial results may constitute a useful approximation to the desired output in soft-real-time situations. In such situations, therefore, a computation will not be discarded even if it cannot finish before a deadline. This method of allowing incomplete computations to produce approximate results is called imprecise computations.

By using this method, there must be a minimum requirement for the partial results corresponding to a certain task. In other words, a minimum part of the task must be accomplished in time in order to make the partial results useful.

A task is accordingly divided into two sub-tasks by a system manager or scheduler. One of the sub-tasks is a mandatory sub-task and must be completed in time to satisfy the minimum requirement. The other sub-task is optional and it is used to refine the partial results. When a task reaches its deadline, its optional sub-task can be terminated and left unfinished. Results obtained so far will be returned as the approximation.

The optional sub-task may be further divided into several sub-sub-tasks. Each time after finishing a sub-sub-task, the system will return its results. When the deadline expires, all partial results of sub-sub-tasks will be combined with the results of the mandatory sub-task to produce final results. It can be seen from the above descriptions that the response time of a task can be guaranteed while the quality of the results is dependent upon the time available for computations.

2.2 Performance polymorphism

The performance polymorphism is also called *multiple version method*. This method is similar to the imprecise computations in that they trade off result quality to meet computation deadlines. However, they are different in specific techniques used.

In using the performance polymorphism, a task need not be divided. It is always accomplished as a whole. In order to guarantee the response time of the task, the system will have a number of different versions of computations for accomplishing the task. Basic functionalities of these computation versions are the same, that is, they all are used for completing the certain task. However, they differ in time and resources used and qualities of results produced.

Normally, a real-time system with the performance polymorphism has at least two versions of computation for a task. There is a primary version and one or more alternative versions. The primary version produces precise results, but it needs longer processing time. The alternative versions need less computation time, but they produce imprecise results (i.e., approximations). In running such a system, the system manager or scheduler will accept a task, and decide which version of computations should be used to accomplish the task according to the time deadline attached. The response time can be guaranteed, and again the quality of results relies on the available processing time.

Although the above two methods both provide facilities to guarantee response time, they have shortcomings in some aspects in terms of problem-solving. On the one hand, for example, if a task is indivisible and must be treated as a whole, the first method cannot be used. On the other hand, the overheads incurred in the second method may restrict its capability, because additional resources are needed to store multiple versions and the cost of the system scheduling may be high. In some cases, it might be desirable to combine these two methods in the same system so that a stronger real-time capability can be achieved.

3. Real-time AI problem-solving

The two methods for guaranteed response time have been implemented for a number of traditional computing applications such as real-time sorting. In those applications, the execution time of a certain algorithm can be determined in advance because it usually performs its task in an predicatable fashion. Theoretical analyses of algorithms are feasible. During a system development, therefore, specific timing requirements can be attached to different algorithms or to different versions of an algorithm. According to these timing requirements, it can easily be decided which algorithm should be used under certain circumstances.

Given two sorting algorithms p_sort and s_sort, for example, the former one is an algorithm implemented for parallel processing while the latter one is for sequential processing. The respective execution time of each algorithm relies on the number of elements N in a list to be sorted. As the overhead of starting the parallel sort is expected to be quite large, p_sort will use more time than s_sort until N exceeds a threshold value. According to the above analysis, strategies can be developed for choosing an appropriate algorithm based on the number of elements to be sorted.

In AI problem-solving, situations become much more complicated than in traditional computing domains. This is mainly because of the non-deterministic nature of knowledge-based methods.

It is commonly accepted that heuristic search is a fundamental technique in AI problem-solving. Unlike in traditional computing domains, however, there are often no well-structured algorithms in the traditional sense. In other words, it is very difficult to theoretically analyze and predict how a knowledge-based algorithm will perform. Response time cannot normally be predicted. The performance of a knowledge-based system is determined by many factors, such as knowledge/data organization in the system, and strategies in using the knowledge. Relationships among knowledge and among data entities also play a very important role.

Given a knowledge-based system, for example, we may assume that it will possess all necessary knowledge for solving problems in a certain domain. However, the system will not know in advance what exact situations it will encounter. Its problem-solving capability depends solely on its knowledge and information it receives from the external world. This is the non-deterministic nature of AI problem-solving. According to a specific situation, the system can execute all its knowledge, or it may only use a certain part of its knowledge. Thus, the system performance in terms of execution time will vary substantially under different circumstances. Furthermore, even if the system always uses the same part of its knowledge, different strategies applied will also affect the performance. For instance, a different execution order of a set of production rules may produce different performance although the same final result may be obtained.

Some recent research work has attempted to put restrictions on the use of knowledge for search and to limit the expressive power of knowledge representation formalisms, so that the performance of a certain system may be roughly estimated (Paul, Acharya, and Black, 1991). However, results are not very satisfactory. The non-deterministic nature of

AI problem-solving makes it difficult to use those techniques which have been successfully applied in traditional systems. New methods must be found. In the next section, we will present a hierarchical model for real-time domains and relevant issues will be discussed.

4. Using an hierarchical structure to model real-time domains

Because the performance of knowledge-based algorithms is difficult to predict, the techniques of imprecise computations and multiple versions cannot be used directly. In a knowledge-based system, the required kind of system manager or scheduler does not normally exist. The way in which knowledge may be applied cannot be determined in advance. It is not feasible to statically choose which algorithm should be used. Furthermore, a strict time limit will also forbid a system to spend too much time on scheduling rather than on computation.

4.1 An hierarchical model for real-time AI applications

Theoretical analyses of the performance of a knowledge-based system is often very difficult and in many cases infeasible. In practice, it is normal to use experimental methods to determine the upper bound and the lower bound of the system's execution time. If the worst case (i.e., the most difficult situation) can be predicted, the system performance under such a circumstance can then be estimated. Similarly, the performance in the simplest case may also be obtained.

Because the difference of performance between the worst and the best cases can be very large due to the nature of AI problem-solving, the system performance in any cases will have wide variance. The inferencing process in a real-time environment should be determined dynamically instead of being determined statically in advance of execution.

For example, if an algorithm is always chosen based on its performance in the worst case, it may produce unnecessary imprecise results. Assume that there are two algorithms A and B for performing a certain task. In its worst case, A needs only *10* time units to finish but provides an approximate result. B needs *18* time units to complete but can produce a precise result. If for a task there are *13* time units available, then A will be chosen and the result is only an approximation. In fact, such a task may not be the worst case of algorithm B, and therefore, B can finish in *13* time units and produce a more precise result. However, the system is not able to know such a fact in advance so that the precise result can never be obtained.

Motivated by the above considerations, an hierarchical structure for real-time domains has been developed, in which the inference is conducted in a progressive manner and where the best possible result will always be guaranteed given a time limit.

4.2 The structure

By using this hierarchical method, an application domain is divided into several levels of abstraction. At the highest level is the most abstracted description of the domain. Below the highest level, there are a few lower levels which are constructed by adding some more detailed descriptions to their respective higher levels.

For example, assume that an application domain can be represented by a set of six parameters called p_1, p_2, p_3, p_4, p_5, and p_6. Then an hierarchical model may be constructed based on these parameters (Figure 1).

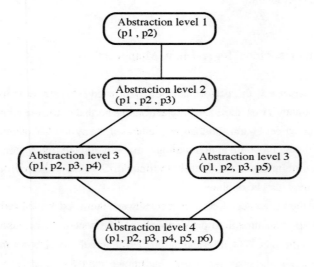

Figure 1. Hierarchical model of a real-time domain

As can be seen from figure 1, p_1 and p_2 construct the highest abstraction level. p_4 and p_5 are added respectively to *level 2* to form two different descriptions that are at the same level. *Level 4* is the lowest level and has the complete set of parameters.

The lowest level has the complete description of the whole application domain. The highest level must have a minimum set of descriptors (e.g., parameters) which is able to represent the domain. How to choose the minimum set of descriptors is domain dependent. We define such a minimum set of descriptors as a ***core*** of the domain. An application domain may have more than one core. In figure 1, for instance, if p_1 and p_3 can form the minimum set, then these two parameters will form another core of the domain. If a core constructs a complete description of a domain (i.e., no other optional descriptors may be added), then the domain is indivisible. Our hierarchical model is not applicable to such situations.

4.3 Problem-solving strategies based on the model

As discussed above, the different abstraction levels organize a domain into a core and several optional sub-domains. Associated with each level, there is corresponding domain knowledge for accomplishing tasks in that particular area. From figure 1, for example, at *level 1* the system is able to perform inferencing tasks based on parameters p_1 and p_2, and will produce relevant results. In other words, the domain and domain knowledge are all organized into the hierarchical structure, and reasoning processes will also be confined at each level.

The core represents minimum requirements of the problem-solving in the domain. Tasks falling into the core area must be completed in any case. Constructions of other abstraction levels add more details to the core, and are optional. Results obtained from these levels are refinements to the minimum requirements.

Although the nature of AI problem-solving is non-deterministic, it is generally the case that the system performance relies on the complexity of the application domain. The system needs less time to produce a result in a simplified domain than in a more complicated one. Thus, the system performance may be defined as a function of the complexity of a domain. Furthermore, the complexity of a domain is dependent upon the descriptors of the domain. More descriptors will present more complicated situations. In our hierarchical model, therefore, the complexity of the domain is increased from higher levels to lower levels. In the worst case, inference at a lower level needs more time than at a higher level. That is, the upper time requirement bound of all reasoning algorithms at a higher level is always smaller than at a lower level.

The techniques of imprecise computations and performance polymorphism are combined in our model to accommodate the requirements of AI problem-solving. If a task at the core level (i.e., the highest level) cannot be finished before its deadline, the system crashes. Otherwise, it will always be able to produce some results which meet the minimum requirements. If more time is available, the system may provide refined results by conducting inference at some lower levels provided that additional information is available. Descriptions of the domain that have brother relationships are at the same level, and they represent alternatives to produce results with the same quality. Which alternative should be chosen depends on the available information and knowledge.

The basic steps of problem-solving in this system can be summarized as follows:
1) When a task is accepted, the domain descriptors for representing the task will be determined.
2) A sub-task is created based on the requirements of the core, i.e., the sub-task is created at the highest abstraction level. Other available information is left redundant at this stage.
3) Only domain knowledge related to the core level is activated. The reasoning process is confined to this level.
4) Results are produced which meet the minimum requirements (that are decided by system developers).
5) Check the current time against the task's deadline. If more time is available, a new sub-task is created at a lower abstraction level based on the additional information which was redundant at previous stages.
6) Computations previously conducted at the higher level will not be repeated. Previous results can be directly used for computations at the lower level.
7) If the new sub-task cannot be finished in time, it is abandoned. Results from the previous sub-task are returned as final results.
8) If there is still more time available after the completion of the sub-task, and the lowest level has not been reached, then repeat from *step 5*.
9) When a sub-task created at the lowest level is completed, the system produces a perfect result in time.

From the above discussions, it can be shown that such a model will be able to guarantee the response time if the execution of a task at the core level can be completed in time. The core execution time represents the minimum timing requirement of the system. The quality of results will always be improved while more time is available. The best result is guaranteed in a certain time period.

5. Implementing the hierarchical model using OOP techniques

The hierarchical model presented above can be conveniently implemented by using OOP techniques. OOP methods offer many advantages over traditional techniques (Nierstrasz, 1989). Its important features, such as data encapsulation, objects as active agents, and inheritance, provide powerful tools for AI systems development. A number of examples are in areas of knowledge representation (Fikes and Kehler, 1985) and real-time expert systems (Ghaly and Prabhakaran, 1991). Some recent research work has equipped OOP languages with the capability of representing time constraints, and successfully applied OOP techniques to building real-time systems (Barry, 1989; Ishikawa, Tokuda, and Mercer, 1990). These provide evidence that OOP techniques are suitable for real-time applications.

In our particular case, the different levels of the hierarchical structure are represented by different *classes*. A class in OOP organizes together all objects with similar attributes and behaviours. Thus, domain descriptors required by a certain abstraction level are defined as class attributes in a class corresponding to that level.

As in the hierarchical model, a lower level will have new domain descriptors in addition to those of its ancestor level. This can be implemented by using inheritance. The ancestor level is defined as a super-class, and the immediate descendant is defined as its sub-class. The whole structure can therefore be recursively defined *via* inheritance.

As an active agent, an object possesses its own methods to manage its behaviour. Domain knowledge is developed according to the domain descriptors at each level. Any knowledge related to other descriptors will not be encoded at an irrelevant level. In other words, no redundant knowledge may exist at any inferencing level. The feature of data encapsulation prevents objects at different levels from accessing each other's internal contents. This guarantees the integrity of each abstraction level.

The hierarchical organization of domain knowledge makes such a system efficient. The inferencing knowledge of a class can only be activated when an instance (i.e., an object) is created for that class. As there is no redundant knowledge in each class, the system is able to focus its attention on the most relevant part of the problem domain, and problem-solving can be effective and efficient.

When the system is in execution, first of all, it will automatically create an instance for the class representing the core level. After creating the instance object, relevant knowledge is accessed and inferencing processes are then activated. The accomplishment of a task will follow the steps presented earlier in this paper. Because an object is capable of inheriting properties from its super-class, contents in a higher level may be transformed to a lower level. Thus, when an instance is created for a sub-class (which corresponds to a sub-task at a lower level), no duplicated computations are necessary. Computations at lower levels are purely for refining results.

6. An illustrating example

In this section, we present an example to illustrate some key points discussed above. The application domain is for signal processing and interpretation, and the KEE (Fikes and Kehler, 1985; KEE, 1989) which is a Knowledge Engineering Environment is used for developing our prototype model.

6.1 Brief overview of the problem domain

In the real-time signal processing and interpretation application, a radar system using diplex-doppler techniques is adopted (Dai et al., 1990). Combined with the radar system, there is a knowledge-based system for processing and interpreting signals obtained from the radar receiver.

The system monitors a protected area in a short range. According to data received, it will decide whether or not there is a moving target in the area. If a target presents, the system will further decide whether the moving object is a human being based on its prior domain knowledge. By assessing the situation against a set of pre-stored criteria, the system is able to determine the danger caused by the presence of the target, and to decide if alarms should be raised.

The basic information that can be extracted from reflected radar signals is as follows: 1) *Presence of a target*; 2) *Moving speed of the target*; 3) *Moving direction of the target*; 4) *Location of the target*; 5) *Radar cross-section of the target*. The available information will help to decide if it is necessary to activate an alarm system. In other words, more information enables the system to make a more appropriate decision and so reduce the possibility of false alarms.

6.2 System development in KEE

A small prototype system is implemented in KEE. Three abstraction levels are constructed for this domain. The task of the core level is to detect the presence of a moving target. The domain descriptor represents the phase difference between transmitted and reflected signals. If there exists a difference, then a moving target presents. Domain knowledge at this level consists of a number of algorithms for checking the phase difference among signals.

As depicted in figure 2, the second level has two alternatives. In addition to the information available at *level 1*, each of them has other descriptors such as speed and direction of the target respectively. Finally, at *level 3* the location and cross-section of the target are presented.

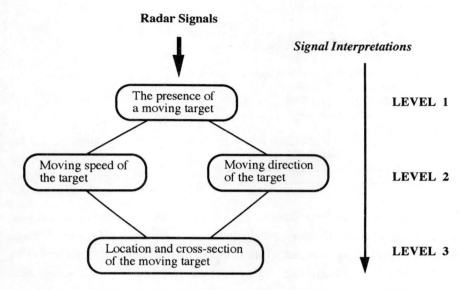

Figure 2. A real-time signal processing and interpretation domain

Each abstraction level or its alternatives is represented as a frame in KEE. A frame consists of data entities for representing instance attributes and a number of methods (i.e., procedures) for performing inference.

For example, the implementation of the lowest level is as follows:

```
_ _ _ _ _ _ _ _ _ _ _ _ _ _ _ _ _ _ _ _
Unit: Third.Inference.Level in knowledge base SPI
      Superclasses:    Second.Inference.Level.Speed,
                       Second.Inference.Level.Direction
      ......
_ _ _ _ _ _ _ _ _ _ _ _ _ _ _ _ _ _ _ _
      ......

Member slot: Location from Third.Inference.Level
      Inheritance: OVERRIDE.VALUES
      ValueClass: NUMBER
      Comment: "Representing the distance between the target and the receiver."
      Value: Unknown

Member slot: Cross.Section from Third.Inference.Level
      Inheritance: OVERRIDE.VALUES
      ValueClass: NUMBER
      Comment: "Representing the size of the target."
      Value: Unknown
```

```
           Member slot: INFERENCE from Third.Inference.Level
               Inheritance: METHOD
               ValueClass: METHODS
               Cardinality.Max: 1
               Cardinality.Min: 1
               Comment: "Inferencing based on all available information."
               Values: Third.Level.Reasoning.Procedures

           Member slot: RESULTS from Third.Inference.Level
               Inheritance: METHOD
               ValueClass: METHODS
               Cardinality.Max: 1
               Cardinality.Min: 1
               Comment: "When deadline expires, return any results that are currently
                        available."
               Values: Return.Results.From.Third.Level
    _ _ _ _ _ _ _ _ _ _ _ _ _ _ _ _ _ _ _ _ _ _ _
```

Inferencing procedures are written as Lisp functions and organized into a similar hierarchical structure which corresponds to the hierarchically organized frames. When an instance of a frame class is created, it will possess all member slots which appeared in the definition of that class. The instance is activated by receiving messages. According to the content of a message, the instance frame will invoke appropriate reasoning procedures to conduct inference. When the deadline of its task expires, the instance frame terminates the reasoning procedures. A special procedure is then invoked to return the current results.

KEE offers an excellent tool for prototyping a demonstration system. Object-oriented techniques provided by the KEE environment can not only model a real-time domain effectively, but also offer an efficient way to organize and apply domain knowledge. Unfortunately, however, the KEE environment is not suitable for developing and delivering a real-time system. This is mainly due to the fact that KEE lacks facilities for representing time constraints. Furthermore, as KEE is implemented in Lisp, the execution speed may cause problems in real-time situations.

7. Concluding remarks

In this paper, we have discussed some important techniques for developing real-time AI systems in which the response time can be guaranteed. A hierarchical model for representing real-time application domains was presented. It has been shown that object-oriented methods can be effectively used to implement such a model.

At the present time, there are many AI systems development tools and environments available. However, these tools generally lack capabilities of representing timing requirements of real-time applications. Therefore, they are inadequate for prototyping and developing real-time systems in many cases.

As current OOP environments become more and more powerful, it is proposed to build a knowledge-based system development tool by combining an OOP language with an object-oriented database system. Explicit facilities for representing timing requirements must be provided by such a tool, and the development of the hierarchical model must be supported. The C++ language is a very good candidate due to its efficient implementation. Although much work needs to done, it is a very promising research direction.

References:

Barry, B. M. (1989). Prototyping a Real-Time Embedded System in Smalltalk. *Proc. of Object-Oriented Programming, Systems, Languages, and Applications (OOPSLA'89)*, October, 255-265.

Dai, H., Anderson, T. J., & Monds, F. C. (1989). Issues of Real-Time Expert Systems. In A. F. Smeaton & G. McDermott (Eds.), *AI and Cognitive Science'89* (pp.179-197). NY: Springer-Verlag.

Dai, H., Anderson, T. J., & Monds, F. C. (1990). A Framework for Real-Time Problem Solving. *Proc. of the 9th European Conference on Artificial Intelligence (ECAI'90)*, 179-185.

Faulk, S. R., & Parnas, D. L.(1988). On Synchronization in Hard-Real-Time Systems. *Communications of the ACM*, **31** (3), 274-287.

Fikes, R., & Kehler, T. (1985). The Role of Frame-Based Representation in Reasoning. *Communications of the ACM*, **28** (9), 904-920.

Ghaly, R., & Prabhakaran, N. (1991). Rule-Based Object-Oriented Approach for Modelling Real-Time Systems. *Information and Software Technology*, **33** (4), 250-258.

Ishikawa, Y., Tokuda, H., & Mercer, C. W. (1990). Object-Oriented Real-Time Language Design: Constructs for Timing Constraints. *Proc. of European Conf. on Object-Oriented Programming / Object-Oriented Programming, Systems, Languages, and Applications (ECOOP/OOPSLA'90)*, 289-298.

KEE (1989). *KEE Reference Manual*, IntelliCorp Inc., USA.

Kenny, K., & Lin, K. J. (1990). Structuring Real-Time Systems with Performance Polymorphism. *Proc. of 11th IEEE Real-Time Systems Symposium*, 238-246.

Kenny, K., & Lin, K. J. (1991). Building Flexible Real-Time Systems Using the Flex Language. *IEEE Computer*, **24** (5), 70-78.

Krishna, C. M., & Lee, Y. H. (Eds.). (1991). *IEEE Computer, Special Edition on Real-Time Systems*, **24** (5).

Laffey, T. J., Cox, P. A., & Schmidt, J. L. (1988). Real-Time Knowledge-Based Systems. *The AI Magazine*, Summer, 27-45.

Liu, J. W. S., Lin, K. J., & Shih, W. K. (1991). Algorithms for Scheduling Imprecise Computations. *IEEE Computer*, **24** (5), 58-68.

Nierstrasz, O. M.(1989). A Survey of Object-Oriented Concepts. In W. Kim and F. H. Lochovsky (Eds.), *Object-Oriented Concepts, Databases, and Applications* (pp.3-21). NY: Addison-Wesley.

Paul, C. J., Acharya, A., & Black, B. (1991). Reducing Problem-Solving Variance to Improve Predictability. *Communications of the ACM*, **34** (8), 81-93.

Argumentation in Weak Theory Domains

Kathleen Freeman, Arthur M. Farley
Department of Computer and Information Science, University of Oregon
Eugene, USA

Abstract

We present argumentation as a method for reasoning in "weak theory domains", i.e., domains where knowledge is incomplete, uncertain, and/or inconsistent. We see these factors as related: methods for reasoning under incomplete knowledge, for example, default reasoning, plausible reasoning, and evidential reasoning, may result in conclusions that are uncertain and/or inconsistent. Knowledge in many domains in which we'd like computers to reason can be expected to be incomplete, and therefore, possibly inconsistent. Also, some domains, e.g., legal reasoning, may be inherently inconsistent. Therefore, it is important to investigate methods for reasoning under inconsistency.

We explore the use of argumentation as a basis for this task. Argumentation is a method for locating, highlighting, and organizing relevant information both in support of and counter to a plausible claim. This information can then serve as a vehicle for comparing the merits of competing claims.

We present aspects of our preliminary investigation of a formal theory of argumentation: (i) identifying a formal theory of argumentation; (ii) implementing the theory in a computer program; (iii) gathering example problems and associated arguments; and (iv) evaluating the theory with respect to the example arguments.

Current work concentrates on the structure and generation of independent arguments for an input claim and its negation. Future work will focus on argumentation as a series of adversarial moves that support and counter a claim.

1 Introduction

The ability to reason is a hallmark of intelligence and therefore an essential skill to be realized by artificial intelligence systems. Reasoning (i.e., making inferences, solving problems, revising beliefs) depends on a reasoner's current knowledge and beliefs. Yet, when we turn our attention to real-world domains, we find that what is known or believed is often incomplete, ambiguous, imprecise, and inconsistent. In such domains, there will always be gaps in what is known due to a lack of information gathering capabilities, the combinatorial complexity of the space of possible situations, the possibility of exceptions to general rules, and the prevalence of changing circumstances.

We call real-world domains about which knowledge is incomplete, inconsistent, or otherwise uncertain, "weak theory domains". Expert systems "tend to be built precisely for domains in which decisions are highly judgmental" (Buchanan & Shortliffe, 1984) and are thus expected to operate in weak theory domains. Various techniques for belief propagation and approaches to fault diagnosis have been developed in an attempt to cope with the uncertainty and unreliability in these domains (see, e.g., Dubois & Prade, 1989; Lea Sombe, 1990). Many of the problems studied in artificial intelligence research on knowledge representation and automated reasoning reflect the consequences of dealing with weak theory domains.

As an example, consider the following "Bermuda" problem, based upon the classic example of Toulmin (1958):

We know that usually anyone born in Bermuda can be assumed to be a British subject. However, someone born in Bermuda to alien parents is not a British subject. Someone with a British passport can also be assumed to be a British subject. Most people who are English speaking and have a Bermudan identification number were born in Bermuda.

Finally, any person with a Bermudan id number is eligible to obtain Bermudan working papers. We have just been introduced to Harry, who speaks English, has a passport that is not a British passport, and shows us his Bermudan working papers. We are wondering whether he is a British subject or not.

The knowledge in this problem is inconsistent, simultaneously supporting both conflicting conclusions. It is incomplete, since information that could help support one conclusion or the other, such as whether or not Harry's parents were aliens, is missing. Also, the knowledge given about the problem is uncertain. Hedges such as "often" and "usually" are indications that a rule is less than certain.

There are other sources of uncertainty, as well. Should the knowledge that Harry has working papers support the conclusion that he has a Bermuda social security number (and thereby support the born in Bermuda assertion)? Or, is the lack of a British passport grounds for claiming that Harry is not a British subject? How much support does either of these give to the conclusion? How should support be combined across the various arguments that support Harry's being a British subject or not being a British subject?

As we see above, support for a claim that is accumulated by reasoning in a weak theory domain may be inconclusive or controversial. Claims supported by *plausible reasoning* (e.g., Polya, 1968) based upon uncertain, incomplete knowledge are said to be *defeasible* (e.g., Pollock, 1987) due to the uncertainty of generalizations and the availability of arguments to the contrary.

If we can not decide the truth of most claims with certainty, how can we proceed with reasoning at all? Our approach is to explore the use of argumentation as a basis for deriving and justifying claims in weak theory domains. A reasoner should be able to justify claims made, both on their supporting merits and with respect to support for alternative claims. Any claim is presumed to be controversial, i.e., in need of support and vulnerable to objections against which it must be defended. This presumption is appropriate for most claims made in weak theory domains. Weak links in the support for a claim should be explicated; links that can be countered by support for alternative claims should be discounted. Finally, there is a need for methods to compare the strength of support among alternative claims.

Our proposal is that argumentation be used as a method for locating, highlighting, and organizing relevant information in support of and counter to a plausible claim. Argumentation can be viewed as a mechanism for generating a class of super-explanations: in addition to providing "logical" support for belief in a claim (i.e., as we would expect to find in a typical explanation), reasoning that counters the claim is determined, as well. Argumentation is seen to be a vehicle for comparing the merits of support for competing claims. Such an ability is crucial for the justification of decision-making by artificial intelligence systems that hope to perform effectively in weak theory domains.

Argumentation has long been viewed as an important reasoning tool outside artificial intelligence (e.g., Rescher, 1977). In artificial intelligence and elsewhere, argumentation has been variously investigated from the standpoints of rhetoric (e.g., Horner, 1988; Kahane, 1988), philosophy (e.g., Rescher, 1977; Toulmin, 1958), discourse analysis (e.g., Flowers, McGuire, & Birnbaum, 1982), legal reasoning (e.g., Ashley, 1989; Levi, 1949; Marshall, 1989; Rissland, 1989), and others. Our work borrows from this previous work, generalizing and refining it so as to focus on a *formal* theory of argumentation that can provide *decision support and justification* based upon *plausible* reasoning.

In the remainder of this paper, we present aspects of our preliminary investigation of argumentation. Our initial goals have been: (i) identify a formal theory of argumentation; (ii) implement the theory in a computer program; (iii) gather example problems and associated arguments that demonstrate various aspects of argumentation; (iv) evaluate the theory with respect to the example arguments. Our results will be of interest to researchers in both artificial intelligence (AI) and argumentation. For AI programs, the ability to generate arguments will be a useful technique in real-world contexts. For argumentation researchers, AI methodology offers a new way for formalizing and evaluating theories of argumentation.

2 A Model of Argument

In our approach to argumentation, we make two significant, simplifying assumptions: (i) of the five aspects of argumentation in classical rhetoric (invention, arrangement, style, memory, and presentation/delivery), we concentrate on invention, i.e., the process of developing a defensible "proof" for an assertion; (ii) of the three types of "proof" in classical rhetoric (logos, ethos, pathos), we deal entirely with logos, i.e., logical reasoning (both informal and formal logic). These restrictions are intended to focus the work on the more formal aspects of argumentation.

Our approach is based on the following, complementary definitions of argument: (i) "the grounds ... on which the merits of an assertion are to depend" (Toulmin, 1958), and (ii) "a method for conducting controversial discussions, with one contender defending a thesis in the face of object[ions] and counterarguments made by an adversary" (Rescher, 1977). There are two senses of argument indicated by these definitions. The first defines argument as a supporting explanation, as in 'she made an argument'. The second concentrates on argument as an activity in which two or more agents engage. We see an argument as a method, or process, as well as a product. Thus, the generation and representation of arguments are both crucial to our theory.

Our theory will develop both senses of argument in successive stages. In the first stage, argument structures for a claim and its negation are generated. These structures summarize the relevant, available knowledge about the claim, its support, and support for its negation. In this paper, we give a description of our theory of argument structure and its generation and presentation by a current implementation. Then, we outline approaches we intend to follow in the next stage, developing a model of the process of argumentation as dialectical, based upon the argument structures described here.

We represent an argument in a modified version of the form suggested by Toulmin in his book *The Uses of Argument* (1958). According to Toulmin, an argument comprises input *data* (i.e., a problem description, evidence, grounds) said to support a *claim* (i.e., conclusion, solution, belief). The authority for taking the step from data to claim is called a *warrant*. The warrant may have *backing*, or justification. The data and the warrant may not be enough to establish the claim with certainty, i.e., the resultant claim may be *qualified* (e.g., "probably"). The claim may be subject to *rebuttals*, special circumstances where the warrant would not hold (i.e., "unless...").

Modifications to this structure are needed to (1) formalize Toulmin's ideas (especially warrants and qualifications); (2) provide a macro structure for arguments, e.g., extended chains of support for claims, multiple backings for claims; and (3) explicate various sources of uncertainty, i.e., arguable points in the domain knowledge.

An argument is represented as a structure of claims, interconnected by "Toulmin argument units" (*taus*). A tau is essentially the data-warrant-claim structure described above. But data and warrants are also viewed as claims, and an argument structure is a hierarchic structure of claims and taus, with taus supporting claims as conclusions and depending on claims as support. Since all the major elements of a tau are claims, we will refer to the tau as a data-warrant-*conclusion* structure to avoid ambiguity.

In addition, we make the following modifications/extensions to the basic Toulmin argument structure. Claims comprise data, i.e., a statement of the claim, a qualification, and a backing. There are two types of backing: *atomic*, as input information from outside the domain of argumentation (i.e., "inartistic proofs", Horner, 1988), and *tau*, where the claim is supported by data and a warrant. A claim may have multiple supporting arguments (backings); the qualification on a claim summarizes the strengths and weaknesses of all the supporting arguments. In addition, a claim has a pointer (possibly nil) to its negation and, thus implicitly to its rebuttals, which are the arguments supporting its negation.

Warrants have a slightly different structure from other claims. In addition to qualification, backing, and rebuttal pointer, warrants have two data fields, *antecedent* and *consequent*. A warrant also has a *type* associated with it which represents information as to the strength of the connection between antecedent and consequent. While a warrant has antecedent and consequent, indicating the normal direction of its application, we can

Table 1

antecedent	consequent	reasoning step	
p	q	modus ponens	(MP)
~q	~p	modus tollens	(MT)
q	p	abduction	(ABD)
~p	~q	contrapositive abduction	(ABC)

use warrants in the opposite direction as well. Given a warrant with antecedent p and consequent q, we define these *reasoning steps* in Table 1.

The last two reasoning steps are fallacies (asserting the consequent and denying the antecedent, respectively) for deductive reasoning, but are often used in plausible reasoning to indicate support for a claim, though not conclusively. Polya (1968) discusses similar "patterns of plausible inference". He calls them "examining a ground" (roughly MP and ABC, above) and "examining a consequence" (roughly MT and ABD, above). Since the ABD and ABC reasoning types are not conclusive, there is a need to attach qualifications to claims. Qualifications also capture the inconclusiveness of uncertain warrants. We next define a formal language for claim qualifications and warrant types. We then specify their interaction with the four types of reasoning steps indicated above in determining resultant claim qualifications.

3 Representing Uncertainty

As a first approximation to the representation of degrees of belief in a claim, we restrict qualification values to be one of the following: *strong (!), usual (!'), more credible (+), less credible (-), weak (X),* and *contingent (?')*. The first five are ranked in decreasing order of degree of belief as given. While we are not adopting a strict probabilistic interpretation here, they can be related to probability values of 1, 1-∂, > .50, > 0.0, and 0; contingent indicates a lack of input data support. This broad-scale quantification of degree of belief limits our ability to capture differences among strengths of some arguments. However, we later demonstrate instances where we can make important distinctions using this highly restricted set of values.

The qualification on a claim is that associated with its strongest supporting argument. The qualifications on (input) data are given as atomic backing at input time and remain unchanged thereafter, unless better support can be derived from tau backing. The strength of any claim drawn from application of a warrant (i.e., a tau backing) is the least of the qualifications associated with the application: the qualifications on data serving as antecedents, the qualification on the warrant, and that derived from the type of warrant and reasoning step applied, as discussed below.

With each warrant we associate a type, reflecting the strength with which its conclusion can be drawn from the given antecedent. The types we specify are: *necessary and sufficient, necessary, sufficient, default,* and *evidential*. In this paper, we focus on the use of the latter three types. *Necessary* can be represented as a *sufficient* warrant with antecedent and conclusion reversed, while *necessary and sufficient* is equivalent to two *sufficient* warrants, one in each direction, between antecedent and conclusion.

Warrants are expected to be written in the direction that accommodates the strongest possible type. For example, for a causal relation such as that between fire and smoke, since fire causes (is sufficient grounds to conclude) smoke, smoke can therefore be said to be evidential grounds for concluding fire. In the warrant that represents this relation, fire should be the antecedent and smoke the consequent, since that is the direction of the stronger relationship.

Table 2

warrant type	reasoning step	tau backing
sufficient	MP, MT	strong/weak
sufficient $\to_s$	ABD, ABC	more/less credible
default	MP, MT	usual/weak
default $\to_d$	ABD, ABC	more/less credible
evidential	MP, MT	more/less credible
evidential $\to_{ev}$	ABD, ABC	more/less credible

Finally, we discuss how the type of warrant and the type of reasoning step are combined to yield the qualification for warrant application aspect of the tau backing. We summarize the interaction between warrant type and reasoning step in Table 2.

As in deductive logic, a *modus ponens* reasoning step with a *sufficient* warrant yields a *strong* qualification on the warrant application. *Modus tollens*, or reasoning deductively from a weak consequent, will lead to a *weak* qualification for a *sufficient* warrant. A *default* warrant differs from an *evidential* warrant in that the default warrant assumes a higher, i.e., *usual* -vs.- *more credible*, qualification on warrant application using *modus ponens*. Reasoning abductively, essentially from a credible or strong conclusion to antecedent, will contribute *more credible* qualifications on claims from the warrant application. The contrapositive of abduction, reasoning from a *less credible* or *weak* antecedent results in a *less credible* consequence.

4 Generating Argument Structures

We next describe a recursive algorithm for generating argument structures, given a claim, a set of domain knowledge (as warrants and data), and a set of inputs (as data). Warrants, for now, are not treated as claims. In Toulmin's terminology, the arguments generated are warrant-using rather than warrant-establishing arguments. For any given claim, an argument is generated in a backward-directed fashion by looking for backing for the claim. A claim has atomic backing if it is already present in the given data base. Tau backing for a claim and its negation is generated by searching for warrants that are relevant to the claim, i.e., containing the claim in the antecedent or conclusion of the warrant. For each relevant warrant, a new tau structure is generated; new claims, from the other "side" of the warrant, are added to a global claims list. Any loop (i.e., a claim being used to support itself in an argument) or contradiction (i.e., the negation of a claim being used to support the claim) is pruned during the argument generation process.

The argument algorithm then calls itself recursively, attempting to find backing for the next claim on the claims list. The process continues until the claims list is empty. The algorithm then restarts the argumentation process for the claim that is the negation of the original claim. Final output is a graphical or textual presentation of the generated argument structures, i.e., claims and taus for both the input claim and its negation.

This algorithm has been implemented in a Scheme program. It has been tested on a number of small examples drawn from the literature in both argumentation and AI. We present program input and graphical output generated by the program for several of these examples in the next section.

5 Examples

We give several examples of argument structure generation that highlight different facets of the theory. The examples include arguments having knowledge that is incomplete, uncertain, or inconsistent, using different inference types, and involving default knowledge. We display several argument structures generated by our current implementation of the theory, highlighting relevant aspects of the arguments presented.

5.1 Example 1

Here we revisit the "Bermuda" argument of the introduction, represented as warrants and input data as follows:

```
(w1 ( (bermuda born) )    -->_d  ( (british subject) ) ! given )
(w2 ( (bermuda born) (alien parents) )  -->_s  ( (not british subject) ) ! given )
(w3 ( (british passport) ) -->_s  ( (british subject) ) ! given )
(w4 ( (english speaking) (working papers) ) --> ev  ( (bermuda born) ) ! given )
(w5 ( (bermudan id#) )   -->_s  ( (working papers) ) ! given )

(d1 (english speaking) ! given )
(d2 (not british passport) ! given )
(d3 (bermudan id#) ! given )
```

Argument structures that summarize the evidence for and against the claim that Harry is a British subject are shown in Figure 1.

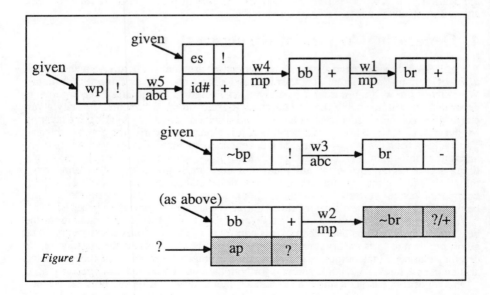

Figure 1

Claims are shown as nodes and warrants are represented as arcs between them. The notation above the arcs indicates the id of the applicable warrant, and notation below the arcs gives the reasoning step type. Notation in the claim nodes gives the statement of the claim and its qualification. Those claim nodes that are contingent, i.e., without supporting input data, are darkened.

In this example, we see that there is some support for each of two mutually exclusive claims, "british subject" and "not british subject". The strongest argument for "british subject", that Harry was born in Bermuda, has a weak link at "bermudan id#". From warrant W5 we see that "working papers" could be concluded from "bermudan id#", but here the warrant is used abductively and the qualification reflects uncertainty resulting from this form of reasoning. That is, there may be other ways of obtaining working papers than by having a bermudan identification number. Since the knowledge base is silent on these other ways - if any - belief in "bermudan id#" may be strengthened. However, it remains that abductive reasoning is less certain than deductive reasoning.

The next argument structure shows that the data that Harry does not have a British passport makes the claim that Harry is a British subject less credible (though not weak, since the reasoning type used in this argument is non-deductive). In the last argument tree we see there is potential direct support for the negative claim (as opposed to the indirect support that accrues to the negation when the claim is shown to be less credible or weak). The support would simultaneously defeat the strongest argument for the claim, by showing that Harry's parents were aliens and thereby rendering irrelevant the argument for the claim based on Harry's being born in Bermuda. Given the current knowledge base, this argument is only a hypothetical one.

5.2 Example 2

Next we consider the "republican-quaker-hawk-dove" example, similar to that discussed by Poole (1989). We present the warrants and input data, as follows:

```
(w1 ( (republican) )   -->  ev   ( (not dove) ) ! given )
(w2 ( (quaker) )   -->  d   ( (dove) ) ! given )
(w3 ( (not dove)   -->  ev   ( (supports star wars) ) ! given )
(w4 ( (quaker) )   -->  s   ( (religious) ) ! given )

(d1 (religious) ! given )
(d2 (republican) ! given )
```

The argument structures generated for dove and ~dove are shown in Figure 2. Again we see some support for both claims. Support for "~dove" is based on evidential support from "republican", while the support of religious supports some belief in "quaker", which then can support "dove" as *more credible*. If we knew that the person was a "quaker" rather than just "religious", as the problem is usually presented, then the qualification for "dove" would be stronger, as *usual* due to the default warrant w2.

Knowledge about star-wars could be used as an additional argument for one or the other of these two claims.

Both of the last two arguments above are inconsistent in that there is no consistent assignment of probabilities to the two outcomes, as satisfying the qualitative qualifications would imply both have probability above 0.50 (Goldszmidt & Pearl, 1991).

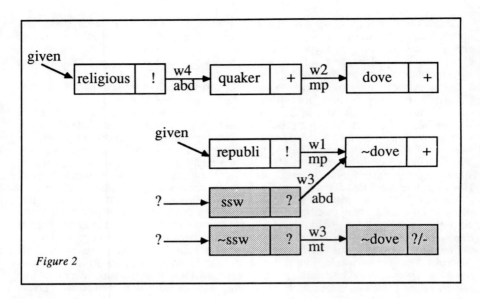

Figure 2

5.3 Example 3

Our version of the standard penguins-birds-flying example, illustrating how argumentation would handle standard default reasoning issues as encountered by inheritance-based representations, is as follows:

(w1 ((bird))   -->$_d$ ((flies)) ! given)
(w2 ((penguin))   -->$_s$ ((not flies)) ! given)
(w3 ((penguin))   -->$_s$ ((bird)) ! given)

(d1 (bird) ! given)

The argument structures generated for flies and ~flies are shown in Figure 3. Given the data as "bird", we find that the strongest argument would be for "flies", with qualification *usual* due to the default warrant w1, while "~flies" would only be deemed *credible*, due to the weak abductive argument that supports penguin. On the other hand, if we had the data "penguin", the claim "~flies" would be established with qualification *strong* due to the sufficient warrant w2, while "flies" would be established with a qualification of *usual*. While the arguments are inconsistent, as discussed above, there is a preference for one conclusion (the normally preferred one of "~flies") over the other.

Here our broad quantification of belief levels is sufficient to draw the expected results (given the stronger claim is taken to be the result) for this kind of implicit default reasoning. However, if warrant w2 were only taken as default information about penguins, we would not be able to distinguish between the two argument strengths. Again, we see the impact of warrant type decisions on argument outcomes.

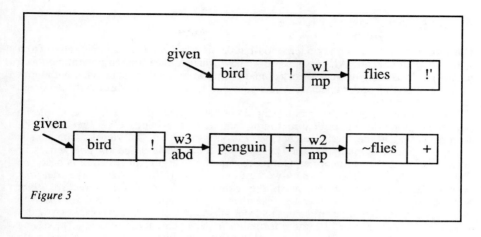

Figure 3

5.4 Example 4

Here we look at a multiple causes situation involving rain, sprinkler, and wet grass (Pearl, 1987). We generate some troublesome results, which motivates the need to address argumentation as an interaction between arguments, using additional types of plausible inference, as we plan to explore in the next phase of our research. The specification is as follows:

(w1 ((rain)) -->$_s$ ((wet grass)) ! given)
(w2 ((sprinkler)) -->$_s$ ((wet grass)) ! given)
(w3 ((wet grass)) -->$_s$ ((wet shoes)) ! given)

(d1 (sprinkler) ! given)

The argument structure generated in support of "rain" is shown in Figure 4. (With the input knowledge, the program is not able to generate any support for "~rain".) Here, the warrants are representing causal relationships. Diagnostic reasoning often reflects weak abductive reasoning over such "causal warrants". Here "rain" is defeasibly supported by "wet grass", which is strongly supported by the "sprinkler" input data. This last part is odd in that "sprinkler", known to be true, provides a causal explanation of "wet grass" and should undercut "rain" rather than support it.

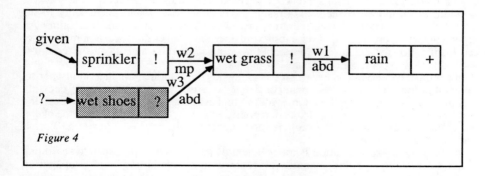

Figure 4

6 Related Research

Our research in modeling argumentation builds on other work in argumentation from both inside and outside AI. Also, since we have chosen to explore argumentation as a method for reasoning in weak theory domains, this work overlaps somewhat with work in AI on reasoning under uncertainty and plausible reasoning. We discuss these related areas in this section.

There has been a great deal of investigation of argument outside AI, for example, in philosophy, rhetoric, and critical reasoning (e.g., Cox & Willard, 1982; Freeley, 1990; Horner, 1988; Kahane, 1988; Rescher, 1977; Toulmin, 1958). Our view of argument as a combination of explanation, or "proof", and dialectic is based on two senses of argument found in this literature. Our representation of argument structure is based on Toulmin's (1958) well known data-warrant-claim model of argument. The idea that argumentation is an important reasoning technique is discussed in (Freeley, 1990; Kahane, 1988; Rescher, 1977), and elsewhere. It is mainly our formal approach to modeling argument that differentiates us from this other work. At least one investigator (Hample, 1982) has called for more formality in argumentation research: "[modeling arguments] will be beneficial in several respects...models yield a precise and clear understanding of the exact character of the theories they serve". One of the main goals of the current work is to employ the computer as a modeling tool to achieve just this effect. The dynamic nature of computer programs supports research in argumentation generation as well as argument analysis.

AI researchers have also undertaken work in argumentation, and recently research in this area seems to be on the rise (AAAI Spring Symposium, 1991). Most AI work in argumentation has been done in contexts other than reasoning per se, e.g., natural language processing (e.g., Flowers et al., 1982), software engineering (e.g., Conklin, 1988; Lee, 1991), or legal reasoning (e.g., Ashley, 1989; Marshall, 1989; Rissland, 1989). Some of this work is mainly concerned with argument analysis and structure (Conklin, 1988; Lee, 1991; Marshall, 1989; Storrs, 1991), while other work is mainly concerned with argument generation (Ashley, 1989; Flowers et al., 1983; Rissland, 1985). Results about argument representation and moves are important for our theory.

Our current research continues this work and adds to it, by (1) its emphasis on both the structure and generation of arguments; (2) use of a domain independent theory of argument; (3) incorporation of two senses of argument, supporting explanation and dialectical process; (4) implementation of the theory in a computer program, and evaluation of the results; and (5) analysis and representation of weak theory domains.

The last point brings this research into the realm of AI work in uncertain reasoning. Since we are investigating argumentation as a general approach to reasoning, it is worthwhile to briefly mention similarities and differences between this approach and other, better established approaches to uncertain reasoning. (Helpful surveys of these approaches are Dubois and Prade (1989) and Lea Sombe (1990).)

Argumentation is a general approach to uncertain reasoning that expects inconsistency, in addition to, indeed perhaps as a result of, incompleteness and uncertainty in the knowledge, and handles it by generating arguments both for and against a claim. An argumentation approach emphasizes justifying decisions, rather than decision making per se. Justification for a claim includes both providing support for a claim and defending it with respect to support for other plausible claims.

In contrast, other work in uncertain reasoning has concentrated on uncertainty per se, rather than on incompleteness or inconsistency. Also, the emphasis is on decision making, e.g., by choosing the claim with the highest certainty, rather than on decision justification. If justification is included, it consists of knowledge that supports the claim, but ignores weak points in the support and support for alternative claims (e.g., Porter, Bareiss, & Holte, 1990).

In our theory, argumentation is independent of the various approaches to reasoning under uncertainty. Some method for representing, combining, and propagating uncertainty is presumed, but it can be numeric or symbolic, probabilistic or not, and so on. For the implementation of our theory, we have invented a simple, symbolic representation of uncertainty and an algorithm for propagating it throughout an argument.

This method captures uncertainty introduced by non-deductive reasoning types, as well as uncertainty in the knowledge itself.

In sum, the current work, along with other AI work in reasoning under uncertainty, assumes the existence of uncertainty in the domain knowledge. Uncertainty is broadly defined to include incomplete and inconsistent knowledge. Argumentation is seen as a useful, general technique for deriving and justifying claims in light of the problems that arise from reasoning in weak theory domains, particularly inconsistency.

7 Conclusions

We have presented a formal theory of argumentation as a method for providing decision support and justification for plausible reasoning in weak theory domains. Two senses of argument, argument as supporting explanation and argument as dialectical process, were identified. The complete theory is to include a structure for representing and an algorithm for generating each of these types of argument. We described a partial implementation of the theory, a program that generates supporting arguments for a given claim and counter-claim. The program uses plausible reasoning (e.g., abductive reasoning) in addition to deductive reasoning (i.e., modus ponens and modus tollens). Hypothetical reasoning is also incorporated to point out incompleteness in the knowledge. A simple method for representing, combining, and propagating the uncertainty introduced by plausible reasoning is given; the same method is used to represent uncertainty in the knowledge base. The algorithm itself can be combined with any general method for handling uncertainty. Inconsistency in the knowledge base is expected, so the program always attempts to generate support for both a claim and its negation. As yet, there is no way to directly compare support for alternative claims. We gave examples of program output for a set of paradigmatic problems from weak theory domains.

Next, we plan to design and implement a formal theory of argument as a dialectical process. The next stage will use the format of a two-sided argument to intertwine the strengths and weaknesses of support for competing claims so they can be directly compared to one another. We plan to identify additional moves (e.g., see Rissland, 1985) for dialectical arguments, as well as heuristics for their selection, in order to support and control the generation of the arguments described in this paper.

For example, as noted in example 4, the pro side could present its argument for rain, that the grass is wet and rain is a cause of wet grass. But the con side can undercut this argument by pointing to an alternative explanation for the wet grass, namely, that sprinklers also cause wet grass, and the sprinkler was indeed on. Pearl (1988) points out that this argument move, "explaining away", should be viewed as another type of plausible inference.

In example 3, though the conclusion flies is highly likely based on the given data, bird, an arguer would look for the weakness in the argument, and seek verification that the bird is not a penguin. Hypothetical lines of reasoning will also be explored, to point out additional knowledge that, if available, could bolster a claim or resolve an apparent inconsistency. Hypothetical arguments can be based upon the contingent (darkened, in our figures) aspects of arguments. These contingent elements could serve as the basis for the generation of information gathering goals, if the argument process were embedded in an entity that solves problems in weak theory domains.

Issues that must be addressed in future work include (i) identifying a structure for dialectical argument; (ii) identifying a process for generating this structure that permits intertwining the tasks of supporting and defending a claim, along with rebutting and undercutting competing claims; (iii) identifying argument moves and heuristic strategies for controlling the argument generation process.

8 Acknowledgements

This research was partially supported by a University of Oregon Ph.D. Fellowship.

References

Ashley, K. (1989). Toward a computational theory of arguing with precedents. *Proceedings of the Second International Conference on Artificial Intelligence and Law,* 93-102.
Buchanan, B. & Shortliffe, E. (1984). *Rule-based expert systems.* Reading, MA: Addison-Wesley.
Conklin, J. (1988). Design rationale and maintainability (Report STP-249-88). Microelectronics and Computer Technology Corporation, Austin, TX.
Cox, J. R. & Willard, C. A. (Eds.). (1982). *Advances in argumentation theory and research.* Carbondale, IL: Southern Illinois University Press.
Dubois, D. & Prade, H. (1989). Handling uncertainty in expert systems - pitfalls, difficulties, remedies. In E. Hollnagel (Ed.), *The reliability of expert systems* (pp. 64-118). West Sussex: Ellis Horwood Limited.
Flowers, M., McGuire, R., & Birnbaum, L. (1982). Adversary arguments and the logic of personal attacks. In W. Lehnert & M. Ringle (Eds.), *Strategies for natural language processing* (pp. 275-294). Hillsdale, NJ: Erlbaum Associates.
Freeley, A. (1990). *Argumentation and debate: Critical thinking for reasoned decision making* (7th ed.). Belmont, CA: Wadsworth Publishing Company.
Goldszmidt, M. & Pearl J. (1991). On the consistency of defeasible databases, *Artificial Intelligence,* 52, 121-149.
Hample, D. (1982). Modeling argument. In J. R. Cox & C. A. Willard (Eds.), *Advances in argumentation theory and research* (pp. 259-284). Carbondale, IL: Southern Illinois University Press.
Horner, W. (1988). *Rhetoric in the classical tradition.* New York, NY: St. Martin's Press.
Kahane, H. (1988). *Logic and contemporary rhetoric* (5th ed.). Belmont, CA: Wadsworth Publishing Co.
Lea Sombe (1990). Reasoning under incomplete information in artificial intelligence. *International Journal of Intelligent Systems,* 5, 323-472.
Lee, J. (1991). DRL: A task-specific argumentation language. *Proceedings of the AAAI Spring Symposium on Argumentation and Belief,* 122-132.
Levi, E. (1949). *An introduction to legal reasoning.* Chicago, IL: University of Chicago Press.
Marshall, C. (1989). Representing the structure of a legal argument. *Proceedings of the Second International Conference on Artificial Intelligence and Law,* 121-127.
Pearl, J. (1987). Embracing causality in formal reasoning. *Proceedings of the Sixth National Conference on Artificial Intelligence,* 369-373.
Pearl, J. (1988). *Probabilistic reasoning in intelligent systems.* San Mateo, CA: Morgan Kaufmann Publishers.
Pollock, J. (1987). Defeasible reasoning. *Cognitive Science,* 11, 481-518.
Polya, G. (1968). *Mathematics and plausible reasoning* (2nd ed.) (vol. II). Princeton, NJ: Princeton University Press.
Poole, D. (1989). Explanation and prediction: an architecture for default and abductive reasoning. *Computational Intelligence,* 5, 97-110.
Porter, B., Bareiss, R., & Holte, R. (1990). Concept learning and heuristic classification in weak theory domains. *Artificial Intelligence,* 45, 229-263.
Proceedings of the AAAI Spring Symposium on Argumentation and Belief. (1991).
Rescher, N. (1977). *Dialectics: A controversy-oriented approach to the theory of knowledge.* Albany, NY: State University of New York Press.
Rissland, E. (1985). Argument moves and hypotheticals. In C. Walter (Ed.), *Computing power and legal reasoning.* St. Paul, MN: West Publishing Co.
Rissland, E. (1989). Dimension-based analysis of hypotheticals from Supreme Court oral argument. *Proceedings of the Second International Conference on Artificial Intelligence and Law,* 111-120.

Storrs, G. (1991). Extensions to Toulmin for capturing real arguments. *Proceedings of the AAAI Spring Symposium on Argumentation and Belief*, 195-204.

Toulmin, S. (1958). *The uses of argument* . Cambridge: Cambridge University Press.

Coordination and Dynamic Organization in the ALM System.

F. Abbruzzese
Dipartimento di Scienze Fisiche, Università di Napoli "Federico II"
Naples, Italy

E. Minicozzi
Dipartimento di Scienze Fisiche, Università di Napoli "Federico II"
Naples, Italy

G. Nardiello,
Dipartimento di Scienze Fisiche, Università di Napoli "Federico II"
Naples, Italy

G. Piazzullo
Istituto per la Ricerca sui Sistemi Informatici Paralleli
Naples, Italy

Abstract

Coordination and organization of independent interacting agents are basic problems when dealing with complex AI applications. Global control, hardwired organization, and by-name interaction, while ensuring coordination, are in conflict with the flexibility and adaptability required by such applications. Distribution of control, dynamic organization, and pattern-directed interaction make better use of available knowledge and fulfil flexibility and adaptability requirements. However, coordination difficulties are increased.

The ALM paradigm, while keeping all advantages of pattern-directed interactions, provides tools for achieving coordination and for building dynamic, flat organizations. ALM is based on a physical metaphor: agents, called Active Entities, interact with "fields" created by other Active Entities and possibly "collide". "Collisions", which constitute a powerful synchronization mechanism, allow Entities to reach any kind of safe agreement; "fields" give to the Entities the "feeling" of the context they are in, so that Entities can specialize in order to operate in that context. Dynamic organizations are built up by grouping Entities into private environments called Computational Islands that may interact as single agents.

1 Introduction

There are basic problems that any proposal of a DAI (Bond & Gasser, 1988a) architecture, characterized by a massive distribution of control, has to address. These problems arise from the conflict between the locality of decisions, behaviours and knowledge and global requirements concerning coordinated actions (Bond & Gasser, 1988b; Gasser, 1991). Further difficulties arise when the distributed system is also an Open System (OS)

(Gasser, 1991; Hewitt, 1985). On one hand, traditional tools for obtaining coordination, like central control, hierarchical architectures, global synchronization, predefinite channels of communication, a-priori Knowledge etc., seem to be in conflict with flexibility and adaptability requirements of OSs. On the other hand, continuous evolution, lack of global control, asynchronicity, incomplete knowledge and delayed information, that characterize OSs, increase coordination difficulties. Constraining the interaction among computational agents improves coordination, but it may affect the flexibility of the overall system. In the Actor System (Agha, 1981; Hewitt, 1977) the possible interactions that an actor may have are constrained by its acquaintances, that are the addresses of all its possible interlocutors. These acquaintances may change as the system evolves, so relations among actors are not hardwired, but may be dynamically defined to meet adaptability requirements. Both dynamic organizations and various mechanism of focused interaction (Bond & Gasser, 1988a) may be implemented, in this way.

However the strict encapsulation of knowledge and of procedures within an actor, and the by-name basis on which interactions take place is in conflict with OSs flexibility requirements. In fact, Open Distributed Systems (ODS) suggest that one should reach an adequate compromise between privacy and availability of knowledge and that one should get rid of names. Patterns directed invocation (Hewitt, 1972) and modifiable encapsulation appear some of the main features that ODSs ask for.

In this paper we show how it is possible to achieve coordination and dynamic organization in the ALM model of computation discussed elsewhere (Abbruzzese & Minicozzi, 1991, in press). ALM is both a computational model and a programming language (Abbruzzese, 1991). In ALM the interaction is pattern directed and the "boundaries" of single agents or group of agents may be dynamically "moved" to regulate what "can be seen" from outside. ALM pattern-directed interaction primitives make knowledge available not only to specific interlocutors but to whichever may need it. Moreover, focused interaction results from agents "feeling" the context they are in and evolving consequently. The synchronization power of an ALM interaction primitive called "Symmetric Influence" enables computational agents to achieve coordination without either global coordinators or a-priori knowledge of the environment. Dynamic organizations are built up by grouping agents into private environments called "Computational Islands" (CIs) that may interact as single agents.

2   Overview of ALM

A computation in ALM is an evolution of interacting Active Entities (in what follows we will use the terms Active Entity, ALM Entity, and Entity as synonymous). Such evolution consists in the continuous death and creation of new Active Entities. An ALM Active Entity is a data structure that evolves according to some prescriptions defined inside it. It is a completely autonomous chunk of knowledge which, inside it, has no explicit reference to other Entities (other agents unique identifiers). Entities interact by means of Influences attached to them that affect the internal structure of other Entities according to compatibility constraints.

Unlike messages, Influences persist or change as the Entities which they belong to persist or change.

An ALM Entity is composed of a Boundary and of a Body. The Boundary has two components: the Influencer and the Filter. The Influencer, in turn, is composed by many Simple-Influencers. Each Simple-Influencer encodes an Influence that each Entity may exerts on the others. This Influence may be either Asymmetric or Symmetric (see later). The Filter is composed of many Simple-Filters. Each Simple-Filter encodes a class of Influences which the Entity is sensitive to. An Influence may affect an Entity only if it is compatible with one of its Simple-Filters.

The Body of an Entity is the component of the Entity that encodes all its possible evolutions as a function of its structure, and, in turn, the structure of an Entity results from the influence possibly perceived. ALM Entities may evolve also in absence of interactions; we call this phenomenon "free-evolution". The result of an ALM Entity evolution is the creation of new Entities and, finally, its death.

An ALM entity defines a private environment of computation. Data, procedures, and control are all local; only the interaction language is supposed to be shared by all Entities. The access to each Entity private environment is protected by the Entity Filter which defines the possible "intrusions" and their destinations.

In the actual implementation of ALM, Simple-Influencers and Simple-Filters are represented by lists of attribute-value pairs. Attributes denote places in the Entity structure, and values may be either constants or constraints ("unbound" values). In this representation a Filter is compatible with an Influencer if one of the Simple-Filters matches one of the Simple-Influencers of the Influencer. Such a matching occurs if all the attributes of the Simple-Filter have corresponding attributes in the Simple-Influencer and the values of this latter match with the values of the former. An Influence of an Entity Ei may "pass through" a Simpler-Filter of an entity Ej, only if Ei is compatible with Ej. If an Influence of Ei "passes through" a Simple-Filter of Ej, the places within Ej, specified by the Simple-Filter attributes with unbound values, are filled with the values of the corresponding attributes in the Influence.

In ALM there are two primitive interactions called respectively Asymmetric Influence and Symmetric Influence. If an Entity Ei has an Asymmetric Simple-Influencer that matches a Simple-Filter of an entity Ej, then Ei may asymmetrically influence Ej. If the Influence of Ei passes through the Filter of Ej then Ej is said to be asymmetrically influenced by Ei. Like a "physical field" Asymmetric Influence may affect several Entities, thus ensuring knowledge availability. Moreover, Asymmetric Influence is ALM main tool for focusing interactions among Entities, as it makes Entities "feel" the context they are in and specialize their behaviour to that context.

An entity Ei can symmetrically influence an Entity Ej if it has a Symmetric Simple-Influencer compatible with the Filter of Ej and, in turn, Ej has a Simple-Influencer compatible with the Filter of Ei. Moreover, if the Influence of Ei "passes through" the Filter of Ej, the influence of Ej has to "pass through" the Filter of Ei. Thus, Symmetric Influence constrains the interaction to be two-way. Like a collision among "physical particles" it makes two Entities affect each other in a single event, thus ensuring coordination.

The following fairness (Francez, 1986) axioms govern interactions within an ALM system:

1. If an entity Ei can influence an entity Ej and this condition persists then Ei will eventually influence Ej;
2. If there is no Entity that can influence an Entity Ej and this condition persists then a free evolution of Ej will eventually take place.

Summing up, each ALM Entity, after being created, affects the others trough its Influence, possibly is subject to the Influence of other Entities, evolves according to its Body, possibly creates new Entities and finally disappears.

3   Coordination and symmetric influences

ALM attains an high flexibility because ALM Entities are not constrained either by a pre-defined organization or by a pre-defined flow of control. Interlocutors are not selected on a by-name basis but by means of patterns that express their features in a conventional language. The use of a conventional language instead of names allows a better use of available knowledge (procedures and partial results) and makes the system robust with respect to changes. Both pattern-directed interaction and persistency of Influences make possible to represent the overall state of computation as a group of quite independent self-explanatory "chunks of knowledge", allowing new incoming data to be easily accommodated in the pre-existing computation. This makes ALM robust with respect to unscheduled events.

However, we notice that systems based on Pattern-directed interaction, suffer of significant coordination problems. In market-like paradigms (Bond & Gasser, 1988a) like Negotiation (Smith & Davis, 1991; Davis & Smith, 1983) such problems are solved by a hierarchical organization of tasks. In this way each group of agents working on the same task has its local coordinator. On the contrary ALM attains coordination without imposing hierarchical constraints and without renouncing to pattern-directed interaction. This result is achieved by exploiting Symmetric Influence that makes two Active Entities reach safe agreements by engaging them in a simultaneous exchange of information. The simultaneity in the information exchange prevents other Active Entities from interfering while the agreement is being reached.

The coordination power of Symmetric Influence can not be obtained by message-passing at acceptable cost when communication is pattern directed and there is not an underlying organization. In order to show this let us consider the following example. After a bloody defeat, troops are trying to come back to their lines. They know that the better thing to do is to group into couples because two soldiers may help each other while groups of more than two soldiers are likely to be discovered by the enemy. Unfortunately, all officers are dead so troops can not rely on the military hierarchy to form the couples (absence of a predefined organization). Moreover soldiers do not know each other names (no knowledge of other agents unique identifiers) and have no upper estimate of their number (no global knowledge of the environment). Now let us suppose that the only available communication mechanism is message-passing; messages are ensured to reach their targets but nothing can be said about their time of arrival. For instance, let us suppose that each soldier has a walkie-talkie and that permission to transmit is eventually given to all soldiers needing it. Under these hypotheses the only thing that each soldier can do is to broadcast a message like this: "My name is ..., I am looking for a companion". Soldiers may answer to these messages proposing themselves as companions. Due to the fact that messages are not issued to specific soldiers but are broadcast, several soldiers may submit a proposal to the same interlocutor. As a consequence proposals are not ensured to be accepted, and each soldier has to try different interlocutors. It is possible to prove that, in the worst case, organizing soldiers into couples requires the exchange of at least $\Omega(N^2)$ messages, where N is the number of soldiers. Thus, no gain in speed with respect to a sequential solution of the problem is ensured because the average number of operations performed by each agent may be $\Omega(N)$. Moreover, it may happens that a soldier will spend all its life looking for a companion because he has no way to realize to be alone.

Formally, the above lower bound holds under the following three hypotheses: 1) the delivery times of messages are unbounded and uncorrelated; 2) the agents that are required to couple are completely autonomous. The first condition says that no bounds can be given for the times needed to delivery messages and that there is an enough weak dependence of the time needed to delivery a message, from the state of the part of the system that is analyzed; a formal definition of the above condition is given in the appendix. The formal definition of complete autonomy is:

Definittion of Complete Autonomy. In a distributed system R we say that the agents of a set E are autonomous with respect to the set of system states S iff for all $s \in S$ the states obtained by modifying s with the subtraction or addition of agents in E are always in S.

Roughly, the above definition says that the states of the agents of E have to be completely uncorrelated, that is, no agent have know of the others.

If we consider not only that the number of soldiers is unknown, but also that new soldiers may continuously arrive, then it can be proved that the interference of the incoming soldiers may force the initial soldiers to exchange an unlimited number of messages. A significative improvement can be obtained only by renouncing to pattern-directed communication and choosing adequately the acquaintances each soldier has. In fact, limiting the possible interlocutors of each soldier, we define an organization that may be exploited to define the couples. As a limit case, the couples can be implicitly defined in the organization.

The high cost of agreement processes, when interaction among agents does not rely on an underlaying organization and is pattern-directed, suggests to define the agreement between two agents as a new interaction primitive: in ALM we have made this choice. In this way the problem can be dealt with at hardware level, where it can be solved more effectively by using ad hoc techniques (Abrruzzese, Nardiello, & Piazzullo, 1992). The agreement needed to form all couples is easily achieved by Symmetric Influence. In fact, if each soldier specifies in its Influencer its name and the fact that it is a soldier, and specifies in its Filter that it needs a soldier with any name, then Symmetric Influence causes couples of soldiers exchange their names in a single event, thus forming a couple. New incoming soldiers can not interfere with this "agreement" process as it requires just one transaction. Moreover, if the number of soldiers does not change, a soldier may realize to be alone exploiting the free-evolution feature. The coordination power of Symmetric Influence is connected to its capability to force two local choices to be coordinated: two active Entities that interact by Symmetric Influence make correlated choices of the Influences that "pass through" their Filters.

Symmetric Influence attains an immediate agreement between two of N entities. In order to cope with agreements among more than two entities we have defined a new interaction operation that generalizes Symmetric Influence to an interaction among an arbitrary number of Active Entities. We have called it Aggregation, because it plays a fundamental role in resource aggregation problems. Aggregation makes an arbitrary number of Active Entities, that satisfy compatibility constraints encoded in their own structure, reach an agreement. Aggregation is required to satisfy the following fairness condition: if Active Entities A1...An satisfy the constraints required for an aggregation to occur and this condition persists, than an aggregation involving all of them will eventually take place. Aggregations, can be implemented at a low computational cost in ALM by using both Symmetric Influence and the free-evolution feature, but discouraging lower bounds hold not only for message-passing but also for the more powerful communication primitives of Linda (Carrero & Gelernter, 1989). More specifically, if N is the number of agents involved in the aggregation algorithm, then a lower bound of $\Omega(N^2)$ holds for the number of communication needed by the algorithm in both message-passing and Linda-like systems.

Finally, ALM free-evolution feature allows the solution of certain problems, that we have called Global Problem, that can be proved to be not solvable with both message- passing and Linda-like systems. Examples of such problems are: 1) counting the number of the completely autonomous agents that satisfies a given condition; 2) electing a unique leader within a set of completely autonomous agents; an instance of the first of these problems is contained in the example described in section 5. Global problems are characterized by the fact that acceptable solutions do not preserve their validity when the environment is enriched with other agents.

4 Dynamic organization and Computational Islands

Organization constrains the interactions among agents and provide a framework of expectations about the behaviour of agents (Bond & Gasser, 1988a). In this way organization reduces the "articulation work" (Bond & Gasser, 1988a) needed for coordinating agents. However, hardwired organization may affect heavily the flexibility of the system. In contrast, in ALM, Computational Islands allow the dynamic definition and modification of many kind of organization.

By means of CIs, ALM computation is organized into several private interacting environments: only Entities belonging to the same CI can directly interact, while Entities belonging to different CIs interact only by means of interface Entities, called Sensors and Effectors. Roughly speaking, CIs factor out features common to the boundary of several independent Entities. They may be seen as high level Entities with Sensors and Effectors as their"boundaries". Sensors and Effectors are ALM entities that regulate communication inward or outward the CIs. More precisely, Sensors of a CI are Entities that do not belong to the CI but may generate Entities belonging to the CI; Effectors of a CI are Entities that belong to the CI and that may generate Entities that do not belong to the CI. Several CIs can perceive events of some other specified CI through the same Sensor and, by the same Effector, a CI can affect several CIs. Since the overall ALM environment is structured as a set of CIs, Sensors of a CI are Effectors of some other CI and Effectors of a CI are Sensors of some other CIs.

A CI is represented by a new components of the Boundary called Island, that, in the actual implementation of ALM, is a list of identifiers. If these identifiers are declared *hereditary*, Entities belonging to the relative CI can be constrained to evolve into Entities belonging to the same CI, otherwise Entities can migrate from a CI to another one. The creation of a CI may be the consequence of many events. For instance, in a distributed problem solving scenario experts needing to interact tightly may agree on creating a CI to protect their work from "unwanted distractions". It may also be the case that a single expert needs to organize its own work as a distributed problem-solving activity, thus creating several CIs and populating them with experts and tasks.

The implementation of dependencies among distinct CIs, requires restricting the access to the CIs, and transforming incoming Entities in order that they satisfy the prescriptions of the CI accepting them. Sensors and Effectors make that possible. Indeed, they work as selective communication channels that transform the information produced in one CI, in such a way that it can be correctly used in another CI. Finally since Sensors and Effectors are ALM Entities, they may dynamically change, fulfilling also reconfigurability requirements.

Unlike hierarchical organizations (Bond & Gasser, 1988a), where the flow of information and control is constrained and defined by the hierarchy, CIs architecture is flat and allows direct interaction among all CIs needing it, via the definition of adequate Sensors and Effectors. Moreover, when CIs do not "know" each of the others, they may interact through publicly known CIs containing some of their Sensors. Publicly known CIs are like Blackboards (Carrero & Gelernter, 1989; Hayes-Roth, 1985; Lesser, Fennel, Erman & Reddy, 1975) in Shared Memory systems, with the relevant difference that they are populated by active modules rather than by passive data.

5 An example

In this section we give an example concerning the evaluation of arithmetic expressions. This example is not intended for demonstration of arithmetic problem-solving but for an illustration of the ALM concepts discussed so far. In the ALM solution both arithmetic expressions and numbers are represented as ALM Entities and the overall evaluation process is represented by interactions among and within adequate CIs. The associativity and commutativity of arithmetic operators are exploited for increasing the parallelism of the computation: data are processed as soon as they are available without caring about the order in which they occur in the expressions, as the order in which operations are carried out does not affect the final result. Finally, the computation is aborted as soon as an error is detected.

The structure of an arithmetic expression implicitly defines constraints on the evaluation process. These constraints govern the creation and the definition of CIs with their Sensors and Effectors. For instance, the correct evaluation of (* (+ 1 2 3) (+ 7 8 9)) implies that the processing of (+ 1 2 3) must be protected from the one of (+ 7 8 9) and that the processing of "*" must follow both of them, hence in this case, three CIs are created. In two of them respectively 1, 2, 3 and 7, 8, 9 interact to yield their sums. In the other one the results furnished by the two previous calculations interact to yield their product. Instead, if the expression to be evaluated is (+ (+ 1 2 3) (+7 8 9)) all numbers will belong to the same CI, as this expression is equivalent to (+ 1 2 3 7 8 9).

Fig. 1 shows how the various CIs are created. There, a publicly known CI contains a database of operators and all expressions that are created during the computation. The Asymmetric Influence of the operators supplies each expression with all the information it needs about its main operator. After that, each expression splits in its arguments that become Entities of a newly created Computational Island. For instance in Fig. 1 "expression looking for operator a" creates the Computational Island CI2, and populates it with its arguments. If the operator is associative and commutative its arguments are inspected to detect situations like the one in (+ (+ 1 2 3) (+7 8 9)), and, if this is the case, the expression arguments split within the CI where they were created without evolving in "expression looking for ..." and without creating new CIs. However, in general, sub-expressions are processed in new CIs, and only the results returned by their evaluations take part in the processing of the CIs they come from. Fig. 1 shows that "expressions " are Effectors in that they create new "expressions looking ...." in the publicly known CI. In Fig. 1 an error is detected during the processing of "expression looking... c". Accordingly, "expression looking... c" creates an "Error" Entity that asymmetrically influences the "Error sensors" of all other CIs. These "Error sensors", in turn, abort the computation in their respective CIs by creating killer Entities.

Let us analyze how numbers are processed within the CIs they belong to. Fig. 2 shows the computation relative to a "+" operator. Numbers to be summed are created by previous computations and each of them appears at an unscheduled time. The end of such flow of Entities is marked by the creation of a "STOP" Entity (we will explain later how this happens). In Fig. 2a Entities representing the numbers couple due to a Symmetric Influence. After that, one element of each couple dies and the other changes into an entity of the same kind representing the sum of the two integers. In turn, the just created entities takes part in the sum process.

When the "STOP" entity is created, its Asymmetric Influence makes the others know that the flow of new entities has been halted, as shown in fig. 2b. After the change Entities continue to sum until just a single Entity remains. At this point, this Entity can not be influenced by other entities, thus it evolves (a free-evolution takes place) into an "Outgoing

Result".

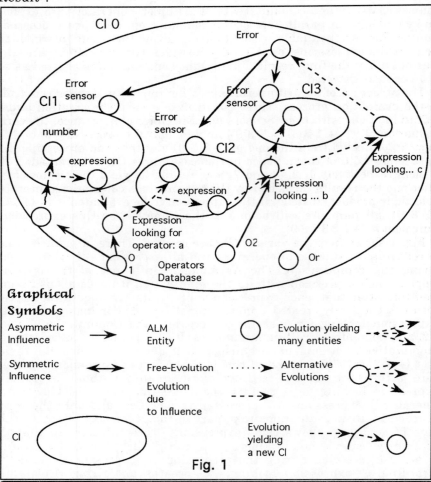

Fig. 1

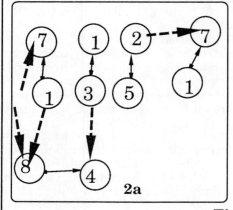

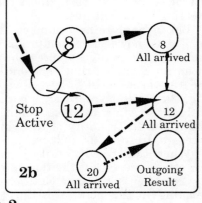

Fig. 2

No Entity waits for a specific partner but only for any other Entity representing a number to be summed. Therefore the flow of control is dynamically determined and the computation can easily accommodate inputs whenever they come. A by-name communication would constrains each number to wait for its pre-defined partner even if it could have fruitfully interacted with some other number already arrived in the CI.
Fig. 3 shows how the CIs exchange partial results and how the "Stop" Entity is created. The "Outgoing Result" Entities cross the limits of their CIs by interacting with Effectors that inform them on the "conventions in use" in their target CIs, thus creating the "Incoming Result" Entities. The "Incoming Result" Entities interact with their previously created "Placeholder" Entities that, when the operator is not associative, inform them of their "role" in the expression (for instance, in the case of a division, if the number is the divisor or the dividend). In this example the "conventions in use" may be identified roughly with the definitions of the arithmetic operation that are being carried out in the target CIs. As a consequence of the interaction "Placeholder" Entities are destroyed. "Placeholder" Entities are created when an "expression" evolves into an "expression looking...".

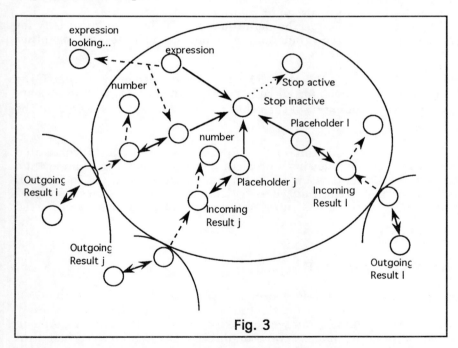

Fig. 3

The absence of both "Placeholder" and "expression" Entities in a CI implies that all the "Incoming Result" Entities have already arrived. The "Stop inactive" Entity is able to "feel" this situation, because its "free-evolution" has been blocked by the Influences of "Placeholder" and "expression" Entities. In fact, the evolution of an Entity may consist just in that Entity remaining unchanged. Therefore, it may be the case that an Entity remains unchanged as long as there are ohter Entities influencing it.

6 Conclusions

In this paper we have shown how ALM faces the coordination and organization problems arising in distributed systems based on pattern-directed interaction.

Symmetric Influence is a powerful tool for coordinating agents, as it allows two pattern-selected Entities exchange information in a single event, achieving agreements at a low computational cost. Computational Islands, with their Sensors and Effectors allow the definition of complex and flexible organizations that may be dynamically created and changed.

7 Appendix

We use the formalism, introduced by Halpern and Moses (1990), to model distributed systems. A distributed system is characterized by the set R of all possible runs. A point (r, t) is a description of the state of the system at time t during run r. A run r' is said to extend a point (r, t) iff the states of the system in the two runs r' and r coincide for all times t'<t.

For simplicity of exposition the definitions that follow are given only for message-passing systems but they can be extended to include other communication primitives.

Definition A1. A system R is said to have an unbounded delivery time if the following conditions hold: for all runs r and all times t, t_1, with $t \leq t_1$, there exists a run r' extending (r, t) such that no message is received in r' in the interval [t, t_1].

Definition A2. We say that a distributed system R has uncorrelated delivery times iff the following statement holds: If at point (r, t) an agent e receives a message c, transmitted by an agent e' at point (r, t'), then for all runs r' in which c is transmitted at (r', t'), and e' has the same history as in r up to time t, there is a run r'' extending (r', t') in which c is received by e at time t.

Acknowledgements

Support for the research reported in this paper was provided by the CNR under contract 90.00406.PF67 of the Progetto Finalizzato Robotica.

ALM paradigm arose from various experiences of the authors. Among these we want to emphasize the stimulating conversations we had with Giuseppe Trautteur. The progress and the experimentation we had with ALM would have been impossible without the work of many people. Among these we are particularly grateful to Giampiero Longobardi, Alberto Genovese, and Massimo Miele, who worked on the implementation of ALM.

References

Abbruzzese, F., (1991).*The ALM Reference Manual* (Research. Rep.). Napoly, Italy: Dipartimento di Scienze Fisiche, Università di Napoli.

Abbruzzese, F., & Minicozzi, E., (in press). ALM: a Parallel Language Aimed at Massive Distribution of Control. Journal of Parallel and Distributed Computing.

Abbruzzese, F., & Minicozzi, E., (1991). Direct Interaction Among Active Data Structures: a Tool for Building AI Systems, In E. Ardizzone, S. Gaglio, F. Sorbello (Eds.), Proceedings of the 2th International Conference of the Italian Association for Artificial Intelligence, (pp. 365-374), Germany: Springer-Verlag.

Abbruzzese F., Nardiello, G., & Piazzullo G. (1992). The Implementation Of New Communication Primitives For DAI Systems (Research Rep.), Napoli,Italy: Dipartimento di Scienze Fisiche, Università di Napoli.

Agha, (1981). Actors: a Model of Concurrent Computation in Distributed Systems. Cambridge: The MIT press.

Bond, A. H., & Gasser, L., (1988a). An Analysis of Problems and Research in the DAI. In A.H. Bond and L. Gasser (Eds.), Readings in:Distributed Artificial Intelligence (pp. 3-35). San Mateo, California : Morgan Kaufmann Publ. Inc.

Bond, A. H., & Gasser, A. H., (Eds.). (1988b). Readings in Distributed Artificial Intelligence. San Mateo, California: Morgan Kaufmann Publ. Inc.

Carriero, N., & Gelernter, D., (1989). How to Write Parallel Programs: a Guide to Perplexed. ACM Computing Surveys , 21, 323-357.

Carriero, N., & Gelernter, D., (1989). Linda in Context. Communications of the ACM, 32, 444-458.

Davis, R., & Smith., R. G., (1983). Negotiation as a Metaphor for Distributed Problem Solving. Artificial Intelligence, 20, 63-109.

Francez, N., (1986). Fairness. New-York: Springer-Verlag.

Gasser, L., (1991). Social Conception of Knowledge and Action: DAI Fondations and Open System Semantics. Artificial Intelligence, 47, 117-138.

Halpern, J. Y., Moses, Y., (1990). Knowledge and Common Knowledge in a Distributed Environment. Journal of the ACM, 37, 549-587.

Hayes-Roth, B., (1985). A Blackboard Architecture for Control. Artificial Intelligence, 26, 251-321.

Hewitt, C., (1972). Description and Theoretical Analysis (Using Schemata) of PLANNER: a Language for Proving Theorems and Manipulating Models in a Robot (AI-TR-258). Cambridge, Ma: MIT Artificial Intelligence Laboratory.

Hewitt, C., (1977). Viewing Control Structures as Patterns of Passing Messages. Artificial Intelligence, 8, 323-364.

Hewitt, C. (1985). The Challenge of Open Systems. Byte, 10, 223-242.

Lesser, V. R., Fennel, R. D., Erman, L. D., & Reddy, D. R., (1975). Organization of the HERSAY II Speech Understanding System. IEEE Transactions on. Acoust.ics, Speech, Signal Processing, 23, 11-24.

Smith, R. G., & Davis, R., (1981). Frameworks for Cooperation in Distributed Problem Solving. IEEE Transactions System, Man, and Cybernetics, 11, 61-70.

Session 2: Cognitive Foundations

The Nature and Development of Reasoning Strategies [1]

Ruth M.J. Byrne
Department of Psychology, Trinity College, University of Dublin,
Dublin, Ireland

Simon J. Handley
School of Psychology, University of Wales College of Cardiff,
Cardiff, Wales, UK

Abstract

Human reasoners can reason from problems they have not encountered before and their creation of new chains of inferences is informative for the construction of artificially intelligent theorem provers that learn to reason in novel ways. We argue that human reasoners generate strategies to guide their sequences of inferences to the solution of novel or complex problems. We propose that human reasoners develop strategies by constructing a new strategic component based on features of a problem, and they assemble a new strategy by combining new and existing strategic components in novel ways. A new strategy can emerge, provoked by a new sort of problem that the existing strategies cannot solve. According to this view, reasoners can spontaneously improve their reasoning skills with practise because they gain expertise not only at implementing the individual parts of a strategy, but also at assembling the parts together efficiently. We report experimental results that support these proposals.

1 Novel inferences

How can a reasoning system -- be it a natural or an artificial system -- improve it's ability to cope with novel and complex inferences that it has never encountered before ?

---

[1] This research was supported by an ESRC grant, reference R-000-23-2491, awarded to Ruth Byrne. We thank Mark Keane and Phil Johnson-Laird for helpful comments, and Alan Milne for writing the program to run the experiment.

The answer to this question lies, we believe, in the development of *meta-inferential strategies*. Reasoners rely on strategies, such as analysing the structure of a problem in advance of any attempt to reason about it, or examining the source of the information in the problem, in order to guide the inferences that they subsequently make.

For example, consider a reasoning system that can make deductive inferences such as the following:

A or B, or both
Not A
Therefore, B

The system would not be able automatically to scale-up to make meta-deductive inferences of the following sort:

A asserts: *A or B or both*
B asserts: *Not A*
Therefore...

unless it contained additional machinery to deal with the source of information, that is, machinery to make an inference from the truth or falsity of an assertor to the truth or falsity of his or her assertion. We propose that this machinery consists in higher-level control strategies that enable reasoners to chart a path through a maze of inferences. We discuss in this paper the nature of these control strategies and their development. Since the human reasoning system manages to cope flexibly with novelty and complexity, we will rely on insights into the nature of human reasoning to guide our proposals about artificial reasoning systems.

Human reasoners construct strategies to solve problems of various sorts (Eysenck & Keane, 1990), and to organise strategies meta-cognitively (Berry, 1983). The study of human reasoning *strategies* has been confined to the choice of a representational format, such as spatial or linguistic strategies (Sternberg, 1980), or to superficial heuristics to pre-empt reasoning (Evans, 1989). The strategies we propose are higher-level control strategies that marshal the basic inferential mechanisms into a coherent structured chain to solve a problem (Kotovsky & Simon, 1989; Newell, 1990).

In the domain of problem solving, researchers have successfully relied on complex problems, such as the Tower of Hanoi, which are relatively remote from everyday experience, and yet tap into the sorts of problem-solving skills necessary for everyday life (Newell, 1990). Likewise, our method has been to examine complex inferences, or *meta-deductions* which may seem artificial, yet allow us to study everyday inferential skills in a controlled situation. We have examined *meta-deductions*, that is, inferences about the truth and falsity of other people's assertions (Smullyan, 1978), such as the following:

Truthtellers always tell the truth, and liars always lie.
Ann asserts: I am a truthteller or Beth is a truthteller, but not both.
Beth asserts: Ann is a truthteller.
Who is a truthteller and who is a liar?

There are a number of ways to work your way through this problem to its solution. Reasoners may begin by supposing that Ann is a truthteller. If so, then her assertion is true, and either she or Beth is a truthteller, but not both. Because they have supposed that Ann is a truthteller, they infer from her assertion that Beth is a liar. If Beth is a liar, then her assertion is a lie, and she says Ann is a truthteller, so Ann must be a liar. The supposition that Ann is a truthteller has led to the conclusion that she is a liar. From the contradiction they can infer that their supposition was wrong: Ann *is* a liar. What about Beth ? Beth said that Ann was a truthteller, so they can infer that she is a liar too.

To solve the problem, reasoners must be able to make deductions from assertions based, for example, on propositional connectives, such as "or", "if", "and", and "not" (Byrne, 1989a; 1989b; Byrne & Johnson-Laird, 1992; Johnson-Laird, Byrne, & Schaeken, *in press*). The study of human reasoning has explored these simple deductions (Braine & O'Brien, 1991; Byrne & Johnson-Laird, 1989; Johnson-Laird & Byrne, 1989; Johnson-Laird, Byrne, & Tabossi, 1989). But, meta-deductions also require a higher level of reasoning. Reasoners must be able to make inferences about the assertors, and reflect on the consequences of their status as truthtellers or liars. They must be able to marshal these inferences into a coherent chain that results in an overall solution about the status of the assertors.

2   Reasoning strategies

We have described some of the strategies that reasoners rely on to make meta-deductions (Johnson-Laird & Byrne, 1990), and we have reported a computational model of these cognitive processes (Johnson-Laird & Byrne, 1991). One strategy that our program relies on is a simple *hypothesise-and-match* strategy. We described this strategy informally when we discussed a possible way to solve the example in the introduction. The structure of the algorithm in our program can be summarised in the following two components:

- **Hypothesise** *that the first assertor is telling the truth.*
    - *Make as many informative inferences as possible from this hypothesis*

- *If it leads to a contradiction, reject the hypothesis and infer that the first assertor is lying. (Otherwise, fail).*
- **Match** *the conclusion about the first assertor with the content of the assertions made by the other assertor(s).*
  - *If they match, infer the new assertor is telling the truth.*
  - *If they mismatch, infer the new assertor is lying. (Otherwise, fail).*

The program relies on many such higher-level strategies. They call on lower-level procedures to construct and describe *models* in order to make inferences based on the propositional connectives, such as "not" and "or". We have discussed elsewhere the nature of the lower-level deductive procedures, and in particular, whether they are based on models, as we suggest, or on inference rules (Johnson-Laird & Byrne, 1991). We will focus here on the higher-level strategies (Byrne & Handley, 1992).

An alternative strategy to solve a meta-deduction such as the example earlier has been proposed by Rips (1989; 1990) and is based on a hypothetico-deductive algorithm. This strategy requires reasoners to hypothesise that the first assertor is telling the truth, and make as many inferences as they can from this hypothesis, and then to hypothesise that the first assertor is telling a lie, and make as many inferences as they can from this alternative hypothesis. For the example problem, reasoners begin in the same way as before, by supposing that Ann is a truthteller, and they reach the contradiction. They infer that their initial hypothesis was wrong, and that Ann is a liar. But, according to Rips's strategy, they do not take a short-cut at this point, and match the conclusion to the content of Beth's assertion, as our simple strategy does. Instead, they must pursue the consequences of the hypothesis that Ann is a liar. If Ann is a liar, then her assertion is a lie: it is false that either Ann or Beth is a truthteller. The negation of the exclusive disjunction is consistent with two states of affairs:

Ann is a truthteller and Beth is a truthteller.
Ann is a liar and Beth is a liar.

Reasoners can rule out the first possibility immediately because they have inferred that Ann is a liar. They infer that Ann is a liar and Beth is a liar as well. Since Beth is a liar, her assertion is false, and so Ann is a liar. They can conclude that Ann and Beth are both liars.

We have implemented such a strategy, called the *full-chain* strategy in our program (Johnson-Laird & Byrne, 1990). The algorithm for the full-chain strategy can be summarised as follows:

- **Hypothesise** *that the first assertor is telling the truth.*
    - *Make as many informative inferences as possible from this hypothesis*
    - *If it leads to a contradiction, reject the hypothesis and infer that the first assertor is lying.*
- **Hypothesise** *that the first assertor is telling a lie.*
    - *Make as many informative inferences as possible from this hypothesis*
    - *If it leads to a contradiction, reject the hypothesis and infer that the first assertor is telling the truth*
- *If a* **consistent assignment** *is reached from both hypotheses make a contingent conclusion; if a contradiction is reached from both hypotheses, the problem is paradoxical.*

The full-chain strategy is a complete strategy that is guaranteed to reach the correct answer for every problem. The simple hypothesise-and-match strategy provides a short-cut through the problem: it obviates the need to make multiple hypotheses, and to keep many alternatives in mind. But, it is a heuristic strategy: it provides the correct answer for some problems, and it fails for others. For example, the strategy fails when no contradiction is reached from the initial hypothesis.

We have shown that human reasoners find problems that can be solved by the simple strategy easier than problems that cannot (Byrne & Handley, 1992; Byrne, Johnson-Laird, & Handley, 1992) even when both sorts of problems can be solved in the same number of steps by the full-chain strategy. These results support our contention that human reasoners rely on simple strategies.

3 The development of strategies

It seems plausible that reasoners construct their simple strategies "on the fly" to suit the particular problems that they encounter, rather than that they come to problems with a set of strategies ready-made in advance. Familiarity with one sort of problem facilitates the development of a suitable strategy (Keane, 1988; 1989; Keane, Ledgeway, & Duff, 1991; Kotovsky & Simon, 1989; Newell, 1990). Reasoners may develop strategies that contain components which are specifically suited to the problem at hand, or they may contain components which are more generally useful.

For example, the hypothesise-and-match strategy that we have described consists of two main components: a *hypothesizing* component, and a *matching* component. The hypothesizing component is a general component which is likely to be useful in the generation of hypotheses in many situations, not only meta-deductive inferences. The matching component is also a general component, which allows reasoners to reason backwards, from information about the truth or falsity of an assertion, to an inference about the truth or falsity of the individual who made the assertion. Such a component is also likely to be useful in many situations aside from meta-deductions.

A new strategy can emerge from several existing ones by re-assembling parts of the existing strategies, provoked by a new sort of problem that the existing strategies cannot solve. A novel problem may also provoke a new component, based on features of the problem. Consider for example, the following problem:

>   *Abe asserts:*   *Ben is a truthteller and Con is a liar.*
>   *Ben asserts:*   *Abe is a truthteller.*
>   *Con asserts:*   *Abe is a truthteller.*

We propose that reasoners who have gained experience with such a problem will develop the following simple strategy (Johnson-Laird & Byrne, 1990). They notice that two of the assertors, Ben and Con, assign the same status to the first assertor, Abe. Thus Ben and Con are either both truthtellers, or both liars. But Abe says only one is a truthteller and the other is a liar. They know immediately that Abe's assertion cannot be true, and they assign the definite status to Abe that he is a liar. They match what they have inferred about Abe to the assertions made by Ben and Con. Since Ben and Con say that Abe is a truthteller, they infer that both Ben and Con are liars.

The algorithm for this *same-assertion-and-match* strategy can be summarised in the following three components:

- *Detect* **same-assertions** *are made by two assertors, who assign a status to a third assertor. (Otherwise, fail).*
    - *If both assign same status, infer both are truthtellers or both are liars.*
    - *If assign different statuses, only one is a truthteller and one is a liar.*
- *Detect the* **third assertors** *assignment of a status to the first two assertors. (Otherwise, fail).*
    - *If third assertor assigns the same status to the two assertors and both inferred to be different status, third assertor is a liar.*
    - *If third assertor assigns a different status to the two assertors and both inferred to be same status, third assertor is a liar. (Otherwise, fail).*
- **Match** *the conclusion about the third assertor with the content of the assertions made by each of the other assertors. (Otherwise, fail).*
    - *If they match, infer the assertor is telling the truth.*
    - *If they mismatch, infer the assertor is lying.*

This strategy contains components which are quite specific to the problem to be solved: the same-assertion part is based closely on one noticeable feature of the problem, that two assertors make the same assertion. Likewise, the second part of the strategy is also based closely on the structure of the problem. These components are well-suited to the problem to be solved, but they are unlikely to be useful for many other sorts of problems. The third part of the strategy is the general matching component we have seen already in operation in the hypothesis-and-match strategy.

The development of a new strategy requires that reasoners not only avail of the individual components -- they must also assemble them together appropriately. If reasoners develop their strategies as they encounter new problems, then they should show an improvement in their ability to get the correct solution to problems, as they gain more experience with the problems. According to this componential view, reasoners should spontaneously improve with practise, because they gain expertise not only at implementing the components, but also at assembling them. We carried out an experiment to examine whether reasoners spontaneously improve with practise on meta-deductions.

We devised two sets of problems, one set based on twelve problems suited to the hypothesise-and-match strategy, and a second set based on twelve problems suited to the same-assertion-and-match strategy. We assigned twenty-four naive subjects to two groups: one group received the hypothesise-and-match problems, and the second group received the same-assertion-and-match problems. Each group received a block of 12 different problems. When they had completed these problems, we gave them a second block of identical problems, so that we could compare their performance on the initial problems, with their performance after practise. The subjects also received other subsequent blocks of problems to examine in detail the development of the strategies, which we discuss elsewhere (Byrne & Handley, 1992).

The problems were presented on a Macintosh Classic computer. The premises appeared on the screen, together with the answer options. An example of the screen presentation is as follows:

Belle asserts: *Jane is a truthteller and Mae is a liar.*
Jane asserts: *Belle is a truthteller.*
Mae asserts: *Belle is a truthteller.*

Belle is a (1) *truthteller*
              (2) *liar*
              (3) *can't tell*

Jane is a   *(1)*   *truthteller*
                (2)   *liar*
                (3)   *can't tell*

Mae is a   *(1)*   *truthteller*
                (2)   *liar*
                (3)   *can't tell*

The subjects task was to decide on the status of each of the assertors. They could highlight an assertor using the arrow-up and arrow-down keys on the keyboard, which allowed them to move from one assertor to another; and they indicated their choice of status for a highlighted assertor by using the "1", "2" and "3" keys. The computer program recorded their choices and their latency to solve each problem.

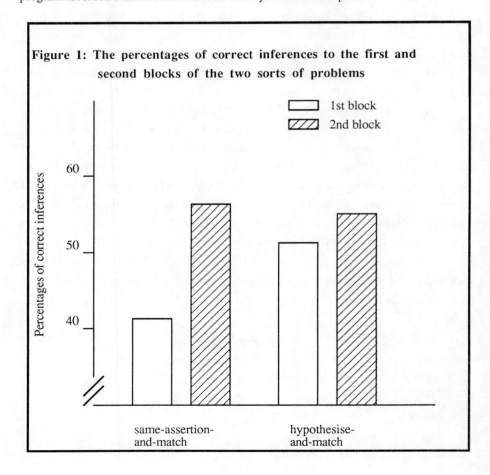

Figure 1: The percentages of correct inferences to the first and second blocks of the two sorts of problems

The results corroborated our predictions that reasoners' performance on meta-deductions spontaneously improves with practise. They made more correct inferences the second time round. The frequency of correct conclusions that the subjects made improved from the first block of problems to the second block of identical problems, for both sorts of problems, as Figure 1 shows (see Byrne and Handley, 1992). The performance of the group of subjects who received the same-assertion-and-match problems improved reliably from the first block (41% correct inferences) to the second block (56% correct). The performance of the group of subjects who received the hypothesise-and-match problems improved only very slightly from the first block (52% correct) to the second block (55%). The two sorts of problems were solved equally often, once reasoners had developed a strategy for dealing with them. As the figure shows, they performed equally well on the second block of hypothesise-and-match problems (55%) and the second block of same-assertion-and-match problems (56%).

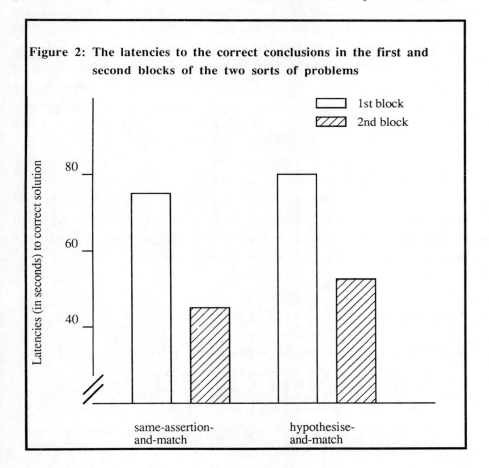

Figure 2: The latencies to the correct conclusions in the first and second blocks of the two sorts of problems

The subjects also became faster at producing the correct conclusion, as Figure 2 shows (see Byrne and Handley, 1992). The group of subjects who received the same-assertion-and-match problems improved reliably from the first block (75 seconds) to the second block (45 seconds). The performance of the group of subjects who received the hypothesise-and-match problems also improved reliably from the first block (80 seconds) to the second block (54 seconds). These results show that reasoners can improve their reasoning skill spontaneously, without feedback. They lend support to our suggestion that reasoning strategies are developed in a piecemeal way.

4 Conclusions

How can we construct an artificial reasoning system that can cope with novel and complex problems which it has not encountered before, and which it has not been programmed to deal with? Human reasoners manage to develop their reasoning skill spontaneously to solve novel and complex problems. We suggest that they do so by relying on a host of simple strategies. These strategies are designed to guide the sequence of inferences through the maze of alternative paths. The strategies consist of components that are constructed "on the fly" and are assembled into a coherent strategy with practise. The components may be general ones, derived from skill with generating hypotheses, for example, or they may be specific ones, derived from detecting unique features of a problem. We have outlined strategies of both sorts, and modelled them computationally. To simulate the flexibility of human reasoning, we suggest that it is crucial to enable new strategies to emerge from existing strategies. If we can gain some insight into constraints on the ways that new strategic components are constructed, and insight into how new and existing components are assembled into a new strategy, then we will gain some insight into the manner in which an artificial reasoning system could be designed to cope genuinely with novel and complex inferences.

References

Berry, D. C. (1983). Metacognitive experience and transfer of logical reasoning. *Quarterly Journal of Experimental Psychology,* **35A,** 39-49.

Braine, M.D.S., & O'Brien, D.P. (1991). A theory of if: a lexical entry, reasoning program, and pragmatic principles. *Psychological Review,* **98,** 182-203.

Byrne, R.M.J. (1989a). Suppressing valid inferences with conditionals. *Cognition,* **31,** 61-83.

Byrne, R.M.J. (1989b). Everyday reasoning with conditional sequences. *Quarterly Journal of Experimental Psychology,* **41A,** 141-166.

Byrne, R.M.J., & Handley, S.J. (1992). *Meta-Deductive Reasoning Strategies.* Manuscript in submission.

Byrne, R.M.J., & Johnson-Laird, P.N. (1989). Spatial reasoning. *Journal of Memory and Learning.* **28**, 564-575.

Byrne, R.M.J., & Johnson-Laird, P.N. (1992). The spontaneous use of propositional connectives. *Quarterly Journal of Experimental Psychology,* **45A**, 89-110.

Byrne, R.M.J., Johnson-Laird, P.N., & Handley, S.J. (1992). Who's telling the truth...cognitive processes in meta-deductive reasoning. In H. Sorensen (Ed.) *AI and Cognitive Science '91.* London: Springer-Verlag.

Eysenck, M.W., & Keane, M.T. (1990). *Cognitive Psychology: A Student's handbook.* Hove, UK & Hillsdale, NJ: Erlbaum.

Evans, J.St.B.T. (1989). *Bias in Human Reasoning: Causes and Consequences.* Hillsdale, NJ: Erlbaum.

Johnson-Laird, P.N., & Byrne, R.M.J. (1989). *Only* reasoning. *Journal of Memory and Language,* **28**, 313-330.

Johnson-Laird, P.N., & Byrne, R.M.J. (1990). Meta-logical puzzles: knights, knaves and Rips. *Cognition,* **36**, 69-84.

Johnson-Laird, P.N., & Byrne, R.M.J. (1991). *Deduction.* Hove, UK and Hillsdale, NJ: Erlbaum.

Johnson-Laird, P.N., Byrne, R.M.J., & Schaeken, W. (1992). Reasoning by model: the case of propositional inference. *Psychological Review.* **99**, 418-439.

Johnson-Laird, P.N., Byrne, R. M.J., & Tabossi, P. (1989). Reasoning by model: the case of multiple quantification. *Psychological Review,* **96**, 658-673.

Keane, M.T. (1988). *Analogical Problem Solving.* Chichester, UK: Ellis Horwood.

Keane, M.T. (1989). Modelling gestalt "insight" problems. In R.M.J. Byrne, & M.T. Keane, (Eds). *Cognitive Science: special issue of the Irish Journal of Psychology.* Dublin, Ireland: PSI.

Keane, M.T., Ledgeway, T., & Duff, S. (1991). Similarity and ordering constraints on analogical mapping. *Proceedings of the 13th Annual Conference of the Cognitive Science Society.* Hillsdale, NJ: Erlbaum.

Kotovsky, K., & Simon, H.A. (1989). What makes some problems really hard: explorations in the problem space of difficulty. *Cognitive Psychology,* **22**, 143-183.

Newell, A. (1990). *Unified Theories of Cognition.* Cambridge, MA: Harvard University Press.

Rips, L.J. (1989). The psychology of knights and knaves. *Cognition,* **31**, 85-116.

Rips, L.J. (1990). Paralogical reasoning: Evans, Johnson-Laird, and Byrne on liar and truth-teller puzzles. *Cognition,* **36**, 291-314.

Smullyan, R.M. (1978). *What is the Name of This Book? The riddle of Dracula and other logical puzzles.* Englewood Cliffs, NJ: Prentice-Hall.

Sternberg, R.J. (1980). Representation and process in linear syllogistic reasoning. *Journal of Experimental Psychology: General,* **109**, 119-159.

Semantic Transparency and Cognitive Modelling of Complex Performance[1]

J.G. Wallace

Information Technology Institute, Swinburne University of Technology
Melbourne, Australia

K. Bluff

Department of Computer Science, Swinburne University of Technology
Melbourne, Australia

Abstract

The simulation of human performance in situations involving interaction with complex technological interfaces highlights some key issues in cognitive modelling. This is illustrated in the behaviour of a pilot engaged in air combat. Modelling a pilot's situation awareness requires an approach that provides for emotional and motivational processes as well as sensori-motor and cognitive processes. This strategic decision raises the question of the type of representation to be employed. A response involves determining the relative roles of symbolic and sub-symbolic representation and, more specifically, the relative merits of rules and neural nets.

The concept of a spectrum of semantic transparency is introduced as a basis for the theoretical analysis of processes underlying human performance in terms relevant to ANN and symbolic representations. This leads to recognition of the inevitability of employing a variety of hybrid forms of representation. It, also, highlights the need for exploration of architectural frameworks to identify those that facilitate articulation and integration of forms of representation and realisation of the potential benefits of their complementarity.

A novel hybrid cognitive architecture is presented which employs interaction between symbolic and sub-symbolic networks to represent the effects of emotional and motivational processes on decision making. This enables a shift from the conventional focus of combat simulation on optimal decision making to the modelling of authentic human decision making with its individual variability and imperfection. The potential of the architecture and hybrid form of representation to resolve general difficulties encountered in the simulation of human performance is discussed.

1 Introduction

The use of simulators in pilot training has clear economic advantages. It is, however, critical that the experience approximates reality. Feedback from pilots who have undergone man-in-the-loop training is far from reassuring on this point. One of many possible examples is provided by AML, a set of software constructed at New Mexico State University, to drive aircraft in flight simulators by selecting manoeuvres. Pilots criticize AML for its "artificiality and predictability". (Goldsmith et al., 1990, p.10).

The reduction of predictability is a major current objective in research on simulators throughout the world. AML's decision-making processes are motivated by performance

---

[1]This research is supported in part by the Aircraft systems Division of the Aeronautical Research Laboratory, Defence Science and Technology Organisation, Australia

rather than psychological processes. This is fairly typical of the approaches adopted elsewhere. Efforts are directed at abstracting enough aspects of human performance for simulation purposes while strenuously avoiding becoming committed to modelling a human operator. The work to be reported is predicated on the view that it is highly improbable that the problems of artificiality and predictability will be solved unless the necessity for incorporating a qualitatively rich, psychological model of the pilot as a source of variability in manoeuvre selection is accepted.

The strategy adopted in our work arises from a comprehensive analysis of the aspects of pilot performance designated by the term 'situation awareness'. A typical definition based on the sequence of behaviour traversed by a pilot engaging in air combat comprises a broad range of sensori-perceptual and cognitive processes. Detection, classification, and identification depend on the general repertoire of sensori-perceptual processes. Inferring the intentions of other agents and assessing the degree of threat involve the application of general reasoning processes to a base of declarative knowledge. Generation, selection and execution of tactical options involve the application of planning and executive processes to a range of procedural knowledge while evaluation of the results hinges on sensori-perceptual monitoring.

This impressive catalogue of processes does not represent the results of a totally comprehensive analysis. A pilot in combat is required to make decisions under constraints of speed and effectiveness in an environment involving a degree of complexity that imposes a continuous, yet variable strain upon the information processing limits involved in cognitive processing. As a consequence, pilot behaviour is determined by a continuous interaction between sensori-perceptual and cognitive processes and emotional and motivational processes prompted by the environment of the performance. A process definition of situation awareness must accommodate the influence of emotion and motivation.

A truly comprehensive analysis of situation awareness must, also, include an unpacking of the state of subjective awareness in processing terms. The functional nature of subjective or conscious awareness is illustrated in variations produced by pilot training. Training is aimed at producing a high degree of overlearning of a range of competencies. Competencies are presented in the form of repertoires of rules to be used at appropriate points on a mission. Pilots are consciously aware of these rules during the learning process. In the interests of speed and effectiveness in combat, performance training and practice are continued until the rules become 'automated' or 'compiled' processes which will be applied in combat without subjective awareness of their selection.

The range of variation in conscious awareness includes a number of other possibilities. Implicit learning can occur without awareness and give rise to implicit knowledge which affects behaviour but is inaccessible to conscious awareness (Berry and Broadbent, 1988). In some cases, automated processes can enter subjective awareness as a result of voluntary or involuntary introspection. This can have disruptive consequences when applied to sensori-motor skills during performance.

2 Semantic Transparency and Representation

Modelling requires an approach that provides for these additional aspects of situation awareness as well as for sensori-motor and cognitive processes. This strategic decision immediately raises the question of the type of representation to be employed. A response involves consideration of the relative roles of symbolic and sub-symbolic representation and, at a more specific level, the relative merits of rules and neural nets.

The answer adopted derives its justification from cognitive psychology buttressed by arguments derived from machine intelligence and pragmatic concern with the feasibility of implementation. The emphasis on analysis and modelling within a psychological frame of reference provides the constraints necessary to achieve an answer.

In the absence of authoritative guidance from neuroscience, psychological analysis of performance must select from a range of approaches to representation on the basis of conflicting accounts of their respective strengths and weaknesses. In cognitive modelling, semantic transparency offers a valuable perspective on the issues involved in determining the type of representation to be employed. For our purposes a suitable definition is provided by Clark (1989) in an extended discussion of Smolensky's views: "A system will be said to be 'semantically transparent' just in case it is possible to describe a neat mapping between a symbolic (conceptual level) semantic description of the system's behaviour and some 'projectible' semantic interpretation of the internally represented 'objects' of its formal computational activity." (p.18). This definition provides a theoretical basis for the formal description of a range of systems varying in their degree of semantic transparency. At one end of the spectrum providing an example of total semantic transparency is a classic symbolic system in which all internal processing proceeds by manipulation of symbols directly and constantly mappable onto features of the system's interaction with its environment. A total absence of semantic transparency would be exhibited by a PDP system in which internal processing proceeded via units devoid of constancy across contexts and any identifiable mapping onto environmental features.

Tempting though it is to assume that all symbolic systems are characterised by a greater degree of semantic transparency than any PDP system this is not the case. The distributions of symbolic and PDP systems on the spectrum of semantic transparency overlap as indicated in Figure 1. The reason for the existence of the overlap lies in the relative strengths and weaknesses of the two types of systems in performance. The following discussion draws heavily on Sun (1991).

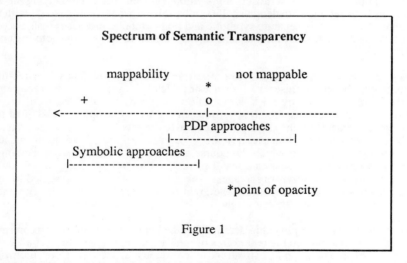

Figure 1

The strengths and weaknesses of symbolic systems are well illustrated in rules as a form of representation. Rules implemented in pure symbolic forms, such as in mathematical logic or Prolog, provide the advantage of well-defined semantics and well studied inference algorithms based on a clearly defined set of axioms. Their inability to handle inexact information, however, leads to particularly severe brittleness in performance.

In contrast, PDP systems have the advantage of being robust, exhibiting generalisation, graceful degradation and a relatively high degree of fault and noise tolerance. In the current state of the art, the downside of pure PDP systems lies in their inability to provide easy access to essential high level cognitive abilities such as rule application, compositionality and explanation which are key features of the performance of symbolic systems. These shortcomings arising from a lack of symbolic capability produce inaccuracy in performance. Pure PDP systems lack precisely specifiable preconditions and actions and rely on an approximate mapping from input to output constructed by minimizing an overall error measure.

Efforts by the proponents of symbolic and PDP systems to compensate for the weaknesses of their preferred mode of representation have led to the proliferation of hybrid systems. These are aimed at obtaining the respective benefits and avoiding the weaknesses of pure symbolic and PDP systems by combining them in a system in a compensatory fashion. This may result, for example, in attempts to combine symbolic and numerical representation by using probabilistic reasoning, fuzzy logic or certainty values in conjunction with rules. Such approaches ameliorate but do not remove the problems of brittleness. More comprehensively, rules can be embedded in a connectionist network architecture complete with weights and transfer functions in an attempt to attain the best of both worlds. (Sun, 1991).

Recognition of the inevitability of employing a variety of forms of representation in hybrid combinations highlights the need for exploration of the types of architectural frameworks that facilitate articulation and integration of combinations and realisation of the potential benefits of their complementarity. This objective in cognitive modelling mirrors aims recently defined by Minsky (1987, 1991) for machine intelligence. He argues that the degree of versatility required in machine intelligence can only be achieved via larger-scale architectures that permit capitalizing on specific strengths of symbolic and non-symbolic modes of presentation, and enable each to be used to offset the deficiencies of the other. Each neat type of symbolic representation of inference must be complemented by scruffier, non-symbolic mechanisms embodying heuristic connections between the knowledge itself (symbolic) and what we hope to do with it. It is not enough to maintain separate symbolic and non-symbolic processes inside separate agencies; we need additional mechanisms to enable each to support the activities of the others.

Analysis of human performance in semantic transparency terms leads inevitability to the adoption of hybrid representations since it typically suggests that the requirements can only be met by drawing on a combination of approaches derived from different points on the semantic transparency spectrum. In our work, the approach adopted to the assessment of semantic transparency centres on the contents of conscious awareness. The degree of semantic transparency of a segment of cognitive performance is determined by the extent to which the course and/or results of internal processing can register in conscious awareness as a result of mapping onto linguistic symbols or non-linguistic symbols such as sensory images. The extent of mapping is assumed to reflect the degree of correspondence between the grain or segmentation of processing and the modularity of linguistic or sensory symbols.

This criterion has been employed in determining the approaches adopted to representing emotion and motivation and the repertoires of rules in our model of situation awareness. Registration of emotion and motivation in conscious awareness from moment to moment typically proceeds in terms of complex combinations of images derived from a variety of senses. Attempts to introspect on these images reveal anything but a neat mapping between a symbolic semantic description of the system's behaviour and a semantic interpretation of internally represented objects subject to formal computational activity. Since neither the behaviours associated with emotions and motives nor the cognitive and neurophysiological processes underlying them are either clearly or comprehensively defined extreme difficulty in establishing correspondence is to be anticipated. As a consequence of their low semantic transparency emotion and motivation are primarily represented as PDP systems. The interaction with the environment and the system's knowledge base is conducted by means of a symbolic-numeric and a numeric-symbolic interface respectively.

As already indicated, pilot training is aimed at producing a high degree of overlearning of a range of competencies presented in the form of repertoires of rules. Pilots are consciously aware of these rules during the learning process and are able to successfully introspect upon them after the completion of training if requested. In combat, however, rules are applied in an 'automated' fashion without subjective awareness of their selection. Schneider and Oliver (1991) present an account of learning in which the acquisition of rules in symbolic form as a result of instruction is the first phase. Thereafter, rules provide the control necessary to modularise and sequence neural nets while they learn the specific associations representing the semantic content of the rules. The control provided by the rules greatly accelerates the learning process in the neural nets. On completion of learning, the neural nets provide a more rapid and robust basis for performance than the symbolic rules.

In view of the continuing high degree of semantic transparency exhibited by the repertoires of rules in learning and performance, they are represented in symbolic form in our model. Speed and robustness are sought through hybridisation and elaboration of the symbolic representation. The symbolic rules are embedded in a complex network structure of nodes representing declarative and procedural knowledge. The nodes are directly interfaced with the neural network representations of emotions and motives and interact with them on a moment by moment basis. The approach adopted to learning employs the same hybrid constituents as Schneider and Oliver but in reversed functional roles since the operation of emotional and motivational neural networks accelerates the acquisition of symbolic rules by highlighting appropriate sequencing and modularisation.

We will now provide an outline of a cognitive model sufficiently extensive to span the full range of processes involved in situation awareness. It, thus, offers a basis for the construction of a psychologically rich model of the pilot as a source of human-like unpredictability in a flight simulator. The model, also, provides an example of the application of a hybrid representation strategy set within an integrating architectural framework.

3 A General Cognitive Architecture

The architecture to be described is shown schematically in Figure 2. It is strongly influenced by the BAIRN system of Wallace et al. (1987) which provides a general theory of human cognition with a particular emphasis on the function of learning.

The central feature of the architecture is a semantic or core network which contains the current state of the system's world model. The semantic network is expressed in a hybrid representation. Each of the nodes in the network contains a symbolic representation of procedural knowledge and related declarative knowledge. Symbolic information rather than activation energy is passed between semantic nodes. Decisions on which semantic nodes occupy, at any moment, a limited number of parallel channels available for semantic processing hinge on both symbolic and non-symbolic inputs to each of the nodes.

To acquire a channel, a node requires both appropriate symbolic information and activation potential. The symbolic information is derived from either other nodes in the semantic network, or as a result of the operation of a number of sensory networks. These sensory networks execute sensori-perceptual processes by employing a hybrid approach in which stimuli registered from the environment are analysed by non-symbolic feature detectors (SFD), The results are processed to produce symbolic data as input for semantic nodes. Selective attention is produced by permitting the needs of semantic nodes for specific symbolic information to provide goals for the sensory network.

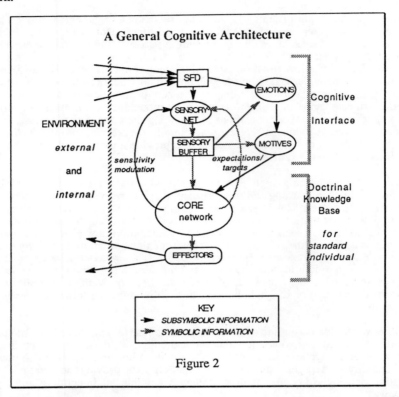

Figure 2

The activation potential necessary for a semantic node to acquire a processing channel is provided in a non-symbolic form by a motivational network. The motivational network, in turn, interacts with an emotional network. The emotional network links sensori-perceptual input in both symbolic and non-symbolic form with physiological actions which define primary emotions such as fear, anger and pleasure. The motivational network responds to the results of the emotional network's processing. Motivational nodes are activated by a combination of non-symbolic input representing physiological states and symbolic input representing situational features in the total environment of the system.

The output of the motivational network is the activation which drives the semantic nodes. Semantic nodes receive activation from one or more sources in the motivational network, as determined by innate or learned associations.

With this brief overview of the architecture as background, a more detailed account will be provided of the three interacting networks that make up the system. As the source of the energy or activation potential that drives the system, the emotional network will be considered first.

3.1 The Emotional Network

Each emotion is a member of a set of emotions defined for the system, $e_n \in E$. An emotion (e_n) is represented by a subset of the possible individual energy/intensity sources, $I_k \in I$, linked in a sub-network. Each intensity level, I_k, is the result of the operation of a distinct function of its input(s) x_i,

$$I_k = F_k(x_i)$$

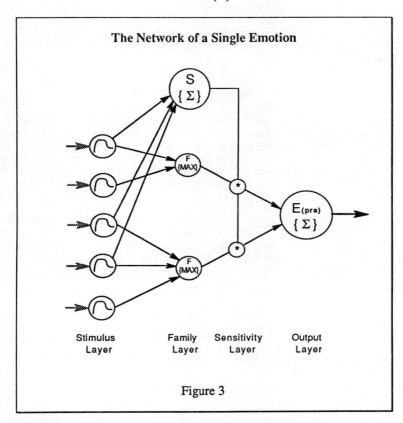

Figure 3

Figure 3 illustrates the sequence of processing within a sub-network. The energy/intensity output from a sub-net (e_n) at time t is determined by four operational features.

(a) The Stimulus Layer comprises intensity sources each registering environmental features and computing a current intensity by means of a distinct function.

(b) The first hidden layer is termed the Family Layer. It reflects a division of the intensity sources into sub-categories, F_n. Each sub-category contains a number of intensity sources grouped due to the conceptual similarity of the environmental features which they register and/or the similarity of the time scale governing changes in the environmental features. The energy/intensity output from each sub-category at time t is the currently greatest intensity output produced by the intensity functions in the sub-category

$$F_n(t) = \max(I_j(t)), \forall I_j \text{ contributing to } F_n.$$

(c) Each F_n enters the Sensitivity Layer, a second hidden layer where it is modulated by the current value of a Sensitivity parameter, S. As Figure 3 indicates, the Sensitivity parameter represents a second effect of a sub-set of the intensity sources. These intensity sources register environmental features that, in addition to contributing directly to the current level of emotional intensity, have a longer lasting effect on the sensitivity or receptivity of the system to input relevant to the specific emotion.

(d) The modulated values issuing from each sub-category at time t, $F_n(t) * S(t)$, are summed to produce a preliminary intensity value for the specific emotion at time t

$$E_{pre}(t) = \sum F_n(t) * S(t)$$

The intensity value is preliminary since the energy/intensity output of each emotion is modulated to take account of the total energy state of the emotional network before its transmission to the motivational network. The total energy state of the emotional network at a time t is reflected in the value of a parameter E_{max}. E_{max} can be viewed as representing the current state of arousal of the system and reflects its psychological and physical/physiological condition.

The processing sequence is indicated in Figure 4. The value of $E_{pre}(t)$ for each emotion is transmitted by its transfer function to a Modulation Layer. The modulation process, M, distributes the total energy currently available, $E_{max}(t)$, between the emotions in proportion to their individual current intensity values, $E_{pre\ n}(t)$. This is achieved by multiplying each intensity value by an Output Modulation parameter, OM, where

$$OM = \frac{E_{max}(t)}{\sum E_{pre\ n}(t)}$$

This yields a final intensity value for each emotion at time t,

$$E_n(t) = E_{pre\ n}(t) * OM(t)$$

This approach to the modulation of emotional intensity values produces variations in the general distribution of energy in the system that are consistent with the performance characteristics associated with varying levels of arousal. More specifically, it produces the relationship between level of arousal and the plasticity or variability of the behaviour of the system depicted in Figure 5.

Levels of arousal, E_{max}, towards the lower and higher ends of the scale produce sub-optimal levels of variability in behaviour. Low arousal is associated with narrowing of the differences in intensity between individual emotions. This increases the probability that the general behaviour of the system will be characterised by rapid switching between sequences of processing or 'trains of thought' yielding an overall 'butterfly minded' effect.

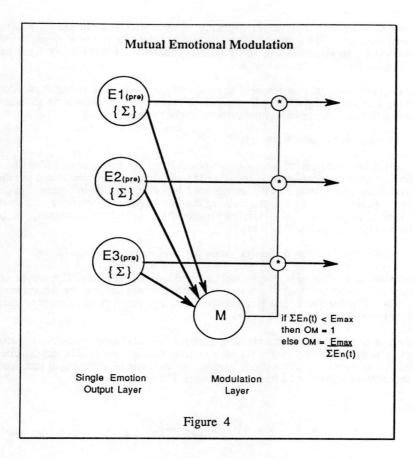

Figure 4

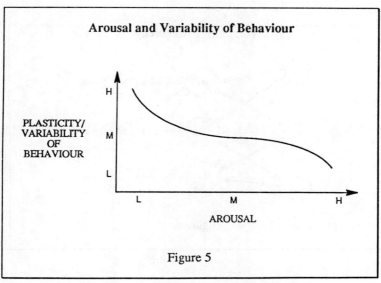

Figure 5

High arousal, in contrast, leads to widening of the differences in intensity between individual emotions. In systemic terms this diminishes the probability of switching between sequences of processing and produces behaviour of a relatively rigid nature and exhibiting tendencies to 'tunnel vision'.

Sustained, effective performance of a demanding type, such as is required of pilots in situation awareness, is most likely to occur if variation between processing sequences is proceeding in the presence of an intermediate level of arousal.

3.2 The Motivational Network

Each motive is a member of a set of motives defined for the system, $m_n \in M$, and receives input via connections to a subset of the emotions represented in the emotional network. As a result, the specific emotions connected and their current levels of activation provide the basis for determining the level of energy/intensity of a motive, m_n, at time t. The actual level of intensity $m_{input\ n}(t)$ is a result of submitting the basic level $m_{pre\ n}(t)$ to a modulation process.

$$m_{input\ n}(t) = m_{pre\ n}(t) * IH_{m\ n}(t)$$

The modulation process, IH_m, implements a motivational hierarchy. The nature of the hierarchy reflects an ordering of primary or innate motives defined for the system and the results of subsequent combinations of primary motives derived by learning processes from the system's experience.

The hierarchy, as Figure 6 demonstrates, operates via inhibitory connections between motives. The extent of the inhibition of a motive, m_n, at time t, $IH_{m\ n}(t)$, is directly proportional to the sum of the unmodulated, or basic, levels of intensity, at that time, of all of the motives higher in the hierarchy than m_n. Thus

$$m_{input\ n}(t) \propto \sum_{k=1}^{n-1} m_{pre\ k}(t)$$

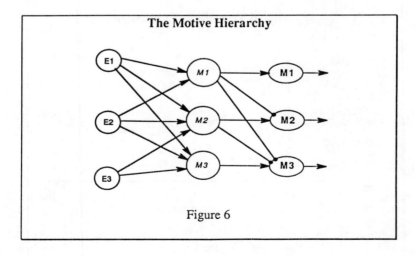

Figure 6

The level of activation output from each motive, $m_n(t)$, for transmission to the semantic (core) network is determined by application of a distinct transfer function, G_n, to $m_{\text{input } n}(t)$, so that

$$m_{\text{output } n}(t) = G_n(m_{\text{input } n}(t))$$

This enables implementation of specific characteristics of intensity variations over time in individual motives.

3.3 The Semantic Network

The activation outputs, $m_{\text{output } n}(t)$, received by each node in the semantic network are translated into activation potential by application of a motivation expected value (mev) profile. Each individual mev_p in the profile represents the maximum contribution that a specific motive can make to the activation potential of a specific node. With this restriction

$$P_{S_n} = \sum_p \Psi(t)$$

$$\text{where } \Psi(t) = m_p(t) : m_p(t) < mev_p$$
$$= mev_p : m_p(t) \geq mev_p$$

A mev_p is a measure of the motivational satisfaction that the system expects to ensue from the operation of a node, and each node has a mev_p responding to each of the individual motives that were involved in its creation. If a node has no mev corresponding to a particular motive the motivational drive emanating from that motive makes no contribution to the node's prepotency. This is because the learning mechanism has not constructed a connection between the motive generator and the node.

The actual prepotency, P_{S_n}, that a node attains is also dependent upon the state of its activation condition sets. Thus at any stage in the system's operation, the results of an activation spread, that is, which node will fire next, depends upon the particular combination that exists of the following factors.

(a) The motivational status of the system.
(b) The mev profile of the recipient nodes.
(c) The output connections from the terminating node.
(d) The mode of activation spread.
(e) The information stored on the activation data queues of the recipient nodes.

The combination of these factors has the potential to make the system's performance an exhibition of rich and varied behaviour, responding to its environment in a combination of data-driven and goal-driven manners.

4 Conclusion

The cognitive architecture outlined above represents a departure from the mainstream of combat simulation. Its focus is not on optional decision making but rather on modelling authentic human decision making, "warts and all". The modular nature of the architecture provides the necessary flexibility. It enables the maintenance of the core network in a steady state while simulating the effects of individual differences in personality by varying the contributions of the motivational and emotional networks. This reflects the reality of the human situation in which symbols are dense in the sense that the same symbols subtend different operational semantics due to differences in the previous history and experience which they represent in individuals.

The flexibility in cognitive modelling enables investigation of the effects of variations in the results of training. Performance differences can be created by varying the declarative and procedural knowledge represented in the core network while maintaining the motivational and emotional profile constant. Alternatively the knowledge represented in the core network can be kept constant while investigating the effects of personality and other differences by manipulating the motivational and emotional networks.

In combat simulation, the cognitive architecture can be used to provide an intelligent, relatively unpredictable and challenging basis for the behaviour of friend or foe units with varying assumptions regarding their training, specific mission instructions and motivational and emotional states.

The modularity and flexible nature of the architecture endows it with considerable generalisability. It could, for example, be readily deployed in simulations of the processes underlying performance in other situations involving interaction with a complex, technological interface and/or decision making under stress. Fire fighting and currency trading are activities falling into this category and worthy of investigation.

On the more general methodological front the architecture illustrates the potential of hybrid representational approaches for mitigation of the operational rigidity and brittleness of totally rule based representation. Interaction with sub-symbolic networks endows symbolic representation with some of the flexibility or fuzziness usually ascribed to connectionist representations. There is, in effect, an implicit fuzzy logic in the processes underlying the variations in performance produced by the system.

The modular structure facilitates investigation of the results of incorporating alternative mechanisms. It is readily possible, for example, to explore the relative merits of adopting different types of neural network structure and operation in implementing the motivational and/or emotional networks.

Finally, the hybrid architecture offers promising approaches to achieving efficiency gains in the operation of rule based knowledge bases. Considerable ingenuity and effort is necessary to set in place the multi-layered detailed symbolic representations required to produce the type of variability characteristic of human performance. Interaction between the symbolic and sub-symbolic networks makes it possible to produce desired results with relatively shallow symbolic, rule based structure and significantly less complexity in individual rule structures.

References

Berry, D.C. & Broadbent, D.E. (1988). Interactive tasks and the implicit-explicit distinction.*British Journal of Psychology*, **79**, 251-272.
Bressler, S.L. (1990). The gamma wave: a cortical information carrier? *Trends in Neuro Sciences*, **13**, 161-162.
Clark, A. (1989). *Microcognition*. Cambridge, MA: The MIT Press.
Duffy, F.H., Bartels P, & Burchfiel, J.L. (1981). Significance probability mapping: an aid in the topographical analysis of brain electrical activity. *Electroencephalography Clinical Neurophysiology*, **51**, 455-462.
Edelman, G.M. (1989). *The Remembered Present*. New York, NY: Basic Books.
Fuster, J.M. (1989). *The Prefrontal Cortex (2nd edition)*. New York, NY: Raven Press.

Gray, C.M., Konig, P., Engel, A.K., & Singer, W. (1989). Oscillatory responses in cat visual cortex exhibit inter-columnar synchronization which reflects global stimulus properties. *Nature*, **338**, 334-337.

Goldsmith, T., McMahon, D., & Schvaneveldt, R. (1990). *Tactical Air Combat Intelligent Trainer (TACIT)*. Memoranda in Computer and Cognitive Science, New Mexico State University.

Kolers, P.A., & Smythe, W.E. (1984). Symbol manipulation: alternatives to the computational view of mind. *Journal of Verbal Learning and Verbal Memory*, **23**, 289-314.

Mesulam, M.M. (1981). A cortical network for directed attention and unilateral neglect. *Annals of Neurology*, **10**, 309-325.

Minsky, M. (1987). *The Society of Mind*. New York, NY: Simon and Schuster.

Minsky, M. (1991). Logical Versus Analogical or Symbolic Versus Connectionist or Neat Versus Scruffy. *AI Magazine*, **Summer '91**, 35-51.

Optican, L.M., Gawne, T.J., Richmond, B.J., & Joseph, P.J. (1991). Unbiased measures of transmitted information and channel capacity from multivariate neuronal data. *Biological Cybernetics*, **65**, 305-310.

Schneider, W., & Oliver, W.L. (1991). An Instructable Connectionist/Control Architecture: Using Rule-Based Instructions to Accomplish Connectionist Learning in a Human Time Scale. In Van Lehn (Ed.) *Architectures for Intelligence*. Hillsdale, NJ: Erlbaum, (in press).

Sharkey, N.E. (1991). Connectionist representation techniques. *Artificial Intelligence Review*, **5**, 143-168.

Silberstein, R.B., Schier, M.A., Pipingas, A., Ciorciari, J., & Wood, S. (1990a). Dynamics of the steady state visually evoked potential topography in a visual vigilance task. *Brain Topography*, **3**, 272-273.

Silberstein, R.B., Schier, M.A., Pipingas, A., Ciorciari, J., Wood, S., & Simpson, D.G. (1990b). Steady state visually evoked potential topography associated with a visual vigilance task. *Brain Topography*, **3**, 337-347.

Smolensky, P. (1987). Connectionist AI and the brain. *Artificial Intelligence Review*, **1**, 95-109.

Smolensky, P. (1988). On the proper treatment of connectionism. *Behavioural and Brain Sciences*, **11**, 1-74.

Sun, R. (1991). *Integrating Rules and Connectionism for Robust Reasoning* (Technical Report CS-91-160). Waltham, MA: Brandeis University, Department of Computer Science.

Wallace, J.G., Klahr, D., & Bluff, K. (1987). A self-modifying production system model of cognitive development. In D. Klahr, P. Langley, & R. Neches (Eds.). *Production System Models of Learning and Development* (pp. 359-435). Cambridge, MA: MIT Press.

Others as Resources: Cognitive Ingredients for Agent Architecture[1]

Maria Miceli, Amedeo Cesta, Rosaria Conte

IP-CNR, National Research Council
Rome, Italy

Abstract

This paper presents some preliminary tools for empowering the architecture of autonomous agents by providing it with a complex of strategies for using dependence-based knowledge to select agents for cooperation. In this perspective others are seen as solutions to one's problems, and may be recategorized as resources.

Three main objectives are pursued in this work. First, to insert a previously developed theory of dependence in a plausible architecture for cognitive agency, showing the role of dependence-based knowledge with regard to the typical functionalities of agency. Second, this work aims to contributing to distributed systems research, by showing how the structure of dependence in a multi-agent world provides criteria for a rational model of interaction and communication. Third, a problem-driven approach to sociality is adopted, where the problem of why social interaction occurs is raised and some answer is found in the agents' problem solving.

The main task this work tries to accomplish is that of endowing an agent with strategic criteria for using the amount of knowledge it has about others, their abilities and goals. Heuristic functions are described which are grounded on our model of dependence relations.

1 Introduction

This paper presents some preliminary ideas for empowering the architecture of autonomous agents, providing it with a powerful interactive tool, namely a complex of strategies for using dependence-based knowledge to select social resources, that is, agents useful for cooperation. In a preceding paper (Castelfranchi, Miceli, & Cesta, 1991) it was shown that social dependence is a resource-dependence in which the resource slot is filled in by a social agent. In this perspective, therefore, others are seen as solutions to one's problems, and may be recategorized as resources.

The present work has a number of objectives. First, it aims to insert our theory of dependence in a *plausible architecture for a cognitive agent (CogAgent)*, showing the

---

[1] This work has been partially supported by CNR under Target Project "Sistemi Informatici e Calcolo Parallelo", grant n.104385/69/9107197 to IP-CNR. The authors participate in the "Project for the Simulation of Social Behavior" at IP-CNR.
Correspondence should be sent to the first author at IP-CNR, Viale Marx 15, I-00137 Rome, Italy (e-mail: pscs@irmkant.bitnet).

role of dependence-based knowledge with regard to the typical functionalities of an intelligent autonomous agent (decision-making, planning, scheduling, etc.).

Secondly, this work intends to *contribute to some extent to developments in distributed systems research*. In distributed systems, the necessity for building totally independent agents is now growingly emphasized (see Reddy and O'Hare (1991) for a review of blackboard systems). In the Contract Net Approach (Smith, 1978) where Contractor agents bid for tasks advertised by Manager nodes, agents were interdependent processes and not fully independent experts. Lesser and Corkill (1981) proposed their Functionally Accurate/Cooperating model to promote total independence between agents. However, both this model and blackboard systems are found to face a number of problems, among which the overhead of communication. A trade-off seems to exist between expert systems computational overload and lack of flexibility and the consequent resort to distribution of control, on one hand, and overheads of communications, on the other (cf. Reddy & O'Hare, 1991). Our theory of dependence suggests that the structure of dependence in a multi agent world does not only allow for the future pattern of interaction among agents to be predicted (this has been shown to some extent in Castelfranchi et al. (1991) and in Conte and Castelfranchi (1992)), but it also provides criteria for a rational model of interaction and communication among cooperating agents. A multi-agent dependence structure, once accessed by agents, reduces and improves communications among agents: dependence-based knowledge provides the agent with reasons for sending messages and ask for help.

Finally our work is intended *to adopt a radically bottom-up, or problem-driven approach to sociality.* In it, sociality is not traced back to varying degrees of overlapping among the agents' mental attitudes --which is the dominant view in AI social studies (Cohen & Levesque, 1990; Grosz & Sidner, 1990; Pollack, 1990; Rao, Georgeff, & Sonenberg, 1991). Rather, here the problem of "why social (inter)action?" is raised and some answer is found in the agents' problem solving. Social agents are seen as one providing solutions to problems encountered by another. In fact, the capacity for re-categorizing others as resources for one's goals appears as a powerful tool of *CogAgent*, greatly increasing his adaptive behavior.

2 The *CogAgent* architecture: Description and status

We are currently developing an agent architecture for testing cognitive modelling hypotheses for multi-agent interaction. Fig.1 shows its basic control strategy. All the knowledge referred in the paper is represented by a knowledge representation service named KRAM, developed within our group (D'Aloisi, 1992; D'Aloisi & Castelfranchi, 1992).

Our agent is supposed to have a behavior which is strongly goal driven. Goals are given to the decision making module which at present executes a simple Goal Choice Procedure. We plan to further investigate more sophisticated strategies for this module, grounded on existing relevant work (Ingrand & Georgeff, 1989; Russell & Wefald, 1991).

The goals selected to be executed are delivered to a planner which produces a sequence of actions useful to achieve each of these goals. The planner we are implementing is based on a structured memory of previous plans, and implies re-using concepts (Cesta & Romano, 1992). We will not enter the details of plan creation since this aspect of our system is not especially relevant to the aims of the present work.

Once got a plan to execute, the agent activates a scheduling procedure (by scheduling we mean the activity of situating untimed actions on a time-line, satisfying their resource requests) which analyzes the availability of resources and assigns each action a given execution time-interval. This is the module strictly related to the subject of this paper. The availability of resources is checked in two directions:
  • does the agent have action *a* in his repertoire?
and if negative:
  • if the agent needs external resources are they accessible to him?

This paper starts from here. Actually, we will not handle the former question. We will simply postulate that agents have implicit knowledge about their action repertoire. As to the latter question, we will limit ourselves to explore the agent's search for social resources. This boils down to checking whether any other agent exists in the common world which might give agent *x* some help to get the work done. In the following, therefore, we will endeavor to propose and discuss some search strategies that our *CogAgent* should resort to in order to select those agents he can treat as means to solve his problems, or better, as resources to achieve his goals.

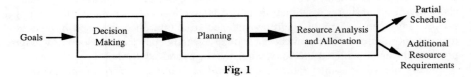

Fig. 1

Two general kinds of goals exist in our architecture:
  • *content* goals, i.e., specific represented states of the world the agent wants to be realized;
  • *meta-goals*, i.e. strategic goals, by means of which the agent builds plans for achieving the content goals; as we will show, the meta-goals include what we call dependence-based criteria for searching "useful" agents to interact with.

3   Preliminaries to the dependence-based agency theory

In Castelfranchi et al. (1991) we gave a formal account of a particular theory of agent behavior. To describe the relevant predicates in our model we use a logical formalization drawn from Cohen and Levesque (1990).

Agents are usually dependent on the existence of resources. We call this type of dependence a *resource dependence* (described by the *R-DEP* predicate):

$D1.$ *(R-DEP x r a p)* $=def$ *(GOAL x p)*
  $\land$ *(RESOURCE r a)*
  $\land$ *((DONE-BY x a)* $\supset$ *(EVENTUALLY p))*

r is then a resource for *x* to achieve his goal that *p,* through the performance of a given action *a*.

A basic definition in our theory is that of *social dependence*:

D2. *(S-DEP x y a p)* =def   *(GOAL x p)*
　　　　　　　　　　　　∧ ¬ *(CANDO x a)*
　　　　　　　　　　　　∧ *(CANDO y a)*
　　　　　　　　　　　　∧ *((DONE-BY y a)* ⊃ *(EVENTUALLY p))*

that is: x depends on y with regard to an act a useful for realizing a state p when p is a goal of x's and x is unable to realize p while y is able to do so. In this context, y's action is a resource for x's achieving his goal.

Resource dependence is likely to produce social dependence. In order to describe this property we introduce the notion of *resource control (CONTROL x r)*. An agent x *controls* a resource r when he is able to do an action by which he allows any other agent to perform any action requiring the resource. If agent x depends on resource r for a given p, and agent y controls r, then agent x depends on agent y for using r. So, in this context social dependence (of x on y) is the joint result of resource dependence (of x on r) and resource control (of y over r):

A1. ∃ a_1 *((R-DEP x r a p)* ∧ *(CONTROL y r))* ⊃ *(S-DEP x y a_1 p)*

So far, just cases of *unilateral* dependence (of x on y) have been described. However, dependence may also be bilateral. There are in fact two possible kinds of *bilateral dependence*:

• *mutual dependence*, which occurs when x and y depend on each other for realizing a *common goal p*, that can be achieved by means of a plan including at least two *different* acts such that x depends on y's doing a_1, and y depends on x's doing a_2:

(S-DEP x y a_1 p) ∧ *(S-DEP y x a_2 p)*

In Conte, Miceli, and Castelfranchi (1991), it was pointed out that *cooperation* is a function of mutual dependence. There is no cooperation in the strictest sense without mutual dependence;

• *reciprocal dependence*, which occurs when x and y depend on each other for realizing different goals, that is, when x depends on y for realizing x's goal that p_1, while y depends on x for realizing y's goal that p_2:

(S-DEP x y a_1 p_1) ∧ *(S-DEP y x a_2 p_2)*

Reciprocal dependence is to *social exchange* what mutual dependence is to cooperation.

Others' benevolence, as we shall see, is an important resource in dependence relationships. In Castelfranchi (1990) and Conte et al. (1991), we argue against the current notion of *benevolence*, suggesting some further refinement of it; however, to our present purposes it is sufficient to refer to a simpler definition, in line with Cohen and Levesque's (1990):

D3. *(BENEVOLENT y x p)* =def *(BEL y (GOAL x p))* ⊃
　　　　　　　　　　　　　　　　　(EVENTUALLY (GOAL y p))

So, if y believes that x has the goal p then also y comes to have the same goal p.

Among the social goals predictable from a dependence relationship, a crucial role is played by the *goal of influencing*. By x's goal of influencing y, *INFL-GOAL*, we mean x's goal that y *has* a certain goal p:

D4. (INFL-GOAL x y p) =def (GOAL x (EVENTUALLY (GOAL y p)))

In our model, from x's assumed dependence on y we derive x's goal of influencing y to perform the action(s) that allow him to achieve his goal:

T1. (BEL x (S-DEP x y a p)) ⊃ *(INFL-GOAL x y (DONE-BY y a))*

4 Dependence-based criteria for searching "useful" agents

Our starting point is x's assumed dependence on someone else with respect to a certain action a in view of a given goal p. More exactly, since it is not necessarily the case that x already knows who is the y it is dependent upon, we will have:

(GOAL x (EVENTUALLY p))
∧ *(BEL x* ¬ *(CANDO x a))*
∧ *(BEL x* ∃y_i *(CANDO y_i a))*
∧ *(BEL x* ∃y_i *((DONE-BY y_i a)* ⊃ *(EVENTUALLY p)))*

Let us provide a very simple example to help us show and clarify each step of our exposition. Let us suppose that x and $y_1, y_2,..., y_n$ are people working in a research center, and that the e-mail server of the computer network is down. Agent x has the goal p to make the e-mail server work (in order to send some message), but he is unable to do the (set of) action(s) a required for making the server start up, and he assumes he is unable to do a. Moreover, x assumes that in the department there are some ys (at least one) who are able to do a.

In our example the Goal p is a *content* goal of x's. Now, let us turn to his dependence-based *meta-goals* or criteria. The problem we face is that of providing strategic criteria for using the amount of knowledge the agent has about others, their abilities and so on. We draw on our dependence analysis for setting up reasonable search functions on the abovementioned knowledge.

The module we are describing is based on a general control function called when an action a exists that the agent is unable to execute. The pseudo-code of the function follows:

    Function *Find-Agent (ActionToBeDone)*
        Let *((PossibleAgents (CollectKnownAgents KnowledgeBase)*
             AbleAgents
             WillingAgents))
        For *CurrentAgent* **in** *PossibleAgents* **do**
            If *(Satisfy CANDO CurrentAgent ActionToBeDone)*
            Then *(Cons CurrentAgent AbleAgents)*
        For *CurrentAgent* **in** *AbleAgents* **do**
            If *(Satisfy WILL CurrentAgent ActionToBeDone)*
            Then *(Cons CurrentAgent WillingAgents)*
        Return *(PickOne WillingAgents)*

The function increasingly restricts lists of agents: a list of the agents known to agent x (*PossibleAgents*), a list of the agents known to x as able to do action a (*AbleAgents*), and a list of the agents known as both able and willing to do action a (*WillingAgents*). To create the last two lists *CogAgent* checks if an agent y satisfies the predicates *(CANDO y a)* in the former case and *(WILL y a)* in the latter. The test is performed by calling the function *Satisfy* which applies whatever heuristic is known to the agent for determining the satisfaction of the predicate (see the following pseudo-code).

> **Function** *Satisfy (Predicate Agent Action)*
> **Let** *((UsefulHeuristics (CollectKnownHeuristics Predicate)))*
> **While** *UsefulHeuristics* **do**
> **Let** *((Heuristic (PickOne UsefulHeuristics)))*
> **If** *(Apply Heuristic Agent)*
> **Then Return** *True*
> **Else** *(Delete Heuristic UsefulHeuristics)*
> **Return** *False*

The agent applies different sets of heuristics for the satisfaction test on the two predicates. These strategies will respectively satisfy the *CANDO* and *WILL* criteria.

In the next sections we describe in detail the heuristic functions we have drawn from our theory of dependence. What is left out of the present discussion are the possible strategies implemented by the function named *PickOne*. Implementing this function in a flexible way could allow us to model agents with very different behaviors.

5 CANDO criterion: Search for an agent who is able to perform the required action

As previously stated, our agent x already knows (on the grounds of previous failure experience, or of some general knowledge about its own action repertoire) he cannot do a. Since at present his architecture does not include learning skills, the only resource he is left with is someone else's performance of a. So, in order not to give up p, x must look for some y who *CANDO a*.

A first problem x will come to face is then: How to find out those ys who *CANDO a*? In other words, the *CANDO* criterion should be operationalized into some sub-criteria for *CANDOs*' detection. Given a certain y_1, x may search for either one or the other of two general kinds of information in order to assess whether y_1 *CANDO a*. We shall call the first a bottom-up, or performance criterion for *CANDOs*' detection, while the second might be called a top-down or categorial criterion.

5.1 Bottom-up or performance criterion

Agent x may search his memory for a belief such as

$(BEL\ x\ (DONE\text{-}BY\ y_1\ a))$

because if an agent has done an action, he is in fact able to do it:

$(DONE\text{-}BY\ y_1\ a) \supset (CANDO\ y_1\ a)$.

Such a belief derives from x's direct observation of y_1's behavior (x already saw y_1 doing a); so, we are assuming x to be endowed with a belief maintenance apparatus. This apparatus should be able to manipulate different labels associated to the beliefs according to the fact that a belief is due to the agent's direct observation, the agent's being told, etc. (Galliers, 1991).

5.2 Top-down or categorial criterion

However reasonable, the former criterion sounds a bit unlikely to occur. More exactly, we believe it is applied when no other information is available, i.e., when the top-down criterion is not applicable. In fact, once faced with their incapacities, people do not always act in a vacuo, blindly looking for those agents they depend on for the fulfillment of their needs. It seems both more realistic and rational to claim that agents often resort to some "frozen" knowledge about who *CANDO* what. By frozen knowledge we mean knowledge already incorporated within a main component of a rational agent architecture: his goal/action categorization, which is mapped to some extent on his categorization of others he may interact with.

To go back to our example, agent x (let us call him John) is not likely to wonder for each of his colleagues whether he saw them resetting the e-mail server. John already knows that a class of people -- say, computer scientists and technicians, and in particular system managers -- are endowed with the *CANDO* in question. So, when faced by a particular colleague y_1, Carol, he is likely to see whether Carol belongs to that class of agents.

This implies that when the top-down criterion is applied, the first query x should make is about the kind of a required: Does a belong to a class A of actions which can be performed by a certain class Y of agents? If so, x will look for those specific ys he knows to belong to that class. Suppose John knows Carol is a computer scientist: this allows him to infer that she can reset the e-mail server. To put things more formally

if (BEL x ((ISA a A) ∧ (CANDO Y A)))
then (SEARCH x (ISA y Y))

This equals to saying that, when faced by a particular y_1, an agent x who follows the top-down criterion will not ask himself whether *(DONE-BY y_1 a)*, but whether *(ISA y_1 Y)*, in order to assess whether *(CANDO y_1 a)*.

To sum up, we might conclude that

(BEL x (CANDO y a)) ⊃
((BEL x (DONE-BY y a)) ∨
(BEL x (ISA y Y) ∧ *(ISA a A)* ∧ *(CANDO Y A)))*

A further specification should be made. For the sake of simplicity, we assume that both the top-down and the bottom-up criterion are autonomously applied by x. However, in a number of cases it is possible that someone else's communication about y's capabilities plays a role in inducing x's belief about *(CANDO y a)*. Suppose a third agent z (Jack) says (or already said) John that Carol is able to reset the e-mail server, either because Jack saw her doing it or because he knows she is a computer scientist. In such cases, the source of x's belief that *(CANDO y a)* is neither x's direct experience of y's performance nor x's own knowledge about classes of actions and agents; rather, the source of x's belief is just z's communication. This of course raises the problem of the authoritativeness of the source of information, in terms both of its expertise and of its trustworthiness.

To go back to our *CANDO* criterion, x will recursively apply it to each of the agents in his world until he finds one agent (say, y_1) that *CANDO a*.

If x does not find any, of course he will end his search and probably give up his goal p. In this case, application of the criterion has just prevented x from embarking in useless and effort-consuming requests for help from people who would not be able to satisfy them.

However, if x's search is successful, that is, if he finds that *(CANDO y_1 a)*, what will he do? Will x automatically turn to y_1 for help?

An agent who can do a certain action is not necessarily an agent who wants to do that action. This is the basic reason why detection of others' *CANDOs* and of one's dependence on them is a necessary but not sufficient condition for obtaining p, that is to say the actual *(DONE-BY y a)* which implies that eventually p becomes true.

For this very reason, the abovementioned *CANDO* criterion is just necessary but not sufficient to assess which are the agents x can effectively turn to. Agent x has to assess not only whether y_1 is competent, but also whether y_1 is willing to *(DONE-BY y_1 a)*. In our example, once assumed that Carol is able to reset the e-mail server, John should wonder whether she is likely to do so, i.e., whether, for one reason or another, she will have the goal to perform such an action. In fact, according to a postulate of rational agenthood, in order to perform any action an agent must want that action:

(DONE-BY y a) $\supset$ *(GOAL y (DONE-BY y a))*.

So, x needs to apply some other criteria to select among the agents who *CANDO a* those who are likely to have *(GOAL y (DONE-BY y a))*.

6 *WILL* criteria: Search for an agent who is willing to do *a*

We shall propose three *WILL* criteria or strategic goals to be pursued by x in his search for an agent y who not only *CANDO a* but also has the goal of performing it.

It is worth observing that we do not assume these criteria to be organized into a fixed sequence. The ordering of application of such criteria may depend on some priorities established by the specific agent x in question according to his own preferences or biases. In other words, search strategies may vary from one agent x_1 to another x_n as far as the ordering of application of these criteria is concerned.

6.1 *WILL* criterion (a): Search for an "exploitable" y

The most straightforward *WILL* criterion is to see whether y already has the goal to perform a, in order to take advantage of y's performance. Suppose John knows that Carol will need to use the e-mail and, being able to reset the server, is likely to do the job: John will just wait for her to do so.

Moreover it is not necessary that y has the same goal p in order to have *(GOAL y (DONE-BY y a))*; action a might be performed by y in view of some other goal q^2. If, for instance, p is "to use the e-mail" and a "to reset the e-mail server", it can be the case that Carol does a not in view of p, but for some other goal. Suppose she does not need

---

[2]In fact, as observed in Conte and Castelfranchi (1992), not all consequences of one's actions are wanted, that is:
$\neg$ *(((DONE-BY y a) $\supset$ p) $\supset$ (DONE-FOR y a p))*
where *DONE-FOR* is defined as follows:
(DONE-FOR y a p)=def (DONE-BY y a)
$\wedge$ *(GOAL y p)*
$\wedge$ *(BEL y ((DONE-BY y a) $\supset$ (EVENTUALLY p)))*
It is possible, then, that *((DONE-BY y a) $\supset$ p))* -- which is what x is interested in, even though $\neg$*(DONE-FOR y a p)*.

the e-mail, but has the goal q of "complying with the director's request". John, who knows about both that request and Carol's goal q, also knows that $q \supset p$. So, he can still wait for Carol to do the job.

6.1.1 Sub-criteria for exploitation

Now, how can x assess whether y has *(GOAL y (DONE-BY y a))*?

Again, we can assume both a *categorial* and a *performance criterion* to be applicable. In the former case, x can see whether y belongs to a class of agents with *(GOAL Y (DONE-BY Y A))*. For instance, John can see whether Carol belongs to the class of (competent) users of the e-mail server. So:

if (BEL x ((ISA a A) ∧ (CANDO Y A) ∧ (GOAL Y (DONE-BY Y A))))
then (SEARCH x (ISA y Y))

In the case of the performance criterion, on the contrary, x should either actually see y performing a or carry out some plan recognition according to which x can believe y is going to perform a. Suppose John sees Carol going to the room where there is the e-mail server after the director's request: John can easily infer that she is going to perform a.

6.2 *WILL* criterion (b): In search of benevolence

In computational models of rational agenthood, agents are shaped as benevolent and cooperative. Indeed, this allows an optimal level of coordination among system components to be achieved. In the last few years, however, the benevolence assumption has been increasingly seen as insufficient.

This is not to deny that benevolence plays a role in social life, if not an exhaustive one. Indeed, people look for benevolence, even though their search is often doomed to fail. In the following, agent x is assumed to search for benevolence; x will try to find benevolent agents among those he believes to depend upon.

There seem to be at least two possible types of benevolence: individualized, or personal, and non individualized benevolence. Let us see them in turn.

6.2.1 Individualized benevolence

This type of benevolence occurs whenever a given agent y exists such that y wants that the goal(s) of a given agent x's is/are achieved. In other terms, y is likely to adopt the goals of a specific agent. To simplify matters, we will not raise the most interesting question whether y is likely to adopt a subset of x's goals or any goal of his. It is reasonable to believe, however, that when benevolence is individualized it usually applies to any goal of the recipient's: if y is benevolent with regard to x as a specific agent (as opposed to an element of a set), y usually wants x to be satisfied and well-adapted. In real life, this usually happens within biological or, more generally, personal relationships.

6.2.2 Non individualized benevolence

This type of benevolence occurs when y (usually a given set of ys) has the goal that a given set of xs achieve a given set of goals. (We will not examine here the case, rather uncommon indeed, of "universalistic", or philanthropic, benevolence.)

In the general form of non individualized benevolence three sets of objects are mentioned: the recipients, their goals, and the benevolent agents. Consequently, if the recipient believes he belongs to a set which receives benevolence with regard to a given set of goals, he has reason to believe there should be at least one benevolent y:

$(BEL\ x\ ((BENEVOLENT\ Y\ X\ P) \land (NON\ EMPTY\ Y) \land$
$(ISA\ x\ X) \land (GOAL\ x\ p) \land (ISA\ p\ P))) \supset$
$(BEL\ x\ \exists y((ISA\ y\ Y) \land (BENEVOLENT\ y\ x\ p)))$

Non individualized benevolence seems to be based either on role tasks or on general norms:

• Indeed, *roles* are often prosocial, that is, they imply a set of beneficiaries by definition non equal to the role players. Sometimes, prosocial roles are even tutorial: in tutorial roles, the role-players are expected to act in the interests of the beneficiaries. In general, however, we say that agent x may be a recipient of non individualized benevolence when x is mentioned as a beneficiary in the role of some ys. In such cases, x is due benevolence, and y is held to give help to x relative to some specified set of goals. In our example, the agent might be held by role to reset the e-mail facility. However, note that role-tasks are usually set up by convention: the same agent might not be held to train her colleagues to use it.

• Sometimes *norms* control when adoption should be given to whom independent of the structure of roles. In some contexts, some agents might be held to adopt some goals of other agents: on the bus, people are expected to get up and leave their seats to disabled; car drivers are held to give precedence to cars coming from the right. A special case is represented by reciprocation. A norm of reciprocation might be seen as a case of non individualized benevolence: y is held to reciprocate, that is, to adopt some goals of those agents who have intentionally benefitted her in the past without being held to do so. If John asks Carol, who besides reset the e-mail can write in Spanish, to send a message to some Spanish-speaking collaborator, John will probably be held to return the favor somehow.

6.2.3 *How does x come to know that y is benevolent?*

How does x find out if there should be someone likely to help him? The following are types of tests that x could apply in his search. (Remind that they are by no means to be taken as necessarily sequenced in order of presentation.)

One type of test resorts to *role-tasks*: x will check the "blackboard" and see whether his goal matches with any of the existing roles. Actually this is not to be easily realized since in a typical blackboard structure only the not yet matched tasks are included. A more realistic way of executing the test would be x's checking his own role repertoire and see whether his goal matches with the goal, rather than the tasks, of some beneficiary role. If he finds one, x will reduce the ys list; if he finds none, he might look for some other non individualized benevolence.

A second type of test calls *norms* into play: x will check his domain knowledge for any norm mentioning a set of beneficiaries. See the procedure described in the preceding test.

A third test checks whether others' *reciprocation* of past favors are likely to occur. This type of test implies that the agent architecture is endowed with a memory (not exceeding a set up time lag) of past interactions. Either the agent knowledge should contain indexes of debts and credits of the last interaction occurred, or an interaction buffer should be checked every time other agents are screened. If some credits are found, x will reduce the y's list.

Finally x may check the interaction buffer for *already encountered benevolence*. Here, an interesting trade-off appears. On one hand, if x finds himself to be highly indebted to someone, he could be discouraged from insisting in that direction since his creditor might deny her help. She could even consider the relation with x no longer convenient and decide to break it. On the other hand, creditors usually are those agents which proved most frequently benevolent in the past, and therefore are good candidates for giving help now. As a balanced solution, let us say that x will select for those agents to whom he is not indebted although he has shared with them the highest number of benevolence episodes in both directions.

6.3 WILL criterion (c): Search for a dependent y

As already stated, WILL criteria are concerned with the search for an agent who is willing to perform the required action. Both exploitation and the search for a benevolent y are quite straight ways for achieving this. In the former case, y already has the goal to perform a of his own, while in the latter, being benevolent, y has the goal that x achieves (some or all of) his goals, including that of having y performing a.

A further criterion exists which, however less straightforward and simple, looks quite common and crucial in natural social interaction: the search for an agent who, besides being able to perform a, is in her turn dependent on x for some other action. In other words, x should find out whether it is possible to move from assumed unilateral dependence of x himself on y to assumed bilateral dependence of each one on the other. Bilateral dependence, in fact, would obviously allow for a more powerful position of x: If it is true that, on the grounds of his assumed dependence on y, x has the goal of influencing y to perform a_1, it should be true that also y, on the grounds of her assumed dependence on x, should have the goal of influencing x to perform some other action a_2. So, x might offer his own performance of a_2 in exchange for y's performance of a_1. Going back to our example, suppose that while Carol is able to reset the e-mail server, John knows ancient Greek very well and he knows that Carol doesn't and is in need of some translation from English into ancient Greek. Other things being equal (i.e. leaving aside possible exploitations or benevolences), John would prefer to resort to Carol rather than to George, who is equally able to reset the e-mail server but does not need anything from John.

As one can see, situations of this sort are the bases for a great part of social interaction. We do not pretend here to provide a detailed analysis of bilateral dependence, which would imply a careful examination of both cooperation and social exchange (as for cooperation, see Conte et al. (1991)). We just limit ourselves to suggest a general outline of the search for bilateral dependence in view of influencing y to perform the required action.

First of all, then, x should see whether in his own action repertoire there is some action he can perform and that y is unable to perform. Of course, such an action should be useful for some goal of y's. So the starting point is:

if (BEL x (S-DEP x y a_1 p))
then (SEARCH x (S-DEP y x a_2 q))

6.3.1 Sub-criteria for y's dependence on x

The usual question, then, is: How can x find out y's dependence on himself? Both experience and categorial knowledge can help also in this case. Agent x might see (or have seen) y trying to perform a_2 (or more generally to pursue q) without success; or x may know y belongs to a class of agents who $\neg(CANDO\ Y\ A_2)$. For instance, Carol as a computer scientist is quite unlikely to have had a classical education including the study of ancient Greek. In such cases then the criterion x should follow would be something like:

if (BEL x ((CANDO x a_2) ∧ (ISA a_2 A_2) ∧ $\neg$(CANDO Y A_2)))
then (SEARCH x (ISA y Y))

6.3.2 The other's dependence as a basis for influencing behavior

Once assessed that y depends on x for some action a_2 in view of a given goal q, x is left with the task of influencing y to perform the action a_1 in view of his own goal p. This is in a sense outside the scope of this work, which is interested in the preliminaries to interaction, rather than in interaction itself. However, we can try to sketch very roughly the basic steps of such an influencing behavior or, more exactly, its basic requirements. These are in fact the very reasons why the search for a dependent y turns into the search for a "useful" y, i.e. for an agent who is likely to do what x wants to.

So, in order to profit by y's dependence, agent x needs that y comes to believe two fundamental facts.

First of all, quite obviously, y's dependence on x. In fact, until y does not know about her dependence on x, y is not likely to look for x's help, and x has nothing to offer in exchange for his own request. If Carol does not know that John is an expert in ancient Greek, she does not know that he can reciprocate her favor.

But apart from such trivial things as making the other know he has the CANDO the other is in search for, x's task might be more complex. There are in fact cases where y does not know she should have and pursue some means-goal r in view of q, and x's skills are a means for r, rather than directly for q. Suppose Carol has a son, Peter, who risks to fail at an examination in Classical Letters. Carol's goal q is just that Peter passes the examination. John, who knows about this and about the means-end relationship between having a good translation of Sappho's poems (r) and passing the examination (q), should make Carol assume this relationship, as a means for making her assume her dependence on him. In fact, once assumed the relationship between r and q, Carol will have also r as a goal of hers, and find she is dependent on x for that.

However, there is a further means-end relationship y should assume in order for x to obtain what he needs, i.e., y's performance of the required action a_1: the means-end relationship between y's doing a_1 and x's doing a_2. Such a relationship is special in nature, in that there might be no intrinsic reasons why y's doing a_1 should be a means for x's doing a_2. Let us go back to the previous example. While a_2 (doing the translation) is a means for q (that Peter passes the exam), and a_1 (resetting the e-mail server) is a means for p (using the e-mail), why on earth should a_2 be a means for p, or a_1 a means for q?

This means-end relation is a sort of artifact, a social construction on x's part. It is John who actually creates the relation between his doing the translation and Carol's resetting the e-mail server. That is the very reason why x needs some sort of "persuasive" power over y, in order to make y believe that such a relation is likely to hold (Castelfranchi et al., 1991). Such a persuasive action in general consists of a communicative act, be it either a more or less explicitly stated promise (x warrants he will perform a_2 if y performs a_1) or a threat (x threatens he will not perform a_2 if y does not perform a_1; or x threatens he will perform some other action a_3 which will thwart some goal of y's).

7 Conclusions

In this paper, ingredients for a dependence-based architecture of a Cognitive Agent have been proposed. In particular, a module for *Resource Analysis and Allocation* has been sketched, which analyzes any plan provided by the planner and identifies any possible *CogAgent*'s incapacities to carry it out. A further facility is designed for selecting either physical or social resources necessary for the agent's achievement of his goal(s). Strategies focused on the *search for help-givers* have been proposed, based on specific indicators of *others' usefulness* (the structure of dependence) and *willingness* to give help. Also some *top-down criteria* (such as norm- and role-based types of knowledge) have been indicated as ways of incorporating *knowledge about what to ask of whom*.

Other architectural suggestions have been pointed out during the analysis: the distinction between content- and meta-goals; a categorization of others possibly indexed by some of their general goals, role-tasks, and other norms; a map of the general role structure of the domain; an agent role repertoire coupling his action repertoire; finally, a memory of the last interactions within given temporal constraints (possibly including an updating of the payoff matrix relative to each dyadic interaction).

Essentially, we have been exploring how some social knowledge can be utilized by means of suitable search strategies. This may cast a new light on *sociality,* showing how it *originates from problem solving*. Indeed, we provided a frame in which to insert and utilize knowledge about dependence, one of the most fundamental aspects of sociality. We have been showing how it can be fruitfully invested to select help-givers out of a network of artificial agents.

Although rather preliminary, this type of work might contribute to some extent to developments in the field of DAI, where overheads of communications and interactional deadlock have been shown to increasingly occur.

Our *CogAgent* is highly *deliberative*, essentially *goal-driven* and *knowledge-based*. These features impose a number of constraints upon his performances and applications. However, our aim is to situate *CogAgent* in a world a bit more realistic than what is implied by the benevolence assumption. Our multi agent world is not benevolent. On the contrary *agents are* expected to be basically *self-interested* although not necessarily hostile. Nonetheless, such a world is not necessarily unpredictable either. It can be analyzed and reasoned upon to take advantage of it.

In short, we endeavored *to enable our agent to overcome difficulties and obstacles, without resorting to reactive architectures,* which in our terms seem to manage brilliantly perception and motion but are perhaps less apt to carry out complex interactional tasks and problem-solving.

Acknowledgements

We thank Cristiano Castelfranchi for his thoughtful suggestions and comments.

References

Castelfranchi, C. (1990). Social power: A point missed in Multi-Agent, DAI and HCI. In Y. Demazeau & J.P. Muller (Eds.), *Decentralized A.I.* (pp. 49-62). Amsterdam, The Netherlands: Elsevier.

Castelfranchi, C., Miceli, M., & Cesta, A. (1991). Dependence relations among autonomous agents. In D.D. Steiner & J.Muller (Eds.), *Pre-Proceedings of the Third European Workshop on Modeling Autonomous Agents in a Multi Agent World* (pp. 150-161). Kaiserslautern, Germany: Deutsches Forschungszentrum fur Kunsliche Intelligenz. (also in: E. Werner & Y. Demazeau (Eds.) (1992), *Decentralized A.I. 3*. Amsterdam, The Netherlands: Elsevier).

Cesta, A. & Romano, G. (1992). Using abstraction-based similarity to retrieve reuse candidates. In J. Hendler (Ed.), *Artificial Intelligence Planning Systems: Proceedings of the First International Conference (AIPS92)* (pp. 269-270). San Mateo, CA: Morgan Kaufmann.

Cohen, P.R. & Levesque, H.J. (1990). Intention is choice with commitment. *Artificial Intelligence*, **42**, 213-261.

Conte, R., Miceli, M., & Castelfranchi, C. (1991). Limits and levels of cooperation: Disentangling various types of prosocial interaction. In Y. Demazeau & J.P. Muller (Eds.), *Decentralized A.I. 2* (pp. 147-157). Amsterdam, The Netherlands: Elsevier.

Conte, R. & Castelfranchi, C. (1992). Mind is not enough. Pre-cognitive bases of social interaction. In N. Gilbert (Ed.), *Proceedings of the Simulating Societies Symposium* (pp. 93-110). Guildford, UK: University of Surrey.

D'Aloisi, D. (1992). A terminological language for representing complex knowledge. In F. Belli & F.J. Radermacher (Eds.), *Industrial and Engineering Applications of Artificial Intelligence and Expert Systems (Proceedings of the Fifth International Conference IEA/AIE)* (pp. 515-524). Berlin, Germany: Springer-Verlag.

D'Aloisi, D. & Castelfranchi, C. (1992). Propositional and terminological knowledge representations. *Working notes of the AAAI Spring Symposium on Propositional Knowledge Representation* (pp. 57-66). Stanford, CA: American Association for Artificial Intelligence (an extended version will appear on the *Journal of Experimental and Theoretical Artificial Intelligence*, **5** (3), 1993).

Galliers, J.R. (1991). Modelling autonomous belief revision in dialogue. In Y.Demazeau & J.P. Muller (Eds.), *Decentralized A.I. 2* (pp. 231-243). Amsterdam, The Netherlands: Elsevier.

Grosz, B.J. & Sidner, C.L. (1990). Plans for discourse. In P.R. Cohen, J. Morgan, M.E. Pollack (Eds.), *Intentions in Communication*. Cambridge, MA: MIT Press.

Ingrand, F.F. & Georgeff, M.P. (1989). Managing deliberation and reasoning in real-time AI systems. *Proceedings of the Eleventh International Joint Conference on Artificial Intelligence (IJCAI-89)*. San Mateo, CA: Morgan Kaufmann.

Lesser, V.R. & Corkill, D.D. (1981). Functionally accurate, co-operative systems. *IEEE Transactions on System, Man and Cybernetics*, **SMC-11**, 81-96.

Pollack, M.E. (1990). Plans as complex mental attitudes. In P.R. Cohen, J. Morgan, M.E. Pollack (Eds.), *Intentions in Communication*. Cambridge, MA: MIT Press.

Rao, A.S., Georgeff, M.P., & Sonenberg, E.A. (1991). Social plans: A preliminary report. In D.D. Steiner & J. Muller (Eds.), *Pre-Proceedings of the Third European Workshop on Modeling Autonomous Agents in a Multi Agent World* (pp.32-46).

Kaiserslautern, Germany: Deutsches Forschungszentrum fur Kunsliche Intelligenz. (also in: E. Werner & Y. Demazeau (Eds.) (1992), *Decentralized A.I. 3*. Amsterdam, The Netherlands: Elsevier).

Reddy, M. & O'Hare, G.M.P. (1991). The blackboard model: A survey of its application. *Artificial Intelligence Review*, **5**, 169-186.

Russell, S.J. & Wefald, E.H. (1991). Principles of metareasoning. *Artificial Intelligence,* **49**, 361-395.

Smith, R. (1978). *A framework for problem solving in distributed processing environment*. (Report No. HPP-78-28). Stanford, CA: Stanford University, Department of Computer Science.

Session 3: Natural Language 1

Word Recognition as a Parsing Problem

Paul McFetridge
Simon Fraser University
Burnaby, Canada

Abstract

The problem of word recognition is represented as a syntactic process of parsing sequences of morphemes into words. The system is based on theories of lexical morphology in which affixes are represented as lexical entries. According to lexical morphology, morphological structure is phrase structural. A strict distinction between phonology (or orthographic variation between input strings and their lexical representation) and morphology is maintained. The phonological component together with the lexicon parses an input string into sequences of morphemes. The morphological component parses the sequence into words. Morphological structure is represented as a directed acyclic graph; unification is the single process for combining morphemes. A chart parser is used to apply a a single rule to sequences of morphemes. Examples from English, Turkish and Arabic are discussed. It is demonstrated that the system can handle both highly agglutinative morphology and nonconcatenative morphology. In addition, the theory proposes a universal property of inflectional affixes: they inherit syntactic features from their stems. This property explains how the contributions of many inflectional affixes are realized on the word and permits a single morphological rule type which handles both derivation and inflection.

Introduction

This paper presents a system for word recognition based, in part, on the family of theories of lexical morphology presented by Lieber (1981), Selkirk(1982) and Williams (1981), among others. A key position taken in these theories is that affixes are lexical items, that morphological structure is phrase structural and that the features of the word are inherited from its constituents.

This position contrasts with the Word and Paradigm model (see Anderson (1982)). A paradigm represents all the morphological relations in which a root can participate. The principle argument for word and paradigm morphology is that there are morphological phenomena such as ablaut, reduplication and the binyanim morphology of Arabic that are not phrase structural or that involve no identifiable affixes. That is, there appears to be morphological operations which are not concatenative. Word and paradigm theories incorporate concatenative morphology as a special case of nonconcatenative morphology. Calder(1989) has gives an implementation of Word and Paradigm morphology.

There are reasons to maintain an interest in lexical morphology. First, if morphology (or much of morphology) is phrase structural, then parsing algorithms designed for syntactic analysis may be applied to morphological analysis. Second, there are aspects of word structure, e.g. compounding, which are clearly phrase structural. Lexical morphology allows compounding, derivation and inflection to be handled by a single operation. Finally, the historical genesis of most affixes is by the conversion of free forms to bound forms; for example, Latin prefixes such as *ad* 'to' in *admit* and *con* 'with' in *commute* were also independent words. In lexical morphology, this is explicable as the change in the value of the feature which distinguishes free from bound forms.

Phonology and Morphology

The word and paradigm theory conflates morphology and phonology. The theory permits morphological process to perform phonological operations. A similar tendency (noted by Kay (1983)) to conflate morphology with the principles which govern alternations between input strings and their lexical representations exists in computational linguistics. The system presented here maintains a strict distinction between morphology and phonology. Phonology is the set of principles which relate surface strings with lexical representations. Morphology is the set of principles which governs the combination of morphemes into words. Because of this distinction, the subsystem responsible for generating lexical representations of morphemes could be revised or replaced without affecting the morphological subsystem responsible for combining morphemes into words.

Increasingly, the value of a declarative linguistic analysis independent of implementation is being recognized (see Shieber (1986)). The question of whether the representation of phonology (or its orthographic equivalent) is best procedural or declarative is moot. However, by maintaining the distinction between phonology and morphology it is possible to provide a declarative analysis of morphological structure independent of phonological representation.

In the following section, the parsing subsystems are described. The case studies of next section illustrate how a variety of morphological phenomena can be successfully handled by lexical morphology. An appendix contains a small lexicon for Turkish which together with the rule in figure 1 illustrates many of the points discussed in the text.

System Configuration

The system consists of two separate modules, each operating on its own set of principles. The first module is a lexical parser which takes as input a string. The string may contain an arbitrary number of words of arbitrary morphological complexity. The lexical parser segments the string into a list of possible morphemes. Currently, possible segmentations are restricted to those which span the entire string. As research on handling ill-formed and unknown words proceeds, this restriction will be lifted so that segments of the string may be left unparsed.

After the string has been segmented in morphemes, the set of morphemes are combined into words according to a morphological grammar. The grammar formalism represents both lexical items and morphological rules as directed acyclic graphs (see Shieber (1986)). The lexicon is structured as a set of inheritance networks as in Flickinger, *et al* (1985). Individual lexical items are marked as instances of inheritance paths. For example, in Turkish if a lexical item is an instance of *infl* — i.e., if it is an inflectional affix — then it is also an instance of *affix*. A morpheme such as *imiz* 'our' is marked as an instance of *poss* which itself is an instance of *infl*.

Lexical Parser

The goal of the lexical parser is to segment an input string into morphemes without regard to their combinatorial properties. This level of analysis must take into account orthographic variations (analogous to phonological variations in speech) between the input string and the lexical entries of the morphemes. For example, if *love* is entered in the lexicon with the final *e*, then the lexical parser must insert the *e* when parsing *loving*. The lexical parser uses a finite state transducer to "undo" the effects of the spelling system (analogous to "undoing" the effects of the phonology of the language). The transducer is based on "Two-level morphology" (Koskenniemi (1983)). However, whereas two-level morphology conflates morphology and phonology (or orthographic alternation), this system maintains a clear distinction between them. The lexical parser is not responsible for assigning morphological structure.

Since the mapping from the orthographic (or phonetic) form to the lexical form is one-to-many, the transducer will produce many forms which do not correspond to possible morphemes. Following a suggestion by Kay (1983), explosive ambiguity is eliminated by tying the transducer directly to the lexicon. The lexicon is implemented as a character tree. Each character produced by the transducer permits the parser to traverse the corresponding branch of the character tree. If there is no corresponding branch, the path through the network that the transducer has been traversing fails. The parser continues until either it reaches the end of the string or it fails.

The transducer responsible for mapping phonetic (or orthographic) forms onto lexical representations may write to more than one output tape. Kay's (1987) analysis of the noncatenative morphology of Arabic explores the use of multiple output tapes, each corresponding to a morpheme. This procedure has been exploited here so that the lexical parser can analyze a word or stem into a set of morphemes which begin and end at the same points.

Morphological Chart Parser

Morphemes selected by the lexical parser are parsed into words by a chart parser constructed largely on the principles given by Thompson (1981). A significant extension from the standard chart parser is the introduction of parameterization on the type of morphology parsed. Prefixation and suffixation are parsed by the fundamental rule of chart parsing: when the end of an active edge meets the beginning of an inactive edge and the inactive edge satisfies the conditions on the active edge's extension, a new edge is created beginning at the vertex at which the active edge begins and ending at the vertex at which the inactive edge ends.

The fundamental rule fails to correctly parse the interdigitated morphology of the Arabic verb root. The Arabic verb root is analysed as composed of three morphemes which share beginning and end points. When this configuration is parsed, the fundamental rule is altered so that:

1. the *beginning* of an active edge must meet the beginning of an inactive edge;
2. the inactive edge must meet the conditions on the active edge's extension;
3. the inactive edge, if not the only edge in the active edge's extension, ends at the same vertex as the active edge.

When conditions 1–3 are met, a new edge is created beginning at the vertex at which the active edge begins and ending at the vertex at which the inactive edge ends.

The effect of this alteration is that the first edge which meets the conditions on an active edge's extension establishes the span in the chart which all subsequent edges must span if they meet the active edge's extension.

The fundamental rule is parameterized; a switch determines which version is applied. The switch can be held constant over the entire parse as one might want for English and Turkish or can be set by individual morphemes so that Arabic verb stems are parsed with the new version of the fundamental rule, but Arabic affixation is parsed by the standard fundamental rule.

Case Studies

English

Affixation, both inflection and derivation, is controlled by a single rule given in figure 1. The rule is concerned only with constituent structure; separate principles control the order of morphemes within the word. The rule checks for a constituent marked as a stem followed or preceded by a constituent marked as an affix. Of particular importance is the structure sharing induced by the rule (and indicated by $|x|$ where x is a digit). The rule asserts that the word will inherit by the *infl* and *syn* values of the affix. The former is used to mark whether an affix is inflectional or derivation. All English affixes subcategorize for stems which are marked $[\text{infl} -]$; this accounts for the fact that English does not permit affixation outside an inflectional affix which is marked $[\text{infl} + \quad]$. Inheriting the *infl* feature of the affix serves to declare whether a form has been inflected.

Of some import is the inheritance of the affix's syntactic features. It is demonstrated how this property explains differences between inflectional and derivational affixes. Later, it is argued that this property can be used to account for ordering relations among morphemes and for multiple instances of inflectional affixes in relativized nominals in Turkish.

$$\begin{bmatrix} W \begin{bmatrix} \text{mor} & [\text{infl} \,|1|] \\ \text{syn} & |2| \end{bmatrix} \\ S \begin{bmatrix} \text{mor} & [\text{cat stem}] \end{bmatrix} \\ \text{af} \begin{bmatrix} \text{mor} & [\text{infl} \,|1|] \\ \text{syn} & |2| \end{bmatrix} \end{bmatrix}$$

Figure 1: Affixation

Affixes subcategorize for the stem to which it is added; both the morphological and syntactic features of the stem must match the affix's subcategorization. For example, the affixes in figure 3 subcategorize for uninflected stems of particular syntactic category.

Category Changing and Preserving Affixes

A property which distinguishes inflectional and derivational morphology is that while the latter may change grammatical category, the former never does. This property is captured in the lexical entries of the affixes together with the affixation rule in figure 1. The affixation rule declares that the syntactic features of the word are inherited from the affix. Category changing affixes will be marked accordingly with syntactic features in their lexical entries. Abbreviated lexical entries for the prefix *en* and the suffix *ment* are given in figure 3 as examples. In each entry, subcategorization is realized as an *arg* feature which is itself one of the features representing the morphological properties of the affix. The *arg* feature specifies properties which a stem must have if the affix is added to it. These prefixes also have a *syn* feature which by the rule in figure 1 will be inherited as the *syn* feature of the word of which they are a constituent.

In contrast with derivational affixes, inflectional affixes never change syntactic category. This fact is explained by an inheritance property of inflectional affixes: all inflectional affixes inherit the syntactic features of the stem to which it is added. In a lexical hierarchy, all affixes which are marked $[\text{infl} + \quad]$ will also have the form in figure 4.

$$\begin{bmatrix} \text{mor} & [\text{arg} \,[\text{syn} \,|1|]] \\ \text{syn} & |1| \end{bmatrix}$$

Figure 3: Inheritance by inflectional affixes

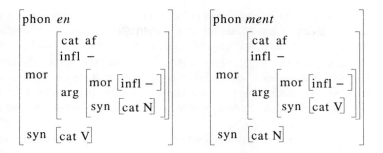

Figure 2: Lexical Entries for *en* and *ment*

For example, the English affix *ing* inherits the syntactic features of the verb to which it is added and contributes a feature representing aspectual information.

$$\begin{bmatrix} \text{phon } ing \\ \text{mor} \begin{bmatrix} \text{arg} \begin{bmatrix} \text{mor } [\text{infl} -] \\ \text{syn } |1| [\text{cat V}] \end{bmatrix} \\ \text{infl} + \end{bmatrix} \\ \text{syn } |1| [\text{asp prog}] \end{bmatrix}$$

Figure 4: Lexical entry for *ing*

The inheritance property of affixes solves a long standing problem in morphological theory: how are features of the word determined by the combination of the features of the stem and affix. Selkirk(1982) and Williams(1981) have proposed that morphological structures are headed and that the syntactic features of the word are inherited from the head constituent. A critical problem with the notion of morphological head is providing a definition which determines for any particular structure which constituent is the head. Williams' definition of the head as the right hand constituent has counterexamples (e.g. $\begin{bmatrix} \text{en} & [\text{rage}]_N \end{bmatrix}_V$). Selkirk proposes that the morpheme which is not rightmost may nonethelessbe the head and hence determine syntactic category providing no other morpheme to its right is marked for syntactic category. Again, this principle fails to explain cases where prefixes determine syntactic category. Finally, Lieber(1981) has proposed that the features of the word are the features of the affix (as in the approach outlined here). If the features of the affix do no also specify syntactic category, then those from the stem are inherited by the word. As stated, Lieber's approach is computationally complex, requiring that features of the word be check for requisite features which if not present are retrieved from the stem.

The approach proposed here eliminates the need for morphological heads and identifies the affix as the sole determinant of syntactic features. It requires no other process than unification to determine the features of the word.

The property of inheriting syntactic information from the stem by inflectional morphemes also ensures that, in morphological structures that contain more that one inflectional affix, the contribution of each affix is reflected on the word. Although this situation does not occur in English, it is common in languages like Turkish. When an inflectional affix is added to a stem, the syntactic features of the stem including those contributed by other inflectional affixes are inherited by the affix and in turn by the word as a whole.

This system of inheritance also blocks the addition of inflectional affixes with conflicting features. For example, assuming that the Latin first person present indicative morpheme -$ō$ (in e.g. $amō$ 'I love') is marked as [-past], it is predicted that it will not occur in structures which are marked [+past], as in the perfect $amāvī$ and the imperfect $amābam$.

Finally, the property of inheriting syntactic features of the stem explains why inflectional affixes do not change syntactic category: an affix cannot both inherit the syntactic features of its stem and be marked for a different set of syntactic features.

Inflection and Derivation

There exists an ordering relation between derivational and inflectional affixes: the former typically appears inside the latter. If a word has been inflected, no further derivation is possible. This ordering relation is accounted for by the rule in figure 1 by inheritance of the *infl* feature. In English, this is a binary feature; inflectional affixes are marked '+', derivational affixes '-'. Derivational affixes subcategorize for stems which are marked $[\text{infl} -]$. Since inflected words will have inherited $[\text{infl} +]$ from their affix, the addition of an inflectional affix blocks further derivation.

Turkish

Turkish is a paradigmatic example of an agglutinating language. It provides an opportunity to illustrate how ordering relations among morphemes can be determined by features on the affixes and to illustrate how the property of inheriting the syntactic features of the stem universal to inflectional affixes is also responsible for accumulating the contributions of each affix on the word.

Hankamer (1989) has argued that morphological parsing – at least for Turkish – must be root-driven as contrasted with "affix-stripping" models. The general procedure is to discover the root, extract from its lexical entry a template of what can follow and use this template to direct parsing. Since Turkish has almost no prefixes[1], left to right parsing will ensure that the root is located first. Hankamer argues that to isolate affixes without the benefit of the template results in an explosion of false analyses. For example, any final low vowel could be the dative suffix attached to a nominal stem and any final high vowel could be the accusative suffix. Thus, any word which ends in a vowel must be considered as a potential noun ending in a case marker.

The approach developed here is neither a root-driven nor affix-stripping algorithm. The combinatorial explosion envisioned by Hankamer does not occur because the lexical parser segments a string into a set of morphemes only if that set spans the entire string. If a final vowel is not a case marker, no configuration of morphemes which include the final vowel as a case marker is possible.

Rather than express the relative order of affixes as a template associated with the root, lexical morphology exploits the subcategorization restrictions on the morphemes themselves The analysis is illustrated with Turkish nouns. These have the structure ROOT + number + (poss) + case Both number and case may have zero exponents (the singular number and the absolute case). True suffixation is handled by the same rule as for English (figure 1). Unlike English, the *infl* feature in Turkish is not binary valued,

---

1. According to Lewis (1967), the only regular prefixation in Turkish is to intensify adjectives and adverbs.

but feature valued. Each affix contributes a feature indicating its presence. For example, *ler* the plural morpheme contributes [mor [infl [num +]]]. Number morphemes subcategorize for stems which are marked $\left[\text{infl} \begin{bmatrix} \text{num} - \\ \text{poss} - \\ \text{case} - \end{bmatrix}\right]$. Possessive morphemes subcategorize for stems which are marked as $\left[\text{infl} \begin{bmatrix} \text{num} + \\ \text{poss} - \\ \text{case} - \end{bmatrix}\right]$. Case morphemes subcategorize for stems which are marked $\text{infl} \begin{bmatrix} \text{num} + \\ \text{case} - \end{bmatrix}$. The contributions to *infl* of previous morphemes are maintained by permitting affixes to inherit *infl* features from their stem. Thus, when a possessor suffix is added to a noun stem marked $[\text{num} +]$, the resulting stem is also marked $[\text{num} +]$.

Since inflectional affixes contribute syntactic features, the inheritance of syntactic features by subsequent inflectional affixes will ensure that all syntactic features contributed by inflectional affixes will 'percolate' to the level of the word. This is illustrated by *evlerimizde* ev+ler+imiz+de 'in our houses'. The addition of *ler* creates a stem marked as (num +) in its *infl* features and $[\text{num pl}]$ in its *syn* feature. The addition of *imiz* contributes the *syn* feature $[\text{poss 1pl}]$ and the *syn* features of the stem to which it is added, so that the *syn* features of the new stem are $\begin{bmatrix} \text{cat} & \text{n} \\ \text{poss} & \text{1pl} \\ \text{num} & \text{pl} \end{bmatrix}$. The addition of the locative *de* contributes the feature $[\text{case locative}]$ and the *syn* features of the stem to which it is added.

Turkish has a relativizing suffix *ki*, which is added after a word has been inflected. As a derivation affix, *ki* does not inherit syntactic features from the stem, but contributes its own. The theory predicts that since *ki* is marked $[\text{infl} -]$ and contributes its own syntactic features adding will have the effect of resetting the INFL features so that inflection can begin again. In fact, this is precisely what happens: cf. *evlerimizdekilerin* ev+ler+imiz+de+ki+ler+in 'of the ones in our houses'.

Arabic

	Perfective Active	Perfective Passive	
I	katab	kutib	*write*
II	kattab	kuttib	*cause to write*
III	kaatab	kuutib	*correspond*
VIII	ktatab	ktutib	*be registered*

Table 1: Arabic verb roots

The Arabic verb stem, of which a very small sample is given in table 1, is formed on a root consisting of consonants to which are intercalated vowels marking voice and aspect. In addition, the pair of consonantal root and vowel sequence can co-occur in different prosodic combinations (called *binyanim*) four of which are listed in table 1.

McCarthy (1981) has argued that each of these — the root, the vowel sequence and the binyan — are separate morphemes arrayed on different tiers as in figure 5. On this analysis, the word *katab* consists of three morphemes: the root *ktb*, the vowel *a* and the prosodic template CVCVC. McCarthy also argues for derivational prefixes; in binyan VIII in table 1 the derivational prefix *t* has metathesized with the initial *k*. In the analysis presented here, these are not analysed as prefixes; rather, the notion of 'prelinking' introduced by Marantz(1982) is exploited. Marantz argues that in reduplicative phenomena, elements in a prosodic template may be lexically associated to segmental content. On this view, binyan VIII may be represented as CtVCVC; the second element of the binyan is lexically specified as *t*.

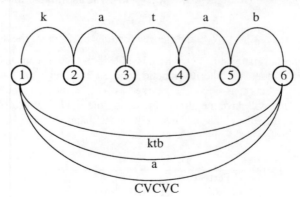

Figure 5: *katab*

The task of the lexical parser is to isolate three morphemes which span the same section of the input string. As previously described, the lexical parser maintains a set of output tapes in parallel. The lexical parser proceeds only if each member of the set is able to traverse the character tree; if one member fails, the entire set fails. If each tape in a set leads to a lexical item, the parser will have isolated three morphemes which span the same section of the string. For example, the string *katab* is analyzed into the set ("CVCVC" "ktb" "a") each member of which begins at vertex 1 and ends at vertex 6.

Figure 6: Chart parsing *katab*

The chart into which the individual characters and component morphemes of *katab* have been entered is illustrated in figure 6.

Although the morphological structure of Arabic initially appears radically different from English or Turkish, structures like those in figure 5 are created by the same morphological rule responsible for English and Turkish morphology, that in figure 1. The single difference between Arabic and Turkish is the change to the fundamental rule of chart parsing. Whereas in Turkish morphemes are arrayed sequentially, in Arabic each morpheme must begin and end at the same point as the others in the verb stem. This difference is captured, not in the morphology, but in the parser. The morphological rules of English, Arabic and Turkish are identical.

Conclusions

The system described here is capable of handling a variety of morphological phenomena with a small set of linguistic principles, particularly the principle that inflectional affixes inherit syntactic features from their stems. It has been demonstrated that a single rule is capable of parsing derivational and inflectional structures, concatenative and nonconcatenative, in a variety of languages. The parsing technology has been extended in a way which is minor to implement but has considerable consequences: the extensions permit the principles responsible for parsing concatenative morphology to parse the kind of nonconcatenative morphology found in Arabic.

The system maintains a distinction between phonology (or orthographic variation) and morphology. The morphological component is strictly declarative. Currently, the implementation of the phonological module is procedural. However, because the architecture maintains the distinction between phonology and morphology, it is possible to remain agnostic about whether the phonological component should be declarative.

There are many issues left to address. Among the inventory of nonconcatenative morphological phenomena that have not been addressed are ablaut and reduplication. Spencer(1988) has suggested that ablaut may be handled by the same principles as those operating in Arabic. Thus, *foot* may be analysed as consisting of the root *ft* together with the singular morpheme *oo*. A possible objection to this is that the semantic connection found among various instantiations of the Arabic consonantal root is not present among words with the consonants *ft*. Marantz(1982) has demonstrated that reduplication does not require transformational power but can be analysed as the affixation of a prosodic template which is associated with material from the stem. It is not yet clear how this can be incorporated into the system.

The system is implemented in Common Lisp.

Acknowledgments

The author is a member of the Institute for Robotics and Intelligent Systems (IRIS) and wishes to acknowledge the support of the Networks of Centres of Excellence Program of the Government of Canada, the Natural Sciences and Engineering Research Council, and the participation of PRECARN Associates, Inc. Acknowledgment is also due to the Centre for Systems Science of Simon Fraser University and the Canadian Cable Labs Fund of Rogers Cablevision for their material and financial support.

References

Anderson, S. R. (1982). Where's morphology? *Linguistic Inquiry*, **13**, 571–612.

Calder, J. (1989). Paradigmatic morphology. *Proceedings of the Fourth Conference of the European Chapter of the Association for Computational Linguistics*, Manchester, 58–65.

Flickinger, D. P., Pollard, C. J. & Wasow, T. (1985). Structure-sharing in lexical representation. *Proceedings of ACL, 23rd Annual Meeting*, 262-267.

Hankamer J.(1989) Morphological parsing and the lexicon. in W. Marslen-Wilson. (Ed.), *Lexical Representation and Process* (pp. 392–408). Cambridge, MA: MIT Press.

Kay, M. (1983). When meta-rules are not meta-rules. In K. Sparck-Jones & Y.A. Wilks

(Eds.), *Automatic Natural Language Parsing* (pp. 94-116). Chichester: Ellis Horwood or Wiley.

Kay, M. (1987). Nonconcatenative finite-state morphology. *Proceedings, Third European Conference of ACL*, 2-10.

Koskenniemi, K. (1983). *Two-level Morphology: a general computational model for word-form recognition and production*. Helsinki: University of Helsinki.

Lieber, R. (1981). *On the Organization of the Lexicon*. Bloomington, Ind.: Indiana University Linguistics Club

Lewis, G. L. (1967) *Turkish Grammar*. Oxford: Clarendon Press.

Marantz, A. (1982). Re reduplication. *Linguistic Inquiry*, **14**, 483–545.

McCarthy, J. (1981). A prosodic theory of nonconcatenative morphology. *Linguistic Inquiry*, **12**, 373–418.

Selkirk, E. (1982). *The Syntax of Words*. Cambridge, MA: The MIT Press.

Shieber, S. (1986). *An Introduction to Unification-Based Approaches to Grammar*. Stanford, Calif: Center for the Study of Language and Information.

Spencer, A. (1988). Lexical rules and lexical representations. *Linguistics*, **24**, 619–640.

Thompson, H. S. (1981). Chart parsing and rule schemata in PSG. *Proceedings of ACL, 19th Annual Meeting*, 167-172.

Williams, E. (1981). On the Notions of "Lexically Related" and "Head of a Word. *Linguistic Inquiry*, **2**: 245–274.

Using Actions to Generate Sentences

Paddy Matthews

Department of Computer Science, University College,
Dublin, Ireland.*

Abstract

In this paper, a method for the generation of text from expert system output is presented which is intermediate in complexity between a simple template-based approach and the more complex approaches based on User Interface Management Systems (UIMS). The system described forms part of a Multi-Media Interface for a Process Control exemplar. The approach outlined in this paper uses the concept of *actions*, functions contained within a template each of which has the responsibility of generating one particular syntactic category of output (e.g. nouns, verbs, etc.). The order of generation of these actions may be varied in order to produce a range of possible syntactic structures to be output. This variation is determined by the value of variables indicating such things as negation, subordinate clauses, tense, etc. which are set during the process of choosing an appropriate template to express the meaning of the Process Control output. The actions are discussed in some detail and the relationship between the actions and the syntactic categories which they correspond to is discussed, with regard to the two languages for which generation has been implemented (English and Dutch). We also discuss possible refinements to the system to make it more sophisticated and user-friendly.

1 Introduction

The work described in this paper was undertaken as part of the ESPRIT Project PROMISE. As part of our task, we were required to construct Natural Language Generation modules for two Process Control exemplars (systems simulating the operation of a chemical plant and a nuclear power station). The two modules were to produce output in two separate languages (English and Dutch). The system would take a piece of output from the Process Control system describing an emergency situation which had arisen and listing possible methods of correcting the situation, and would produce from it a representation of the output as Natural Language text.which would preserve the meaning as far as possible. This output would be sent to the Presentation System in tagged-data format, which would allow it to be presented to the user in the form of text, speech, etc. as deemed appropriate by the Data Manager module. The overall system architecture is described in the diagram on the next page (Fig. 1).

Two main approaches had been used previously in the generation of output from expert systems. The first involved a "fill-in-the-blank" approach, where the names of system variables were simply slotted into blank spaces in a template which was then output. Morphological processing with this approach was minimal. Examples of this approach in recent systems include (Wick and Slagle, 1989), and, at a more advanced level, (Bench-Capon , Lowes & Mc.Enery, 1990). This approach is suitable if we merely wish to present a set of results and do not wish to go into more detailed explanations as to how these results were produced. However, the flexibility of output produced by this method is minimal, and the quality of explanations provided to the user suffers from the lack of understanding of the output by the system itself.

---

* E-mail: in%"matthews@ccvax.ucd.ie". Current address: School of Management Science, Dublin Insititute of Technology, Rathmines, Dublin 6, Ireland.

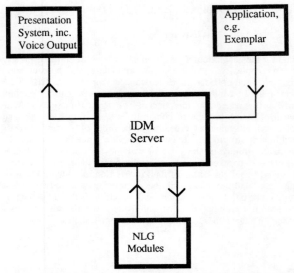

Fig 1 Flow of data between components of system

The second approach used is one involving knowledge about discourse structure, plans, goals, discourse history, etc. Examples include (Allgayer et al., 1989), (Löwgren, 1989) and (Moore and Swartout, 1989). This approach can give extremely impressive and sophisticated output, particularly in systems involving interactive dialogue between the user and the system, but systems using this approach can be extremely complex and time-consuming to construct, and in cases where such a level of sophistication is unnecessary, they represent a waste of time and effort.

We wished to be able to provide a system which could provide a certain degree of syntactic flexibility (partly due to certain features of Dutch syntax which will be explained later) but which, because of our limited budget in terms of man-months, could be constructed fairly quickly. It was decided to design a system which would try to combine some of the simplicity of the template-based approach with the syntactic flexibility of the more sophisticated approaches.

2 Architecture of the NLG Module

The overall architecture of the NLG module can be represented by the diagram on the next page (Fig. 2):

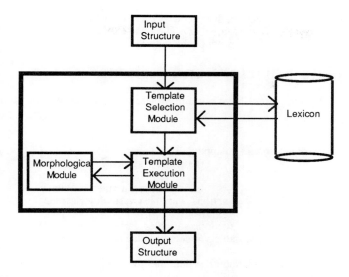

Fig. 2 Architecture for Natural Language Generation Module

The module consists of four smaller modules: a Template Selection Module, a Lexicon, a Template Execution Module, and a small-scale Morphological Module.

The Template Selection Module takes as input the information provided by the Process Control System which is to be output as text. The module operates by means of a set of discrimination nets. These discrimination nets check the input to the NLG module for the presence or absence of certain features (these might be either features such as the syntactic structure of the input, e.g. the presence of certain slots within the structure, or features such as the value of the filler of a slot within the structure), and based on the results of these tests, will perform certain actions, such as carrying out further tests, setting the value of variables indicating such things as tense, number, etc. In certain circumstances, the discrimination nets may operate in a recursive fashion, with parts of the input being sent back to the topmost discrimination net for analysis in their own right (e.g. complex expressions such as "If <clause>, then <clause>"). Based on the cumulative results of a set of tests, an appropriate template is chosen from the Lexicon, and is sent to the Template Execution Module. During the operation of this module, the functions in the template are executed. Each template consists of a function detailing which parts of the input to the NLG module will be dealt with by each action in the template (a mapping function) and the actions themselves, which produce the output. During the process of the execution of the template, any verbs in the output are passed on by means of one of the actions to the Morphological Module, which makes any adjustments necessary for tense, grammatical number, etc. The actions in the template may also take fragments of the process control output structure, and pass them into the discrimination nets at a particular node, e.g. for subordinate clauses. This step is undertaken in order to provide for the maximum flexibility of our templates in terms of the semantic range of output, and reduces the number of templates necessary in our lexicon, as we can classify certain types of output, e.g. if-then structures, as a single template but with various clauses filling the slots within the if-then structure. The Lexicon consists of two types of information: firstly, the templates, used to generate sentences or clauses, and secondly a data dictionary, consisting of Process Control System variables, together with their descriptions in the target language and any other grammatical information which would be necessary to correctly output the descriptions, e.g. grammatical gender.

The architecture of the system is discussed in more detail in (Matthews & Mok, 1992) and (Matthews, Mok & Kavanagh, 1992). The rest of this paper will concentrate on the actions and how they operate to produce the output.

3 Template Structure and Actions

All templates in our system have a common structure, which can be summarized as follows:

>(template-name (mapping
>   (structure-fragment1 variable1)
>   .....
>   (structure-fragment$n$ variable$n$))
>   ((action1 variable$x$ variable$y$ variable$z$)
>   ....
>   (action$p$ variable$q$ variable$r$ variable$t$))
>   ((global-variable$a$:= default-value$a$)
>   ....
>   (global-variable$z$:= default-value$z$)))

where template-name is the name of the template,
> mapping represents a set of relationships between a set of fragments of the structure structure-fragment$x$, where $x=1..n$ and a corresponding set of template variables variable$x$, where $x = 1..n$,

(action$n$ x y z) represents a procedure with template variables x, y and z, where x, y and z are template variables which have had their values set during the mapping stage, and global-variable$x$ is a variable such as **subord**, **neg**, etc., which is reset to its default value when the execution of the template is complete (a list of these variables will be given later in this section).

The mechanism for mapping parts of the structure onto template variables has been implemented as a number of separate functions, one of which is called depending on the structure and the number of template variables. However, in order to ease the comprehensibility of the examples in this chapter, all of these varying functions will be treated as if they were a single function. Therefore, in the example templates in the rest of this paper, a brief description of the possible syntax of the possible structure will be given, indicating the position of those fragments of the structure which are mapped onto template variables, i.e. high(x) indicates that the part of the high() structure between the brackets has been mapped onto the template variable x.

The actions are implemented as functions which take in variables or parts of the structure and produce appropriate text fragments. These actions may produce individual words, recursively send fragments of the structure to the discrimination net, build new structures from existing fragments, obtain the gender or produce the text attached to an item in the lexicon, or, given a verb and values for **tense**, **pers** and possibly other variables, produce an appropriately inflected form of the verb. In this way, the morphological module discussed in earlier documents has been integrated within the main module.

The list of actions held in common by both the Dutch and English generators are as follows:

(output_word *string*): This function takes *string* (a word or words enclosed in double quotes) and outputs it directly, followed by a space.

(**output_verbform** *verb tense person*): This function takes a verb (in infinitive form), and values for person and tense, and produces the appropriate form of the verb. Its operation will be covered in greater detail in the next section.

(**build_structure** *list-of-items*): This function takes the contents of *list-of-items* (structure fragments or individual characters or pieces of text) and builds them into a structure.

(**output_as_text** *structure node-number (variable-settings)*): This function takes a structure (either a template variable or the result of calling *build-structure*), and sends it to the discrimination net at node *node-number*. Certain global variables (such as **subord**) may have their values reset if they are included in *variable-settings*.

(**output_number** *number*): This function simply outputs a number *number*.

The following actions are used only by the Dutch generator. This is because they are used to deal with grammatical phenomena which are particular to Dutch, e.g. separable particles and gender of nouns.

(**output_separable_particle** *string*): This function is used when a separable particle, e.g. 'op' in 'opkomen' ('to come on'), is to be output directly before the inflected form of the verb and is therefore fused with it, as in a subordinate clause, e.g. "Omdat het alarm *op*kwam (Because the alarm came on)" vs. "Het alarm kwam *op* (The alarm came on)". It outputs *string* without a following space.

(**output_as_dutch** *lexicon-entry preceding*): This function takes a lexicon entry, e.g. ALM(19) (a Process Control System variable), and outputs the contents of the description field of the lexicon entry (its description in Dutch), possibly preceded by a definite or indefinite article, depending on the value of *preceding*. *Preceding* can have one of four values: defart, indefart, negart or noart. If *preceding* has the value defart, then 'het' is output before the contents of the description field if the value of the gender field of the lexicon entry has the value n (for neuter gender), and 'de' is output if the value of the gender field is c (for common gender). If *preceding* has the value indefart, then 'een' is output, and if it has the value negart, then 'geen' is output. Finally, if *preceding* has the value noart, then nothing is output before the description field.

(**get_gender** *lexicon-entry*): This function returns the value of the gender field for *lexicon-entry*.

The following action is used only by the English generator:

(**output_as_english** *lexicon-entry preceding*): This function takes a lexicon entr, e.g. ALM(19), and outputs the contents of the description field of the lexicon entry, possibly preceded by a definite or indefinite article, depending on the value of *preceding*. *Preceding* can have one of four values: defart, indefart, negart or noart. If *preceding* has the value defart, then 'the' is output before the contents of the description field. If *preceding* has the value indefart, then 'an' or 'a' is output before the contents of the description field depending on whether the contents of the description field start with a vowel or consonant, respectively, and if it has the value negart, then 'no' is output. Finally, if *preceding* has the value noart, then nothing is output before the description field.

The list of variables set by the discrimination net is as follows:

overall-structure: This represents the entire structure whose meaning is to be output by the discrimination net, and which will be output by the system in the case of failure to choose an appropriate template.

current-structure: This represents the structure currently being worked on by the discrimination net. It may be the same as **overall-structure**, or it may be a part of **overall-structure**, as will often happen when the discrimination net is being called recursively during the execution of the templates. At the beginning of the execution of the discrimination net, the two variables will be identical.

subord: This can have one of two values, TRUE or FALSE. It indicates whether the text to be output is to be expressed as a subordinate clause, e.g. following 'omdat' or 'dat' or 'die', or as an independent clause. Default value is FALSE. This variable is used only in Dutch. This is because Dutch requires a different word order in subordinate clauses. Whereas the inflected verb is normally placed in the second position in a clause, in subordinate clauses the inflected verb is moved to the end. In order to generate Dutch sentences correctly, we need to know which utterances are in main clauses, and which are in subordinate clauses.

tense: For the purposes of this system, this variable will have one of three values for the system: PRES (present tense), PAST (past definite tense, e.g. "Het alarm dat 3 minuten geleden *opkwam* is aktief" - English: "The alarm which came on 3 minutes ago is active".), or IMPV (imperative, e.g. "*Verwacht* een te hoog niveau in Vat D-220" - English: "Expect an excessive level in Vessel D-220".). It indicates the tense of the clause to be output. Default value is PRES. Other compound tenses in both English and Dutch can be built up using these three tenses, together with present and past participles, but other languages will have additional tenses.

pers: This indicates the person of the verb in the sentence/clause. For the purposes of this system, only two values, *sing2* (second person singular, used with imperative) and *sing3* (third person singular, used in all other instances) are used. Default value is *sing3*.

neg: This variable will indicate whether the clause/sentence is to be negated or not. It can have the values TRUE (i.e. the clause is to be negated) or FALSE. The default value is FALSE.

4 Example Templates

We will now illustrate the use of actions in the production of output by means of selected templates in both Dutch and English. The templates themselves are actually coded in C, but we represent them by a Lisp-based pseudocode.

Our first example is a quite complex template used in the Dutch generator, which includes such features as subordinate clauses, and changing word order.

```
(ZIJN-AKTIEF2
    (timerel (x,y))
    ((output_as_dutch x noart)                    ; 'Alarm a'
     (if y != () then ((if (get-gender x) = n then (output_word "dat")
                                          else (output_word "die"))
                      (output_as_text y ((subord := TRUE))))
     (if subord = TRUE then ((output_word "aktief")
                             (output_verbform zijn tense pers))
                       else ((output_verbform zijn tense pers)
                             (output_word "aktief"))))
    ((subord := FALSE)))
```

Here the structure is of the type timerel(event, time) and will produce output of the type "Alarm X dat op time Y opkwam is aktief (Alarm X, which came on at time Y, is active)". The subject of the event (the alarm) is output first. A subordinate clause indicating the time at which the event occurred or its frequency may also be added, if the value of y is not (). This subordinate clause is preceded by the relative pronoun, which may be either 'dat' (neuter gender) or 'die' (common gender), depending upon the grammatical gender of the item being referred to (for the purposes of this module, we are assuming that all items are singular) and is expressed using subordinate clause word order. In subordinate clauses, the inflected verb moves to the end of the clause, e.g. "Het alarm *is* aktief" vs. "omdat het alarm aktief *is*" (subordinate). This is checked for in the main clause once the subordinate clause has been generated.

The next three examples are all from the Dutch generator. The first of them is a simple intransitive verb 'blokkeren' (to block). "X blokkeert" (X is blocking). We do not need to worry about negation or subordinate word order here.

```
(BLOKKEREN
    (blocked(x))
    ((output_as_dutch x noart)
     (output_verbform blokkeren tense pers))
    ())
```

Our next example is somewhat more complex, as we have to deal with negation and subordinate clauses. It is the verb 'draaien' (to turn).

```
(DRAAIEN
    ((x,y))            ; x is of type DO(a) and y is either T or F.
                       ; Depending on the value of y, the system
                       ; variable neg will already have had its value
                       ; set to TRUE or FALSE
    ((output_as_dutch x noart)
     (if subord = TRUE then
        ((if neg = TRUE then
            (output_word "niet"))
         (output_verbform draaien tense pers))
      else ((output_verbform draaien tense pers)
            (if neg = TRUE then (output_word "niet")))))
    ((subord := FALSE)))
```

Our next example involves a separable verb, 'toenemen' (to increase). Here we have to deal with the problem of the separable part of the verb, 'toe'. In an independent clause, the separable part of the verb moves to the end of the clause, e.g. "Het niveau in vat D-220 nam twee minuten geleden *toe*" ("The level in vessel D-220 increased 2 minutes ago"). In a subordinate clause, on the other hand, the inflected verb moves beyond it to the end of the sentence, e.g. "Het niveau dat twee minuten geleden *toe* nam" --> "Het niveau dat twee minuten geleden *toe*nam" (The level which increased two minutes ago), and the separable part of the verb rejoins the main, inflected part of the verb. We do not, however, have to deal with negation in this example.

```
(TOENEMEN
    (increased (x))
    ((output_as_dutch x defart)
     (if subord = TRUE then
        ((output_separable_particle "toe")
         (output_verbform nemen tense pers))
      else ((output_verbform nemen tense pers)
            (output_word "toe"))))
    ((subord := FALSE)))
```

Our next example is of the form *subject* <zijn> *adverbial-complement*. The structure is of the type (a GT b, c), where a and b are expert system variables and c has the possible values T or F. The output should be of the type "a is (niet) groter dan b" in an independent clause or "a (niet) groter is dan b" in a subordinate clause.

```
(GROTER-DAN
    ((x GT y, z))
    ((output_as_dutch x defart)
     (if subord = TRUE then
         ((if neg = TRUE then
             (output_word "niet"))
          (output_word "groter")
          (output_verbform zijn tense pers)
          (output_word "dan")
          (output_as_dutch y defart))
      else ((output_verbform zijn tense pers)
          (if neg = TRUE then
              (output_word "niet"))
          (output_word "groter dan")
          (output_as_dutch y defart))))
    ((subord := FALSE)))
```

Our final two examples come from the English generator. Because of the relatively constant word order of English, they are much less complex.

The first template deals with a sentence of the type *Fault* so *action*. The action part of the sentence should be expressed using the imperative mood and the 2nd person singular. Both *fault* and *action* are clauses in their own right.

```
(SO
    ((so (fault action)))
    ((output_as_text fault 1)
     (output_word "so")
     (output_as_text action 1 (tense := IMPV pers := SING2)))
    ()
)
```

Finally, we have an example of a template which outputs a simple sentence such as 'Operate x'. The structure generated from is of the format (op x).

```
(OPERATE2
    ((op x))
    ((output_verbform operate tense pers)
     (output_as_english x noart))
```

5 Actions and Grammatical Categories

The correspondence between some of our actions and grammatical categories is quite close. For instance **output_verbform** produces an inflected verb form, and **output_as_dutch** and **output_as_english** produce noun phrases. Grammatical gender is taken care of in the Dutch generator within the lexicon entry. Questions of syntax in main and subordinate clauses are dealt with by the templates and the **subord** variable in the Dutch generator. Some questions, such as that of adjectives, remain. At the moment, adjectives are simply dealt with as part of the string which makes up the description field of Process Control System variable entries in the lexicon. A better solution would be to hold them separately within the description field, e.g. instead of

(VAT-20, groot vat, n)

we would have

(VAT-20, vat (adj groot), n)

This is because adjectives are inflected in Dutch depending upon the gender and number of the noun they qualify, but also depending on whether they are preceded by an indefinite or definite article. The production of articles and qualifiers is current;y dealt with by **output_as_english/_dutch**. The only pronouns currently dealt with by the system are referring pronouns preceding a subordinate clause. In order to generate other types of pronouns, we would need to have a record of items which had previously been referred to in the dialogue, in order to allow us to decide which things could be pronominalised without causing ambiguity, and this would involve some sort of a discourse module and a much more sophisticated model for our generator.

6 Conclusions

We have shown in this document how actions are used to change the order of sentence constituents in the output. This enables us to construct a greater variety of sentences from each individual template. At present, our task is made easier by the fact that we are dealing with only three tenses in Dutch, the present, the imperative and the imperfect, but similar structures would work for other, more complex tenses in Dutch such as the future and perfect tenses where an inflected verb form has two parts (the auxiliary verb and the inflected form of the verb itself) which can occur in widely separated parts of the sentence.

One possibility would be to abstract the current set of templates into ones for, e.g. an intransitive verb, an intransitive separable verb, a transitive verb with a subordinate clause, etc. One of these more abstract templates could be called from each template attached to a particular verb. This would reduce the repetition of code in the current templates. It might also be possible to extend the system outlined above so that connected paragraphs of text, with phenomena such as pronominalization, passivization, etc. could be generated.

The system outlined has been implemented in C as a demonstration system for exemplars for two real-time industrial applications (a chemical plant and a nuclear power station). There are currently about 40 possible templates, enabling us to deal with variations upon 7 basic situations. We believe that this system, intermediate between a simple "fill-in-the-blanks" approach and a dialogue-based system, can be used to increase the user-friendliness of many process control systems.

7 Acknowledgements

The author wishes to thank his former colleagues, Huk Mok, Ita Kavanagh, Gabriel McDermott and Philip Li Yim Yang Tai, for their contribution to the implementation of this design, and wishes to acknowledge the financial contribution made by the Commission of the European Communities to the research upon which this paper is based, and to express his thanks to UCD's project partners in DOW Benelux, Scottish Power, EXIS and the University of Leuven for their evaluation of the finished products.

8 Bibliography

Allgayer, J., Harbusch, K., Kobsa, A., Reddig, C., Reithinger, N., and Schmauks, D. (1989). XTRA: a natural-language access system to expert systems. *International Journal of Man-Machine Studies*, **31** (2), 161-195.

Bench-Capon, T.J.M, Lowes, D., & Mc.Enery, A.M. (1990). Using Toulmin's Argument Formalism to Explain Rule-Based Programs. Paper presented at the 5th Explanation Workshop, University of Manchester, April 1990.

Löwgren, J. (1989). An architecture for expert system user interface design and management. In: *UIST: Proceedings of the ACM SIGGRAPH Symposium on User Interface Software and Technology, Williamsburg, USA, 13-15 Nov. 1989*, (pp. 43-52). New York, NY, USA: ACM.

Matthews, P., and Mok, H. (1992). Transforming Expert System Output into Natural Language in a Process Control Environment. In *SAC'92: Proceedings of the 1992 ACM/SIGAPP Symposium on Applied Computing, Kansas City, Mar. 1-3 1992*, **1**, (pp. 116-122). New York, NY, USA: ACM.

Matthews, P., Mok, H., & Kavanagh, I. (1992). Project PROMISE Final Deliverable No. 19 - Task 1.13 (Natural Language Generation). June 1992. UCD Computer Science Department Technical Report TR-92-15.

Moore, J.D., & Swartout, W.R. (1989). A reactive approach to explanation. In *IJCAI-89: Proceedings of the 11th International Joint Conference on Artificial Intelligence, 20-25 Aug. 1989*, **2**, (pp. 1504-1510). Palo Alto, CA, USA: Morgan Kaufmann.

Wick, M.R., & Slagle, J.R. (1989). An Explanation Facility for Today's Expert Systems. *IEEE Expert*, **4** (1), 26-36.

Terminology and Extra-linguistic Knowledge[1]

Jennifer Pearson

National Centre for Language Technology, Dublin City University
Dublin, Ireland[2].

Abstract

The paper describes a project currently being undertaken within the framework of Eurotra 10. The project involves designing a reusable terminological resource for NLP and other purposes. The resource is designed to capture the extra-linguistic knowledge contained in terminology definitions. While the project also involves implementation in the new Eurotra formalism, this paper focusses on the concept of extra-linguistic knowledge and terminology and on the design of the terminological resource.

1. Lexical Resources and NLP Systems

In the early days of development of natural language processing systems, researchers focussed on the design of the system rather than on the creation of resources as input to the system. For example, little attention was paid to the creation of dictionaries. Traditionally, most researchers tended to rely on small hand-created dictionaries for the purposes of demonstrating the implementability of their systems.

More recently, with the development of large-scale NLP systems for real world applications there has been a growing awareness of the need for large-scale lexical and grammatical resources. As the creation of such resources involves major investment in terms of time and money, researchers have been considering the possibilities of re-using existing lexical material. Now that machine readable dictionaries have become available, this has become a major research activity and most of the large NLP research centres have research teams engaged in investigating the possibility of extracting lexical material from existing machine readable dictionaries. In 1991, the European Commission commissioned a study (known as Eurotra-7) on the feasibility of reusing lexical and terminological resources.

The Longman's Dictionary of Contemporary English, more commonly known as

---

[1] The author would like to acknowledge the financial assistance provided by the European Commission for this project. The other partners involved in this project are CRP-CU, Luxembourg, ILTEC, Portugal, INCOM, Federal Republic of Germany.
[2] e-mail: pearsonj@dcu.ie

LDOCE, a dictionary for non-native learners of the English language, Webster's Seventh New Collegiate Dictionary and the Collins English Dictionary, more commonly known as COBUILD, and other general language dictionaries have all been the subject of research by teams at IBM, New York, at the Virginia Polytechnic Institute and State University in Blackburg, the Linguistics Research Centre at the University of Texas, the University of Birmingham, to name but a few of the institutes who have recognized the potential of such material.

2. General language and sublanguage

Previous experience with the Eurotra project has shown that the objective of developing a large scale multilingual machine translation system is problematic, for a number of reasons. Natural language is highly complex and ambiguous and any possiblity of designing a system which would cope with most complexities of natural language must remain remote for the foreseeable future. On the other hand, there is widespread recognition that systems designed for a specific purpose are far more likely to be viable. Take for example, the TAUM-METEO system designed to translate weather bulletins or the system developed at ISSCO (Geneva) to translate avalanche bulletins. These systems are built on the premise that within general language, or overlapping with general language, there are distinct languages with a well defined grammar and a well defined vocabulary.

Some proponents (e.g. Sager, 1982) of the sublanguage approach would argue that the notion of sublanguage applies to the grammar of a particular *subject field*. Thus, according to this approach, the grammar used in medical texts will be different to the grammar used in engineering texts. Other proponents (e.g. Biber, 1988) will argue that grammar is defined by the *text type* rather than the subject field to which the text refers. Thus, an instruction manual will display similar syntactic characteristics regardless of whether the manual refers to the installation of a photocopier, a washing-machine or a computer. While recognizing that the lexical component of a sublanguage, particularly when the term *sublanguage* is being used to describe the language of a single subject field, is also restricted, sublanguage researchers have tended by and large to focus on defining and describing the *grammar* of the sublanguage rather than on the *lexical* component. This is not surprising as it merely reflects what was happening in other NLP projects until recently.

Yet, to a large extent, the whole area of sublanguage research for NLP purposes derives from applied linguists' research into languages for special purposes or LSP, as it is more commonly known. Interestingly, research into LSP started from the premise that within general language, there are subsets of language with their own *special vocabulary* and that these subsets could be defined by an inventory of the vocabulary used in them. Hence, the concept of a language for scientific purposes, a language for business purposes and hence the study of terms, the field of terminology. Initial research in the field of LSP thus focussed on defining the *vocabulary* of a given subject field and interest in the *grammar* of any given subject field came only much later. When grammarians became interested in sublanguage, the focus shifted from *vocabulary* to *syntax*, and it is only recently that the pendulum has begun to swing back towards the exploitation of lexical information.

3. Lexical resources

In a recent quarterly report from the Linguistics Research Centre at the University of Texas, Winfred Lehmann (Lehmann, 1991), the author deals with the topic of the lexicon in natural language processing and expresses her dismay at the lack of attention paid to the lexicon in past work on language processing, in favour of concentration on grammar and parsing. The author emphasizes the importance of the lexicon by stating that

> "...as a fundamental prerequisite to any processing of information both the meaning of the selected item and its inherent potential relationships to other items must be determined. This requirement applies whether one selects concepts or words (possibly analyzed further into morphemes or lexical features) as primitives. More specifically, natural language processing based on concepts - the so called knowledge-based systems - must include for any concept the analogue to grammatical classification that is required for processing based on words."

For the purposes of this project, we have classified lexical resources into two very general categories, namely general language dictionaries and specialised dictionaries. The first category can be further subdivided into a number of categories which include monolingual dictionaries, bi- or multilingual dictionaries, general encyclopaedic dictionaries. Likewise there are different types of specialised terminological resources.
Our project is particularly interested in the nature of monolingual lexical resources in general and in experiments with retrieval of information from machine readable versions of these resources, in particular. Below we give a brief overview of those aspects of lexical resources which have a bearing on our work.

3.1 General Lexical Resources

3.1.1 Purpose
Traditionally, the purpose of monolingual general dictionaries was, and still is, to

convey to the user the *meaning* of the word or phrase which has been looked up and its *usage* in language.

These dictionaries were designed for human use. It was never envisaged that at some future date attempts might be made to automatically extract the information contained therein. While all dictionaries are structured in some sense whether in relation to the alphabetical ordering of all entries or in relation to the structure or order of material contained in each individual entry, dictionary writers in the past also relied on the user's ability to navigate through the dictionary, on the user's knowledge of the world and on the user's ability to cross-reference and consult other entries where necessary. For example, a user seeking an explanation of the word bridge in the context of *musical instruments* will simply ignore any reference to the meaning of bridge in the context of *civil engineering* or *dentistry*. The user will fill in gaps where necessary and will be able to judge from the context of the definition whether or not it is useful for his or her purposes. Not so, a machine.

3.1.2 Organisation

All dictionaries are *structured* in the general sense of the word, whether in relation to the alphabetical ordering of entries or in relation to the structure or order of material contained in each dictionary entry. Yet, traditional lexicographers also rely on the user's ability to navigate through a dictionary, on the user's knowledge of the world and on the user's ability to cross-reference and consult other entries, where necessary. For example, a user seeking an explanation of the word bridge in the context of *musical instruments* will simply ignore any reference to the meaning of bridge in the context of *civil engineering* or *dentistry*. The user will fill in gaps where necessary and will be able to judge how useful a definition is for his or her own purposes.

3.1.3 Content of Dictionary Entries

Typically, a dictionary entry will contain a number of fields which may include such features as the grammatical category of the entry, a definition of the entry with reference to the broad subject field to which it belongs, an indication of synonyms, antonyms, linguistic *usage* of the entry, with phrasal information, lexical co-occurrences.

3.1.4 Defining Mechanisms - Lexical Relations

The *meaning* of the dictionary entry will be conveyed initially either a) by defining the entry in relation to its *intension* (e.g. table - a flat, horizontal slab or board supported by one or more legs, Collins Modern English Dictionary), b) by *relating* the entry to another and/or by *distinguishing* it from a superordinate entry (e.g. seat - a piece of furniture, chair -a seat with a back, typically having four legs and often having arms,

Collins Modern English Dictionary) or c) by specifying that the entry is *part of* a superordinate entry (e.g. leg - support of chair, table, bed etc, Concise Oxford Dictionary). The latter two definitions are examples of *hypernym* and *ontological* relations.

While the subject of lexical relations has recently become an important research topic with the elaboration of complex sets of lexical relations such as Nutter's (1989) Fundamentallay Semantic Relations and those devised by Mel'çuk for his Explanatory Combinatorial Dictionary (Apresyan et al., 1969), the *hypernym* and *ontological* relations are those most commonly used to introduce an entry in general language dictionaries. Many other relations such as *material, function, mode of operation* are also used as extensions to the head of the defining phrase but, as we shall see, not in an easily tractable or systematic fashion.

3.1.4.1 The Intensional definition

The aggregate of the characteristics of a concept constitutes its intension. According to Felber (1984, p.116), the intension of, for example, airplane will consist of the following charactertistics: *heavier-than-air aircraft, power driven, supporting surfaces, which remain fixed under a given condition of flight.*(ibid) The definition indicates the composition and nature of an airplane. The intensional approach is valid as long as the concepts referred to in the definition are self-explanatory or referred to elsewhere in the dictionary. It would even be acceptable for this type of definition to be circular, e.g. builder - a person who builds houses or other buildings as a job , seamstress - a woman who sews or makes clothes as her job (*Collins Modern English Dictionary*, 1986). While these latter definitions tell the human user little about the entries or about the concepts underlying the entries, and would therefore be considered inadequate by conventional terminologists, they may have their place in a semantic network as long as the underlying concept is defined elsewhere.

3.1.4.2 The Hypernym definition

Veronis and Ide (1991) define a hypernym as follows: Y is a hypernym of X if 'this is an X' entails but is not entailed by 'this is a Y'. In this case, an entry will be defined, not in terms of its own properties but in relation to a similar or related concept. Thus, for example, (i) airplane - a vehicle with wings and one or more engines that enable it to fly through the air (ibid) or (ii) glider - an aircraft that does not have an engine but flies by floating on air currents (ibid). While both of these definitions refer to the intension of the entry, they also relate the concept to another concept in the same hierarchy. The entry Vehicle is 'any conveyance in or by which people are transported, especially one with wheels'. Aircraft is 'any machine capable of flying by means of buoyancy or aerodynamic forces' (ibid). If the definition indicates a *hypernym* relation, the subordinate concept automatically inherits the properties of the superordinate concept. Thus, in addition to having all of the attributes of vehicle, an airplane is powered and has wings. There is no need to reiterate the properties which have been inherited. As we will demonstrate, this notion of inheritance is very important for our project.

3.1.5 Tractability

Although lexicography is by definition concerned with the organisation of knowledge and the explanation of words, it does not necessarily follow that a dictionary which can be accessed quite easily by a human user will be equally accessible to a machine.

Most large general language dictionaries will have been compiled by a number of lexicographers. While it is customary to construct dictionaries on the basis of a set of guiding principles, inconsistencies inevitably arise, the most obvious of these being in relation to i) naming conventions, ii) the language used in dictionary entries and iii) the ordering of material within dictionary entries.

3.1.5.1 Naming conventions

A common example of variations in naming conventions is in the use of hypernyms. As discussed in 3.1.4.2, a typical dictionary entry for a concrete noun may commence with a reference to the concept superordinate to the concept being defined (*hypernym* relation). Take for example (fig. 1) the entries knife, fork and spoon in the Collins Dictionary of the English Language. knife is defined as a *cutting instrument,* fork is defined as *a small, usually metal implement,* and spoon is defined as *a utensil having a shallow concave part.* A human user will be able to infer from each of the definitions that any one of these entries could equally have been described as instrument, *a mechanical implement or tool,* utensil, *an implement, tool or container* or implement, *a piece of equipment.* If, however, this information were to be extracted automatically on the basis of hypernyms, the above would all be put into three different categories on the basis of the hypernym used to define them.

3.1.5.2 Use of language

Our investigation of experiments with machine readable dictionaries has revealed how difficult it is to convert general language dictionary entires into a machine tractable format and one of the problems has been that the language used in dictionary entries is not always used in an unambiguous and coherent fashion.

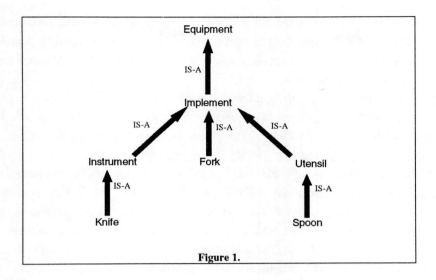

Figure 1.

This type of inconsistency has caused serious problems for automatic extraction of material. The constraint of having to **reuse** material already available means that one has to compensate for inconsistences, fill in gaps where necessary etc. Consequently, much of the effort to date has involved converting material which was never designed to be tractable into a tractable form.

Thus, it is not possible to postulate that the same link words will always be used to introduce a certain type of *field* or that link words will be associated exclusively with one *type* of field.

3.1.5.3 Organisation of material
While the contents of dictionary entries are generally organised in a principled way, the number of fields will vary and it is not possible to postulate that the dictionary entries, for example, for two deverbal nouns (i.e. nouns derived from verbs) will always have the same number of fields or even the same type of fields.

4. Specialised dictionaries and terminological resources

Broadly speaking, there are two types of specialised dictionary or terminological resource i) the monolingual specialised dictionary, occasionally with target language equivalents, and ii) the bi-or multilingual specialised dictionary. As the latter generally only offer the equivalent of a source term in a target language with, occasionally, reference to linguistic usage and very rarely reference to the meaning of an entry, we have focussed our attention on the former category.

4.1 Purpose of terminological resources
Unlike general language dictionaries where an entry is situated within a very broad

context which may result in multiple readings, terminological resources contain only the vocabulary of the subject field to which they refer. They were originally devised as a means of explaining the concepts of a subject field, enabling the user to access and understand the subject field.

One of the earliest proponents of a terminological approach to language was Eugen Wüster, an electrical engineer, who believed that a terminological approach to language could be used for a number of purposes. First of all, he argued in his <u>General Theory of Terminology and Terminological lexicography</u>, published posthumously in 1979 that the terminology of a particular subject field represented the knowledge of that particular field. Thus, if one could describe the terms of a subject field in such a way that it would be possible to infer the organisation of the subject field from the descriptions, one would in fact have a representation of the field. The advantages of such an approach would be that the subject field would, ideally, be accessible to all, whether experts in the field, student translators, documentalists etc.

The underlying premise of Wüster's General Theory of Terminology (GTT) is that terms are labels for clearly defined concepts. Within technical, scientific and technological sublanguages, terminology works in much the same way as a system of proper names works in general language. A term is a label which refers to a concept, and in the best of all possible worlds a term refers uniquely to one and only one concept within a given subject field, i.e. there is exactly one term x which refers to the concept Z and to no other concept, and there is no term y such that y refers to Z and y is not identical with x. Furthermore, as these concepts can be clearly defined, it should be possible to identify the links between concepts and create conceptual hierarchies. The two links most commonly used by terminologists to describe relations were the IS_A and PART_OF links. Thus, within the field of telecommunications, for example, a *transmitting earth station* IS_A *earth station* IS_A station. Within the same field, an *antenna* is PART_OF a *satellite*. With these two links alone, traditional terminologists managed to build sizeable representations of the terminology of particular subject fields. All concepts defined in relation to a superordinate concept automatically inherited the characteristics of that superordinate concept with the result that it was only necessary to specify the distinguishing characteristics.

It is certainly true to say that traditional terminologists were unwittingly creating what we now term *knowledge representation structures* but their objective, like the objective of dictionary writers in the past, was to make material accessible to the human user. It would appear, however, that, given their structured approach to defining concepts, it would be relatively easy to extract material from such resources, if they were available. Unfortunately, this is not the case. There are indeed many specialised dictionaries which deal only with the concepts of one particular subject field but, as we will see, these have the same shortcomings as dictionaries for general language and were not constructed according to terminological principles. There are some small databanks, most notably DANTERM at the Copenhagen School of Economics, which rely on terminological principles but there is no large-scale implementation which could be used for retrieval purposes.

4.2 Content of the terminological entry

The semantic content of entries in monolingual specialised dictionaries is, in some respects, similar to the content of general language dictionary entries. There are, however, a number of differences which are worthy of note. i) Given that terminologies are concerned exclusively with one subject field, it would be very unusual to find more than one reading of a term, except in the case of a term which refers to two different grammatical categories but to one and the same concept. ii) In the entry, no reference will be made to the broad subject field in which the concept is used. This is contained in the title of the terminological resource (e.g. Terminology of Telecommunications). The subject field references contained in specialised terminological resources will refer to narrower subject fields than the scope of the resource itself. Thus, in the Penguin Dictionary of Telecommunications , the entry for signal reads - in *telephony* and in other public networks, an instruction forming part of the process in setting up, maintaining or clearing a call; in the general case, any electrical pulses transmitted in a network to represent message information, or control information in handling the process of communication (Graham, 1983). iii) It is unusual to find information on linguistic usage in terminological resources.

4.3 Lexical relations

An examination of the structure of entries in specialised dictionaries has revealed that the *hypernym* and *ontological* relations are the most common means of introducing a terminological entry. Such relations are generally described as IS_A and PART_OF relations within the field of knowledge representation. From our research, it appears that the interpretation of the hypernym relation can vary quite considerably. In some cases the hypernym which appears as the head of the definition text will be the concept which is immediately superordinate to the concept being defined. Thus, *transmitting earth station* IS_A *earth station*. In other cases, the head of the definition text will again be a superordinate concept but at a much higher level of abstraction which, instead of giving information about the immediate conceptual context of the entry, will define the entry in terms of a class whose properties will apply to any number of concepts from different conceptual hierarchies. Thus, *electron gun* IS_A device, *short haul modem* IS_A device. With these latter examples, one might reasonably have expected the heads of the defining phrases to be *gun* and *modem* respectively, thereby allowing the subordinate concept to inherit relevant characteristics from the superordinate concept. It is likely that the defining language used in the dictionary imposed this very abstract attachment and did not allow for attachments at various points on the hierarchical scale.

In addition to the relations specified in the head of the defining phrase, terminological entries will typically refer to other relations such as *purpose* in relation to methods or pieces of equipment, *manner* in relation to processes etc. Yet, as with general language dictionaries it is not easy to envisage how the entries could easily be processed by machine. The problems are similar - inconsistency of naming conventions, inconsistent ordering of material within the entry, controlled defining vocabulary. The fact that the material does not appear to be tractable in fact matters little to this project because, as ET-7 has shown, there are no terminological resources which are suitable for such processing. Instead, the purpose of our investigation of terminological definitions was to ascertain what type of defining language and what sets of lexical relations were being used.

5. Design of a re-usable terminological resource - the ET10 project

Our experience with terminology and sublanguage research has convinced us that it should be possible, by using the terminology of a subject field and by building on terminological principles, to create a knowledge-based module for the provision of extra-linguistic information to a machine translation system. Given that the Eurotra-7 study had shown that existing terminological resources could not be re-used in this context, we were in the very fortunate position of being able to start from the beginning with the design of our resource and of being able to draw on the findings and recommendations of researchers already engaged in designing tools for automatic extraction of lexical information.

Briefly, our objectives were as follows:

- to produce specifications for building a terminological *resource* which could be re-used in a number of applications;
- to devise a formal *structure* for the form of definitions from which natural language definitions could be derived automatically
- to devise a set of *relations* for describing conceptual links
- to use the above specifications to create a small-scale knowledge representation structure

Our project was undertaken within the framework of a wider EC initiative, known as ET10. To ensure comparability of results, all 6 consortia working within ET10 are required to work with the same corpus, namely the ITU Handbook on Satellite Communications, and with the same formalism, namely the ALEP framework which has been developed on the basis of the ET6/1 project. This is a typed feature formalism which has recently been developed and is currently in the implementation stage. All consortia are required to implement the results of their research in the new formalism. This is also one of the broader objectives of our work.

6. Formulation of guidelines for terminological entries

Unlike the creators of large general lexical for NLP purposes who, given the existence of machine readable dictionaries, are endeavouring to **reuse** these in order to avoid unnecessary duplication of effort, we found ourselves in the fortunate position of having no such resources at our disposal. The focus of work with MRD's is on reuse of **existing** material. Consequently, research results document the achievements, which types of relations have been found, which ones are useful and tractable. Yet, no reference is made to what we shall call 'missing' data, i.e. the sort of information which might have been considered useful but is not contained in the resources which they are investigating.

Within ET10/66, however, we are primarily concerned with producing specifications for creating a **reusable** resource. While we are indeed bound to use the ALEP facilities, we perceive our brief as being broader than this. We are aiming to produce recommendations for formulating term definitions. Our intention is to devise a defining strategy which will situate the concept in its context and will specify/constrain both the semantic and the syntactic co-occurrence restrictions.

Our brief, therefore, was to specify guidelines for the **form** and **content** of the definitions. Our first objective was to devise a set of guidelines which specify the type of information which term definitions must contain. This must be exactly that information which is useful in a lexicon which is to be used 1) in a machine translation system and 2) by a human user. Our second objective was to devise a set of guidelines which specify the **form** those definitions must take. This must be exactly that **form** which facilitates the access and extraction of the useful information by computer.

Our task then was to specify the set of fields for which information may be given in each definition. This is analogous to defining the set of slots in a frame structure and specifying the legal values for the fillers of those slots.

6.1 Selection of lexical relations

Following our examination of existing dictionaries, both general and specialised, and our investigation of existing sets of lexical relations (e.g. Nutter, Mel'çuk), we opted to base our lexical relations on an existing set of relations. While recognizing that existing sets have been devised for general language and might therefore be found wanting in many respects, it seemed more appropriate to adopt this approach than to attempt to devise a set of relations on the basis of existing specialised terminological resources. We drew up a draft set of relations based on Nutter's and Mel'çuk's relations and on our own requirements. These were all incorporated into a format which we have called the *term definition form* (cf Appendix III). We also devised a three-tier classification system whereby class 1 refers to the topmost node or broad subject field, class 2 to a subset of class 1 and class3 to one of the following: *process, property, equipment, method, product.*

6.2 Term definition form - 1st draft

The information which we wished to capture falls into three categories. First, we wished to situate the concept in its broad semantic context by describing the pertinent conceptual relations. Secondly, we wished to capture other linguistic and semantic information pertaining to each concept and to related concepts. Finally, we wished to specify linguistic/semantic co-occurrence restrictions where applicable.

To illustrate our approach, let us look at the term definition form under each of the above three categories using the concrete example of <u>frequency division multiplex</u>, a telecommunications term.

TERM DEFINITION FORM

(1st category information)

TERM:	*frequency division multiplex*
CLASS1:	*telecommunications*
CLASS2:	*transmission*
CLASS3:	*process*
DEFINITION:	
IS_A:	*multiplex*
BELONGS_TO:	*process*
AGGREGATE:	*transmission system*
IS_PART_OF:	-
PRECEDES (event):	*transmission*
FOLLOWS:	-
UNIT_OF_MEASUREMENT:	-
MADE_OF:	-
HAS_PROPERTY:	-
USED_TO:	*combine message channels*
USED_BY:	*frequency division multiplexer*

LOCATED:	-
CONTAINED_IN:	-
PRODUCED_BY:	-
FOR_PURPOSE_OF:	*carrying message channels over a single circuit*
SUFFICIENT_CONDITION_FOR:	*carrying message channels over a single circuit*
NECCESSARY_CONDITION_FOR:	*combining channels separated in frequency*
AFFECTOR:	*multiplexer*
AFFECTED:	*message channels*
ACT/ACTOR:	*multiplexer*
ACT/OBJECT:	*message channel*
ACT/LOCATION:	-
ACT/MANNER:	-

(2nd category information)

SISTER_OF:	*time division multiplex, code division multiplex*
EQUALS:	-
ABBREVIATED_FORM:	*fdm*
SHORT_FOR:	-
ALSO_KNOWN_AS:	-
OPPOSITE_OF:	-

(3rd category information)
COLLOCATIONS

6.2.1 Summary of findings

On the basis of the information contained in the first category, we know that frequency division multiplex is a process used within telecommunications whereby a multiplexer combines message channels so that they can be carried over a single frequency. Frequency division multiplex is an integral part of the transmission system and it takes place prior to the transmission of data.

The slots contained in the first category can be further subdivided into those which describe the intensional characteristics of a concept (its composition, properties, unit of measurement etc) and those which situate the concept in relation to other concepts.

6.2.1.1 The IS_A slot

The filler of the IS_A slot always refers to the node immediately superordinate to the concept being defined, thereby allowing the concept being defined to inherit the characteristics of the superordinate concept. In this case, as frequency division multiplex IS_A multiplex, frequency division multiplex automatically inherits whatever attributes (or properties) are appropriate for multiplex (IS_A inheritance principle). Where the concept being defined is in fact immediately subordinate to the topmost node in a hierarchy, the IS_A slot remains blank and the concept attaches directly to the filler of the class3 slot, which we perceive to be the topmost node of each hierarchy. In the case of <u>multiplex</u>, the IS_A slot remains blank and <u>multiplex</u> is attached directly to *process*.

6.2.1.2 The EQUALS, ABBREVIATED_FORM, SHORT_FOR, ALSO_KNOWN_AS slots

The EQUALS slot may only be filled by absolute synonyms. The ABBREVIATED_FORM slot is self-explanatory. The SHORT_FOR slot will contain the full form of a term in cases where the filler of the TERM slot is an abbreviation. The ALSO_KNOWN_AS slot is designed to capture what we would describe is non-standard versions of a term (elliptical and other versions).

The reason for specifying these slots was that it was anticipated that if the system should encounter any one of these slots, it would allow them to inherit the knowledge contained in the slots which come under the first category. Thus, *fdm* (the filler for the ABBREVIATED-FORM slot of <u>frequency division multiplex</u>) will inherit all of the 1st category information of <u>frequency division multiplex</u>.

6.3 Revised term definition form

We were aware of course that, for any given term or concept, this initial definition form was an overspecification - in most instances some or many of the slots would be irrelevant and therefore remain empty. Our objective at this time though, was to allow for as much information as possible to be included in the definition of any term. Our telecommunications consultant used the definition form in the definition of ten sample terms from the corpus. These were

coding	frequency division multiplex
mobile	modem
modulation	satellite
signal	spotbeam
threshold level	transmission rate

These terms and definitions then became our working set. We wanted to examine the definitions to see how the slots were being used, if more slots were required or if some could be discarded. We also wanted to examine the definitional information to see how useful it was with regard to the construction of a domain ontology or conceptual representation. At this stage too, we began to pay more attention to the requirement that our definitional information be ultimately represented in the typed-feature structure based formalism of the ALEP framework. The definition of a type specification in the ALEP framework requires that one specifies the *appropriate attributes* of any given concept. This would require that in the specification of a type system to represent our definition forms, we should specify only the appropriate slots for any given concept or type of concept. Having set out such an organisation or structure for the representation of concepts in the sample domain, and bearing implementation requirements in mind, we decided to restructure our definition form with the specific objective of minimising redundancy and overspecification. We wanted to have a definition form whereby only the slots which were directly relevant to the definition of a particular concept had to be filled in. This was of course possible in the initial definition form by simply ignoring irrelevant slots, but we felt that there were too few of the slots in the form relevant in any one definition instance. In keeping with the idea of defining appropriate attributes for types in the Alep formalism, and facilitated by the division of concepts into *classes* at the highest level of the hierarchical representation, we decided to specify instead of one single general-purpose definition form, a set of definition forms - one for each class of concepts - where each form would only contain those slots which were necessary for the definition of a concept of that particular class. We therefore designed a definition form for each of the top-level concepts (concept classes) defined in Figure 2. These concept classes correspond exactly to the set of legal **class3** values in the classification system for the domain of telecommunications (TCOMM).

Given this detailed specification of the definition forms for each class of concept, the specification of restrictions on the information that can be provided in certain slots, an abstract definition of the conceptual organisation of the test domain and a hierarchical

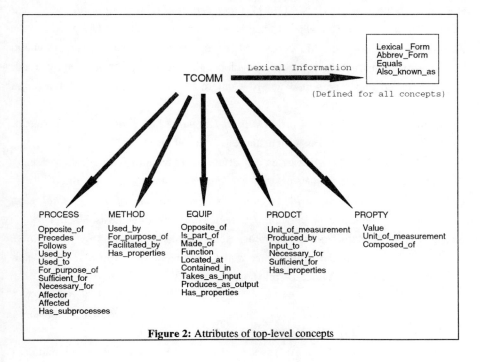

Figure 2: Attributes of top-level concepts

knowledge representation formalism to reflect that organisation we have now constructed a larger sample concept hierarchy. We are now using this information to check that the fields which we have defined captue and represent the complex relationships and interdependencies which exist between the concepts defined by the terms.

References

Apresyan, Y.D., Mel'çuk, I.A, & Zolkovsky, A.K., (1969). Semantics and lexicography: Towards a new type of Unilingual Dictionary. In F. Kiefer (Ed.), *Studies in Syntax and Semantics* (pp. 1-33). Dordrecht, Netherlands: Reidel.

Biber, D. (1988). *Variation across Speech and Writing*. Cambridge, UK: Cambridge University Press.

Collins Modern English Dictionary (1986). London, UK: Collins.

Collins Cobuild English Language Dictionary (1987). London, UK: Collins.

Darian, S. (1982). The role of definitions in scientific and technical writing: forms, functions and properties. In *Proceedings of the 3rd European Symposium on LSP* (pp. 27-48). Copenhagen, Denmark: Copenhagen LSP Centre.

Felber, H. (1984). *Terminology Manual*. Paris, France: UNESCO.

Graham, J. (1983). *Penguin Dictionary of Telecommunications*. London, UK: Penguin Books.

Heid, U., & McNaught, J. (Eds.), (1991). *Final Report of ET-7: Reusability of Lexical and Terminological Resources*. Luxembourg, Luxembourg: CEC, DG-XIII.

Kurohashi, S., Murakami, M., Nagao, M. et al. (1992). A method of automatic hypertext construction from an encyclopoedic dictionary of a specific field. Unpublished paper presented at 3rd Conference on Applied Natural Language Processing, Trento, Italy.

Lehmann, W. (1991). *The Lexicon in Natural Language Processing* (Report). Austin, Texas: University of Texas, Linguistic Research Centre.

Melby, A.K., (1991). Des causes et des effets de l'asymetrie partielle des reseaux semantiques lies aux langues naturelles. *Cahiers de Lexicologie*, **58**, 5-43.

Nkwenti-Azeh, B. (1989). *An Investigation into the Structure of the Terminological Information contained in Special Language Definitions*. PhD thesis, UMIST.

Nutter, T. (1989). Representing knowledge about words. In *Proceedings of 2nd Irish Conference on Artificial Intelligence and Cognitive Science* (pp. 313-328). New York, NY: Springer-Verlag (Workshops in Computing).

Sager, N. (1982). Syntactic formatting of science information. In R. Kittredge & J. Lehrberger (Eds.), *Sublanguage, Studies of Language in Restricted Semantic Domains* (pp. 9-26). De Gruyter.

Veronis, J., & Ide N. (1991). An assessment of semantic information automatically extracted from machine readable dictionaries. In *Proceedings of the Fifth Conference of the European Chapter of the Assocation for Computational Linguistics* (pp.227-232). Assocation for Computational Linguistics.

Wüster, E. (1979). *Einführung in die Allgemeine Terminologielehre und Terminologische Lexikographie.* Vienna, Austria/New York, NY: Springer-Verlag.

Session 4: Learning and Expert Systems

A KNOWLEDGE-BASED AUTONOMOUS VEHICLE SYSTEM FOR EMERGENCY MANAGEMENT SUPPORT

John F. Gilmore

Georgia Tech Research Institute
Atlanta, USA

Abstract

The Federal Emergency Management Agency (FEMA) is tasked with allocating resources in responding to natural (e.g., earthquakes, volcanos, floods) and man-made (nuclear accidents, fires, hazardous waste) disasters. Data collection during these events must be performed in a timely manner to support the real-time decision support structure employed by FEMA. This paper describes an knowledge-based system for the command and control of multiple unmanned autonomous vehicles (UAV) performing surveillance in support of a viable emergency management decision-making system. The goal of the Emergency Management Expert Response thru Goal-driven Entities (EMERGE) is to assist a FEMA UAV operator by monitoring the status of individual UAVs and alerting FEMA personnel to existing or potential trouble areas. A knowledge-based system developed to assist human operators in the control of multiple air vehicles, EMERGE monitors the status of individual UAVs, alerting the EMERGE operator to existing or potential trouble areas. Assisting the operator by focusing his attention on tasks requiring resolution rather than system monitoring activities, EMERGE provides a mechanism by which multiple aerial vehicle surveillance can successfully support data collection and disaster response efforts.

1 Introduction

Unmanned Aerial Vehicle technology has made several major advances over the last five years. With progress in developing single vehicle hardware systems escalating, a multitude of multiple unmanned aerial vehicle applications are being explored. Applications involving several relay and observation vehicles controlled by a single operator are being viewed as an effective future application of UAV technology. However, the most basic operations will readily increase the workload of the operator. Assisting the operator by focusing his attention on tasks requiring resolution, rather than system monitoring activities, provides a mechanism by which multiple autonomous vehicle operations can be achieved.

The goal of the Emergency Management Expert Response thru Goal-driven Events (EMERGE) system is to assist an emergency management operator by monitoring the status of individual UAVs and alerting the operator to existing or potential trouble areas.

Originally developed by Gilmore, Soniat du Fossat, and Roth (1989) as a knowledge-based military reconnaissance system, EMERGE performs these functions in support of a human operator tasked with making all operational decisions.

The EMERGE system consists of a ground station manned by a single operator responsible for the control and monitoring of an unlimited number of UAVs. The UAVs are divided into two classes, relay and observation. Relay vehicles provide a communication link between the ground station and an observation UAV. A single relay is capable of supporting several observation vehicles on long distance, long duration flights. An infinite number of observation vehicles may be employed at any given time. If communication between observation vehicles and the ground station will be unobstructed by terrain, relay vehicles will not be required.

The functionality of EMERGE is presented herein with emphasis on the diagnostic, coordination, and control aspects addressed by the system. To accomplish this, the paper is divided into four sections. First, an EMERGE system description provides a definition of the functional capabilities within the system. Second, the EMERGE system architecture is defined to establish the blackboard system developed in the initial system prototype and its supporting planning, diagnostics, coordination, hazard avoidance, and modeling knowledge sources. Third, results of EMERGE during the execution of a simulated emergency are shown to demonstrate the current level of system performance. Finally, a discussion of future efforts addressing system integration and performance enhancements is presented.

2 EMERGE System Definition

Unmanned aerial vehicles have evolved from experimental prototype systems into fully operational aircraft capable of day and night incident observation. As the utilization of UAVs increases, so does the workload of the human operator. When multiple vehicle control is dictated, the operator can no longer function at an acceptable level of performance and requires an automated assistant. An intelligent associate system working in support of the operator will maximize the number of vehicles each operator can effectively manage during an emergency situation. As timely and comprehensive information, about current conditions after an wide spread disaster (e.g., a hurricane or an earthquake), will provide greater relief to those affected; a knowledge-based associate system clearly will result in the saving of human lives.

EMERGE is a knowledge-based system developed to assist human operators in the control of multiple air vehicles. The EMERGE system consists of a ground station manned by a single operator tasked with the control and monitoring of multiple UAVs. These air vehicles are divided into two classes: observation and relay. Each observation vehicle is capable of:

- flying to EMERGE-designated points-of-interest (POI),

- generating routes that allow the vehicle to reach each POI,

- maintaining a loiter orbit at a selected point,

- receiving ground station and/or relay vehicle commands, and

- avoiding hazardous areas through reactive replanning.

Relay vehicles are for short distance, long duration flights in which the primary objective is to maintain a communication link between deep flying observation vehicles and the ground station operator. On any given EMERGE mission, a minimum of zero relay datalink vehicles may be utilized, but the system has an infinite upper bound. The actual maximum number of vehicles is determined by the skill of each operator. Each relay vehicle is capable of the above observation functions as well as receiving *and* retransmitting ground station-to-observation commands.

Relay vehicles are capable of communicating and interacting with an infinite number of observation vehicles. The number of observation vehicles employed for a given application will be a function of the terrain environment. For example, multiple observation vehicles operating in mountainous terrain may cause a communication problem for a relay vehicle tasked with maintaining a radio link line-of-sight (LOS).

In EMERGE applications that do not require the use of relay air vehicles, such as short range missions, the ground station is capable of directly controlling an infinite number of observation vehicles. Launch and recovery of observation and relay air vehicles is under the jurisdiction of a separate local controller co-located at the launch recovery point. EMERGE's multiple vehicle control capability exists from the time the air vehicle is established on a stable climb to operational altitude until it reaches the recovery approach point.

Georgia Tech's focus in developing the EMERGE system has been on air vehicle control and monitoring. In light of this, several assumptions exist in the EMERGE system prototype. These assumptions include the following:

- onboard navigation combined with ground-based aids are adequate for datalink antenna pointing.

- air vehicles possess inertial navigation equipment of sufficient accuracy to locate the vehicles for precise antenna pointing.

- datalink command and status information are transmitted as short bursts of digital data.

- datalink information is comprised of uplink commands or downlink status or operational parameters.

- ground station and air vehicle antenna pointing at desired units can be achieved accurately and rapidly enough to allow sampling of all data from each vehicle on a timely basis.

- air vehicles' sensors will be infrared, visual, or radar, with a possibility of multiple sensor configurations.

The EMERGE system addresses these UAV functional requirements and baseline assumptions through the use of a blackboard architecture capable of representing application knowledge, reasoning in the domain, and resolving system conflicts.

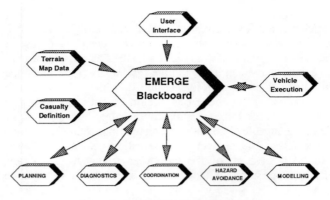

Figure 1. EMERGE System Architecture

3 EMERGE System Architecture

EMERGE is comprised of five functional components integrated by a blackboard system (Nii, 1986) as shown in Figure 1. The EMERGE blackboard is a global knowledge structure initially possessing information on terrain in the application area and predetermined system casualties for evaluating performance during application execution. Reasoning and decision-making occur in the blackboard through the activation of five independent knowledge sources for:

- *planning* to generate flight paths for each vehicle based upon their individual assigned tasks,

- *diagnostics* to monitor vehicle performance through the detection of hardware faults and malfunctions,

- *coordination* to allocate vehicle tasks based on UAV sensor configurations and re-allocate system resources when hardware malfunctions occur,

- *hazard avoidance* to identify hazardous areas (e.g., volcanic ash, power lines, flooding) that may affect a vehicles ability to complete its tasks, and

- *modeling* to verify and validate that proposed routes fall within acceptable vehicle flight parameters.

A blackboard system (Jaggannathan, 1989) provides an architecture capable of supporting parallel expert systems, each focused on a separate area of expertise, but contributing to overall system goals. The diversity of subsystems envisioned for integration in future UAVs indicates the clear requirement of a blackboard architecture if knowledge-based autonomous systems are to continue to evolve. The core of the EMERGE system is the GEST blackboard architecture (Gilmore & Roth, 1990) consisting of the central blackboard data structure, the independent experts, or knowledge sources, and the control module as shown in Figure 2.

The central blackboard data structure is the hierarchical solution space for the problem solving process. In the EMERGE system, the central blackboard data structure maintains

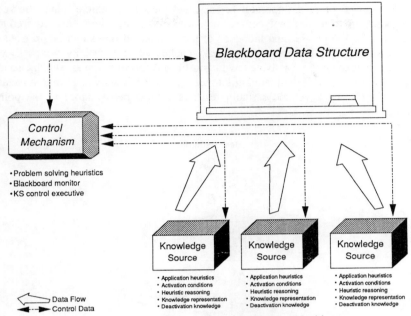

Figure 2. Blackboard System Architecture.

the global information posted by planning, diagnostic, coordination, hazard avoidance, and modeling knowledge sources that contribute the information required to formulate problem solutions. This information is accessed by the control module to verify that knowledge source preconditions (e.g., activate the diagnostic, coordination, and control knowledge source if a casualty occurs) have been fulfilled.

The EMERGE blackboard can support an infinite number of knowledge sources that specialize in solving subproblems that contribute to the overall UAV application goals. Knowledge sources contain application heuristics and algorithms in a knowledge representation tailored to their area of expertise.

For example, the planning knowledge source utilizes search algorithms and pilot-derived landmark recognition heuristics to produce a possible UAV route. Global information inferenced by a knowledge source is posted to the blackboard as depicted by the data flow paths shown in Figure 2. Knowledge of activation conditions, as well as deactivation conditions, is also maintained in the knowledge sources (e.g., if a conflict free route exists, deactivate the planning knowledge source).

The control mechanism is comprised of problem solving heuristics, rather than application heuristics. Control is tasked with monitoring the status of the blackboard to ascertain if a knowledge source can contribute to the problem solution space, and if so, when the knowledge source should be activated. This control alleviates communication loads in a streamlined UAV processing environment whereas having all of the knowledge sources activated would severely impact system performance. In this role as the knowledge source control executive, control utilizes meta-level knowledge where possible.

The EMERGE system processing flow (Figure 3) is initiated when the user has identified the terrain map for the application, generated the vehicle's initial plans, placed hazards at predetermined terrain locations, and constructed time dependent vehicle casualty schedules. Once the operator has selected the initial vehicle route, the vehicle modeling knowledge source analyzes the route to generate the commands required to fly the aircraft. The UAVs are then launched at the designated times and begin execution of their missions. As the system time clock reaches vehicle casualty times, they are posted to the blackboard and activate the diagnostic and coordination knowledge sources. The replanning portion of the planning knowledge source is activated when a hazard is encountered or when a vehicle casualty restricts vehicle performance to the point of impacting system goals.

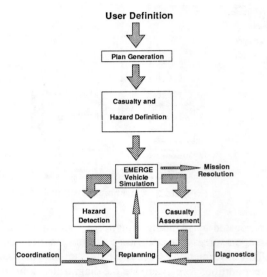

Figure 3. EMERGE System Processing Flow.

The overall goal of the EMERGE program is the knowledge-based control and monitoring of multiple aerial vehicles operating in support of emergency management personnel. The potential of distributed processing in this application domain clearly coincides with the distributed problem solving capabilities of the blackboard architecture. Details on the EMERGE blackboard architecture's planning, diagnostic, coordination, and control and planning knowledge source functions are presented in the following subsections.

3.1 Planning Knowledge Source

The goal of the planning knowledge source is to transform a digital altitude map into an optimal vehicle route based on terrain, hazards, and vehicle knowledge. The Heuristic Autonomous Route Planning Oracle (HARPO) developed by Gilmore and Semeco (1986) addresses the dynamic requirements of the EMERGE operating environment by integrating numeric computations with heuristic search strategies to rapidly adapt to unexpected events. This is accomplished by determining application-specific waypoints, generating their interconnected graph representation, coarsely validating each route segment based on the physical attributes of the vehicle, and searching the route segment candidates to construct an optimal application solution.

Point selection identifies possible waypoints for flight landmarks in the area to be traversed. Waypoints are three-dimensional coordinate point locations from which a number of application-derived paths can be generated (e.g., search routes of picnic areas in a volcano zone). Working with various pilots, several heuristic waypoint generation techniques have been developed with the four most utilized methods being maximum visibility, crossover, quad, and merge points.

A global terrain map graph is created based upon the waypoint candidates generated in point selection. A line of flight computation is made to determine whether a flyable path exists between individual HARPO-generated points-of-interest. Each remaining validated route segment is fused into a search graph with each node pair assigned its traversal criteria value. The criteria value is a weighted function of distance, hazard exposure time, terrain difficulty, and vehicle constraints (e.g., speed, fuel consumption). The weighting equation portion of the criteria function is directly dependent on the application goals and requirements.

The A* heuristic search algorithm is applied to the criteria function search graph to produce the initial vehicle route. The route generated will only be the foundation of the route planning strategy because the volatile EMERGE environment dictates the requirement of a replanning mechanism. A* is used for two reasons. First, it produces an optimal route given actual system constraints and limitations. Second, A* will always identify the optimal route *based* on criteria function values regardless of alterations to the function itself. This last aspect allows the criteria function to be adapted to a specific task or application rather than being restricted to a more general form.

Several important route planning functions are also performed by the planning knowledge source. Replanning is activated when the vehicle recognizes an obstacle (e.g., mountains) or hazard (e.g., telephone towers and lines) that will prevent it from fulfilling its goals. Replanning is accomplished by incorporating the current vehicle location into the search graph, updating the criteria values, and searching for a new route. Some applications require planning to reach a map coordinate or end point and execute a hardwired flight. In end game planning, the vehicle will fly to a pre-planned starting point, follow script-like control inputs, and regain autonomous vehicle control when the vehicle reaches an end point. This capability supports pre-planned reconnaissance requirements such as forest fire area patrolling.

3.2 Hazard Knowledge Source

Practical route planning systems require hazard modeling to [a] identify the type of hazard encountered, and [b] determine the impact of the hazard (e.g., volcano eruption fallout range) in relation to how they effect the vehicle's ability to achieve planned application goals. EMERGE supports generic hazard modeling based upon stationary point sources with varying radial spheres of influence. This model is effective in representing most natural hazards when updated over time to dynamically track hazards such as volcanic fallout. Hazard exposure considerations include hazard distance, impact ranges, and continuous exposure to hazard over time.

EMERGE supports two type of hazards. Briefed hazards are those known prior to the launch of any UAV such as a volcano. This a priori information provides the type of hazard and its approximate location as initial route planning data. Modeled by mapped hazard envelopes in the HARPO system before the mission is flown, flight through these prebriefed hazards can be avoided or minimized depending upon the overall application goals.

Unbriefed hazards are regions of terrain containing a hazard that will be detected as the vehicle executes a route to the goal point (e.g., a forest fire spreading faster than predicted). When initially exposed, the aircraft must immediately replan, updating its search space with the new hazard model information. Replanning may be performed by activating the planning knowledge source if judged to be of an extensive nature, or hazard knowledge source if the envisioned route alterations are judged to be contained within a small portion of the terrain map.

3.3 Diagnostic Knowledge Sources

The diagnostic knowledge source examines the status of individual UAVs to determine if they are operating in an acceptable range permitting the current UAV mission to be completed. Internal to the diagnostic knowledge source are six distinct rule partitions that evaluate engine conditions, flight conditions, navigation conditions, automatic pilot indicators, datalink indicators, and payload indicators.

The engine evaluation focuses on identifying the key engine components and monitoring their effect on the air vehicle. The engine is the most important aspect of the vehicle, as its malfunction, at best, will cause a mission abort, and at worst, the loss of an expensive technology sensitive system. Engine evaluation initially examines the engine oil pressure to determine whether the oil temperature or remaining fuel attributes are within acceptable ranges. Oil temperature variations are analyzed over time with unrecoverable problems resulting in vehicle recall and possible launch of a replacement vehicle. When it is determined that the fuel level is insufficient to complete application goals, the vehicle is withdrawn.

Engine evaluation monitors air speed, engine RPMs, and pitch angle. Air speed instabilities due to landing gear problems result in immediate landing gear retraction. Low vehicle RPM readings are adjusted by opening the engine throttle. Commands to climb or descend are utilized to rectify pitch angle anomalies.

The flight condition evaluation partition makes decisions on mission planning, vehicle recall and recovery, and system self-destruct. Mission planning is activated when altitudinal alterations are required. Vehicle recall results when engine problems indicate a high probability of mission failure. System self-destruct is activated when engine problems occur during war time and the vehicle is not capable of flying out of enemy territory.

Navigation evaluation verifies the vehicle position using inertial system and global positioning system (GPS) data. Rules to generate path correction commands exist, but their associated actions have not been integrated as the inertial and GPS systems are not modeled for feedback.

Automatic pilot evaluation analyzes the power supply current, electronics temperature, yaw gyro spin velocity, elevator activator, pitch gyro spin velocity, aileron activator current, and electronics temperature to assess the operational performance of the automatic pilot.

Datalink evaluation examines measurements of power supply current, antenna servo current, electronics temperature, signal strength, and signal-to-noise ratio in determining datalink status.

Finally, the payload evaluation addresses payload utility in light of supply current, built-in test results, torque motor current, FLIR detection temperature, electronics temperature, and gyro spin rate.

3.4 Coordination Knowledge Source

Resolution of vehicle casualties is accomplished by applying coordination knowledge source heuristics specifically developed to deal with fatal diagnostic problems that jeopardize overall system goals. For example, if a relay vehicle is damaged and must return to base, a decision must be made on whether the mission can still be successfully completed. Under these conditions, coordination knowledge source rules will [a] determine the time required for a replacement UAV to reach the relay communication loiter point, and [b] assess the fuel expended in loitering the relay's observation vehicles. If fuel permits, the knowledge source will recommend [a] the launch of a replacement relay UAV, [b] the recall of the damaged relay UAV, and [c] the placement of the effected observation vehicles in a loiter mode until the replacement relay arrives. EMERGE will automatically execute these actions if the operator does not override the systems recommendations.

EMERGE supports the simulated introduction of casualty malfunctions and anomalies as an aid in evaluating the expert system controller's capability to take the proper corrective actions.

3.5 Modeling Knowledge Source

A realistic autonomous vehicle system must possess a dynamic model of its vehicles to generate and verify that their flight paths are within the operating envelope of each vehicle. For this reason, several coarse airborne vehicle models were originally used to validate flight paths. These models did not possess the flight dynamics required for autonomous control system integration with the route planning system, so a dynamic aircraft model capable of autonomous control system integration was developed to determine the vehicle's maximum rate of turn, and rate of climb or descent as a function of altitude, weight, air speed, and power setting. The vehicle modeling algorithm uses a "cone of maneuverability" to design the flight paths. The size and shape of the cone are functions of the flight conditions (air speed, altitude, weight, power setting) and the structural limitations of the aircraft. As the aircraft's speed increases, so does the radius of turn for the same bank angle. The radius of turn can be decreased, but structural load on the aircraft is increased.

Utilizing a basic six degree-of-freedom (6DOF) model, equations of motion are computed and transformed to a flat earth reference frame with time integration performed to update position, velocity, and attitude. By modifying these equations of motion, this model is capable of representing both fixed-winged (e.g., such as the SKYEYE) and rotary-winged UAVs.

4 Results

The EMERGE user interface consists of a multiple window screen containing information on terrain, vehicle status, hazards, and control decisions as shown in Figure 4. The large center window is a pseudo color representation of digital altitude data for an area of northern Alabama. This window displays the mission scenario representing hazards as concentric circles, mission planning waypoints as small color coded squares, and the active UAV routes as color coordinated lines. The window to the right labeled Mission Control gives the mission time, a set of buttons and the ground truth for up to

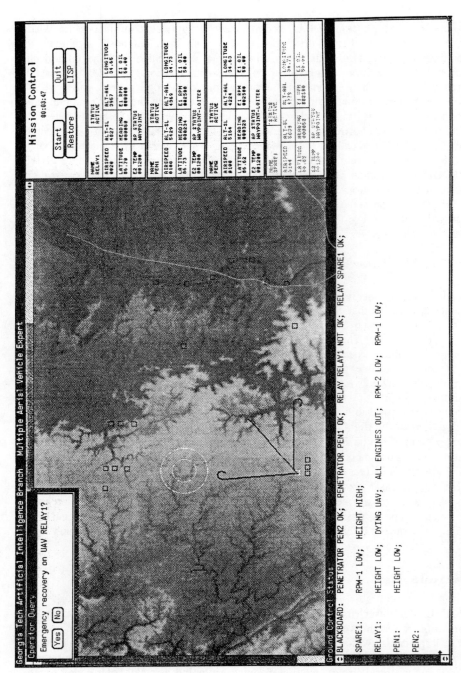

Figure 4. The EMERGE User Interface Demonstrating a Volcano Eruption Rescue Mission

6 UAVs. Each UAV ground truth window displays ground truth data for the UAV and its status color coded to match the UAV track. The bottom window labeled "Ground Control Status" lists relevant facts posted to the blackboard and each UAV Expert System.

Functionality of the EMERGE system can be seen through the use of an example application consisting of four UAVs: one relay (named RELAY1) with two associated observation vehicles (PEN1 and PEN2), and a spare UAV (SPARE1). The application/mission is defined by the user utilizing a variety of menus provided by the EMERGE system. Once the terrain area in which the mission is to be conducted is selected, the user specifies each UAV available for the mission. With the UAVs defined, the user specifies the waypoints and casualties for each UAV.

The intent of the example mission is for PEN1 and PEN2 to patrol camping areas in a state park by flying over waypoints on the terrain specified by the workstation operator. RELAY1 will support the missions of PEN1 and PEN2 by providing the communication links to the Ground Control Station (GCS). SPARE1 is available to replace a failed UAV. The mission specifies a relay engine casualty and a single unbriefed volcano hazard to demonstrate the expert controller's ability to handle mission anomalies.

For this example, PEN1 has been assigned the search of a camping are due north of an active volcano. Seven waypoints designate the area to be searched, while two concentric circles indicate the known volcano location and its unknown (or unbriefed) fallout area. PEN2 has been assigned the search of the river valley to the east of the volcano. RELAY1 is tasked with loitering northeast of the launch point and maintaining a communication link with both observation vehicles. SPARE1 is a generic UAV that can quickly be configured as a relay or observation vehicle.

The status of EMERGE after approximately seven minutes of operation is shown in Figure 5. The UAV paths clockwise from the left are; PEN1, RELAY1, and PEN2. RELAY1's mission was to fly out to its designated waypoint and loiter 15 minutes in anticipation of the communications needs of PEN1 and PEN2, which fly out longer distances to operator designated waypoints for camp ground surveillance activities. To simulate a failure, RELAY1 was programmed to suffer an engine failure three minutes into the mission. This engine failure results in RELAY1 being called home to attempt a safe recovery. The expert system first evaluates the time it will take to configure, launch, and loiter a replacement UAV (SPARE1). The amount of fuel remaining on each observation vehicle if they were to loiter in at their current locations until the new RELAY arrives on station is then examine to determine whether sufficient fuel will remain to complete their searches.

If loitering will adversely effect their search goals (in terms of both fuel consumption and desired search window time), both observation vehicles will be recalled. If fuel and time constraints will still be sufficient, as in this example, the expert system will activate the SPARE1 UAV to replace RELAY1 and complete its mission. These events require EMERGE to reprogram RELAY1 to return home, program SPARE1 to pick up RELAY1's mission where the failure has occurred, and reprogram PEN1 and PEN2 to utilize SPARE1 as a relay and hold position (i.e., the circling pattern of PEN1 and PEN2 shown in Figure 5) until SPARE1 is on station. Eventually, RELAY1's engine problems are severe enough to cause the vehicle to crash as noted by the solid square.

Soon after SPARE1 arrives on station and releases the two observation vehicles to continue their search paths, PEN1 flies into the fallout from the volcano. Once detected, the hazard avoidance knowledge source is activated to assess the hazard and replan the

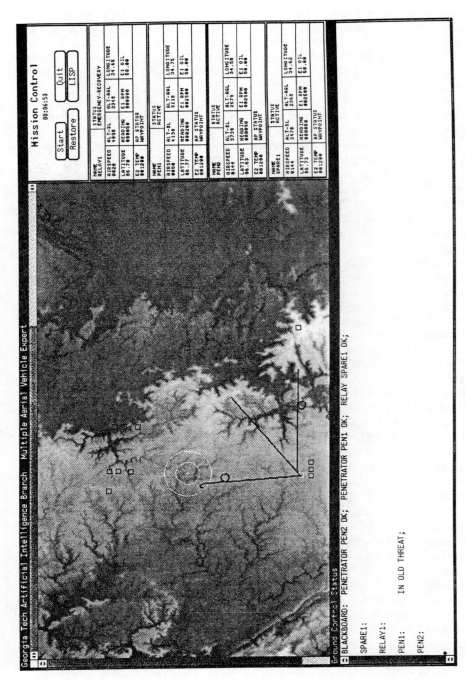

Figure 5. EMERGE Response To A Relay Failure And A Hazardous Volcano Fallout Encounter

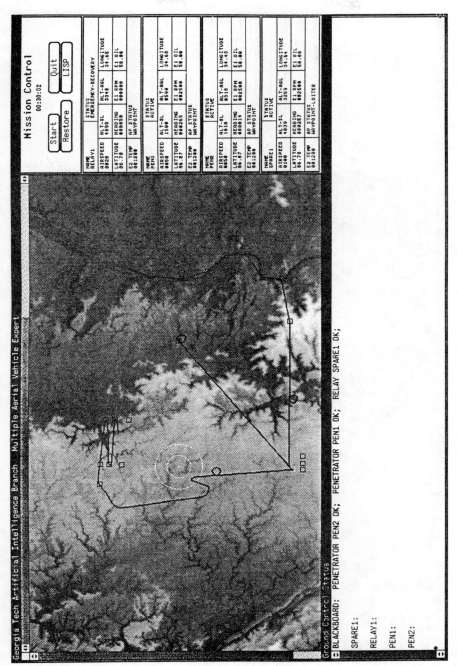

Figure 6. Results Of The EMERGE System After Thirty Minutes Of Autonomous Vehicle Control

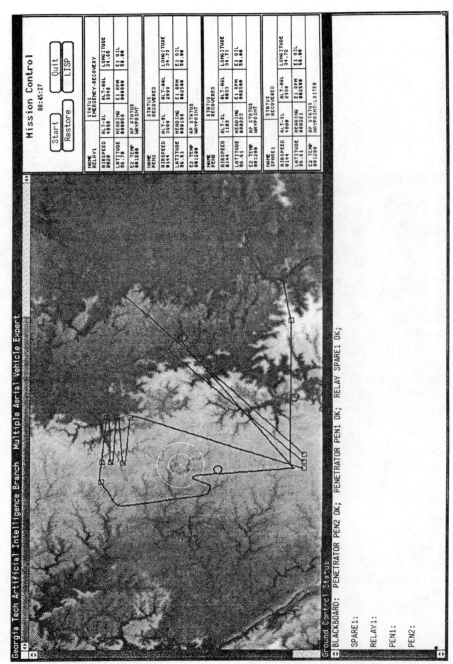

Figure 7. EMERGE System Final Display After The Recover of PEN1, PEN2, and SPARE1 UAVs

vehicle route out of the hazard area as shown in Figure 5. When PEN1 encounters the volcano fallout, its transmits hazard information to the GCS via a telemetry report. EMERGE then reprograms the UAV to take evasive action to exit the projected volcano fallout area. The GCS expert system eliminates mission segments which are within the hazard zone and activates the planning knowledge source to plan around this area resuming the mission at the next safe waypoint. Utilizing this new plan, the GCS and UAVs reprogrammed to avoid the hazard and resume their mission with a minimum of impact.

The vehicles continue on their search routes and complete 80% of their missions by the thirty minute mark as shown in Figure 6. At this time PEN1 is completing its final two camping area searches areas north of the volcano, PEN2 has almost completed its search of the river valley, and SPARE1 is still in a loiter pattern maintaining communication links between the ground control station and each observation vehicles. The completed mission after each UAV has landed at the recovery station is shown in Figure 7.

5 Summary

This paper has described the Emergency Management Expert Response thru Goal-driven Events system currently being used as a testbed for autonomous emergency management and assessment. The EMERGE system is an initial prototype to assist ground station operators in the control and monitoring of multiple relay and penetrator air vehicle. The EMERGE system, as it currently exists, is near real-time with three UAVs and can potentially be real-time on a SUN4 computer for six UAVs with further enhancements and optimization efforts.

The EMERGE system was developed as a research prototype for developing multiple air vehicle control and assessing the feasibility of expert system UAV control. As a first pass prototype, EMERGE contains extensive functionality in the areas of knowledge-based control, route planning, and hazard avoidance. Future efforts may incorporate the support of effects models (earthquake, fallout, nuclear blast), weather reports, and land data (floods, landslides) to expand the surveillance aspects of the system. To be a fully operational system, future system enhancements must also address integrating EMERGE into a ground station simulator, customizing the user interface in light of human factors requirements, and expanding the diagnostic, coordination, and control knowledge base.

6 References

Gilmore, J. F., and Semeco, A. C. (1986). Knowledge-Based Approach Toward Developing an Autonomous Helicopter System. *Optical Engineering,* **25**(3), 893-912.

Gilmore, J. F., Soniat du Fossat, E. C., and Roth, S. P. (1989). *Multiple Aerial Vehicle Expert System.* (Research Report Number 8925). Atlanta, Georgia: Georgia Tech Research Institute, Computer Science Lab.

Gilmore, J. F., and Roth, S. P. (1990). *Generic Expert System Tool Manual 4.0.* Atlanta, Georgia: Georgia Tech Research Institute, Computer Science Lab.

Jaggannathan, V. (1989). *Blackboard Architectures and Applications*. New York, New York: Academic Press Publishing Company.

Nii, H. P. (1986). Blackboard Systems: The Blackboard Model of Problem Solving and the Evolution of Blackboard Architectures - Part I. *AI Magazine,* **7** (2), 38-53.

Nii, H. P. (1986). Blackboard Systems Part II: Blackboard Application Systems, Blackboard Systems from a Knowledge Engineer's Perspective. *AI Magazine,* **7** (3), 82-103.

Learning to Play Connect 4 : A Study in Attribute Definition for ID3

Brendan W. Baird
Department of Computing Science, University of Ulster at Coleraine[1]
N. Ireland

Ray J. Hickey
Department of Computing Science, University of Ulster at Coleraine
N. Ireland

Abstract

The use of algorithms such as ID3 to induce decision trees and rule sets requires that a set of attributes or features be defined with which to describe objects to be classified. This problem is considered in an application to the game of Connect 4 where the task is to learn a set of rules with which a program can play to a reasonable standard. The attributes used evaluate the current position of a game from the point of view of both players and therefore, to a limited extent, implement a defensive as well as an offensive strategy. The attributes characterise moves made by the ultimate winners in a series of games played by novice and moderately good players.

1 Introduction

The selection of attributes to be used in algorithms which learn to classify from examples often presents problems. Attributes are high-level features within a domain which hopefully carry relevant information thus facilitating the induction of a decision tree or rule set using perhaps ID3 (Quinlan, 1986) or CN2 (Clark & Boswell, 1991). Such algorithms have been employed in many domains including medical diagnosis, risk assessment in insurance and game playing.

In this paper we consider the definition of a set of attributes for induction of a rule set to choose moves in the game Connect 4. This rule set would then constitute the advisory module in a program to play the game. No overt look-ahead in the form of a planning strategy or indeed minimax is included in the program (see Anantharaman, Campbell and Hsu (1990) for an account of deep searching as applied to chess). Our aim is to investigate to what extent a reasonable playing capability can be provided by the rules which advise what move to make next on the basis of the current game position as seen through the attributes used.

An alternative approach which is effective in game-playing domains is to make the identification of attributes part of the learning mechanism, i.e to build highly informative attributes from some initial set. This sub-field of learning from examples called

---

[1] Full Postal Address: Department of Computing Science, University of Ulster, Coleraine, Co. L'Derry, N.Ireland BT52 1SA.
EMail: cccf13@uk.ac.ulster.ucvax

constructive induction (Michalski, 1983) has received much attention recently; see, for example, Matheus (1991).

The rules will be obtained from a decision tree induced by ID3 and subsequently pruned using the Niblett-Bratko post-pruning algorithm (Bratko, 1990). This algorithm has been subject to criticism (Mingers, 1989) and has recently been altered (in fact, made more general through the introduction of a new user-defined parameter); see Cestnik and Bratko (1991). It is the original version, however which is used here.

1.1 Connect 4

Described by it's creators as *'the vertical strategy game'*, Connect 4 is a simple game of few rules. Played on a vertical grid of seven columns by six rows, two players (referred to as red and yellow) take turns to drop one of their twenty-one counters into any of the (not yet full) columns in the grid. That counter falls down on top of any previous counters inserted, or to the bottom position if that column is empty.

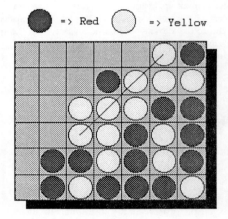

Figure 1 : An example of an end position

The object of the game is to be the first player to get four of their counters in a row, either vertically, horizontally or diagonally. At the end of a game the board may be in a position such as that in figure 1 with yellow winning as shown.

1.1.1 Spot and play 'forks'

A *fork* occurs when a player leaves the board with more than one possible four in a row which could be played on their next move. Thus no matter which of these the opponent blocks, a winning move can be played on the next turn. For example, in figure 2, when red plays column five (as indicated by the red arrow) this leaves either column two or six as a possible winning move. Yellow can only block one of these implying that red wins. Obviously an equivalent defensive strategy would be to identify and block possible forks by the opponent.

For a discussion of the acquisition of a planning strategy to deal with forks in a domain independent way see Epstein (1990).

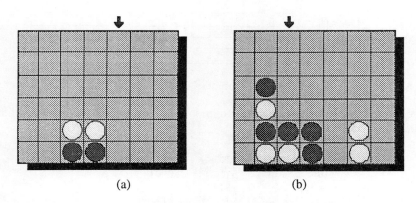

Figure 2 : (a) Red plays a 'fork' to win on next turn
(b) Red plays a 'pin' to win on next turn

1.1.2 Spot and play 'pins'

A *pin* is a situation where, despite the fact that there may only be one possible win on the board at present for a particular player, when the opponent blocks that move they uncover a further winning position for the player. In figure 3 (b), if red plays the second column as indicated on the diagram, then yellow is forced to block a four in a row in column five. However, even after yellow has played column five, red can still win by playing column five as this makes four in a row horizontally on the second row. Again, a player should watch for the opponent also having a pin and block that move. Pins can be greater in magnitude than two moves.

2 Collecting examples

The success of the learning algorithm will be dependent on the examples which are presented to it. We are not aware, however, of any documentation of Connect 4 games. The examples could come from humans in two ways. Firstly, many games could be monitored, recording every move made by the players (some of whom may not be good at the game). Secondly, good players could be used to create examples of particular strategies such as the creation of forks or pins. It is also possible, of course, to combine examples from both these sources.

The first method was adopted with a wide range of players including children and students. A *'winner stays on'* approach to the game playing was used. That is, when someone won a game they became the champion and the next person to play had to challenge them; there were several such parallel 'competitions'.

Each move was recorded by the players on a data collection form as shown in figure 3. The moves are recorded as a series of numbers in the range one to seven, corresponding to the column played. At the end of the game the winner is also recorded on the form. These details are then entered into a text file.

When collecting the data, it was useful to see each player as their colour (red or yellow). However, when writing a program to replay the games it was more useful to think of the

Figure 3 : Format of data collection form

players as player one (p1) and player two (p1). The structure of the text file is outlined in figure 4

file	::=	{game}
game	::=	winner + {move}
winner	::=	p1	p2
move	::=	1	2	3	4	5	6	7

Figure 4 : File structure of example games

Once the games were played and recorded in the appropriate text file a program was written to replay each game by reading the text file.

3 Attribute definition: Capturing features from example games

The goal of learning is to produce a set of rules each one of which advises playing a particular column depending on the current state of the game. The latter is assessed from high-level features or attributes, in the terminology of ID3, which capture salient features of the present position. The choice of particular attributes, called the bias of the algorithm, and how informative they are is of fundamental importance to the quality of learning. An example is a conjunction of attribute-value pairs (the condition) together with a class attribute value (the conclusion). The examples are distilled from the record of the games (described in the previous section); each example is a (typically over-specialised) rule in its own right. The example set as a whole will usually possess noise, the sources of which are, principally, errors in the recording of moves and clashes in examples, i.e. the same condition appearing with different classes.

The example set was built from moves made by the ultimate winner of the game. Thus even though the loser of a game may make good moves before eventually losing, these will not be used. The class attribute was taken to be the column (values 1 to 7) played by the winner at each turn. Thus a game provides one example for each move made throughout it by the player who wins.

The condition part of an example provides an evaluation of the current state of the game. Although knowledge of which move to make is already embedded in the actual moves (as made by the winner) the high-level summary of a position afforded by the condition attributes constitutes a further infusion of expertise. In defining the attributes we sought to restrict ourselves to relevant objective features of a position rather than subjective assessments of say 'strength of the current state for yellow'. Since the attributes were defined after the games were played, the examples do not entirely represent the opinions as regards strategy of the players. The transformation from a raw move to an example is many to one and thus an example is an inference about what a player does when the condition as specified by the attribute values pertains.

We made two attempts at defining the attributes. The first, a naive attempt, was very simple and ignored much that would appear to be relevant about the current state. The second attempt was more ambitious, the aim being to include as much as possible that would appear relevant in assessing the current position from the point of view of both players.

3.1 First definition

Each column was scored as either strong, weak or neutral from the eventual winner's point of view. A column was said to be *strong* if the winner had more counters in it than the loser, *weak* if the loser had more counters in it and *neutral* if both players have the same number of counters in it. Thus there are seven attributes, the ith being the strength to the winner of column i, and each takes values strong, weak and neutral. As indicated above, the class is just the column played by the winner from this position.

As an illustration, given the game position in figure 5, with red (who eventually won) about to play, the evaluation using the attributes defined above is shown in table 1.

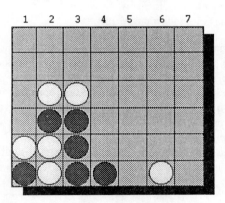

Figure 5 : Example board position - red to play

Table 1 : Scores evaluated from the board in figure 5

Col	1	2	3	4	5	6	7
Value	neutral	weak	strong	strong	neutral	weak	neutral

From this position column six was played. Thus the following example was recorded:

    if   **col1=neutral, col2=weak, col3=strong, col4=strong, col5=neutral, col6=weak, col7=neutral**

    then   **class= col6.**

The evaluation of the moves from all the games produced 531 examples.

3.2 Second definition

Again each column was scored but in this case more detailed consideration was given to the implications of playing a particular column. Since the object of the game is to create 'four in a row' in some direction, the evaluation took into account the extent to which this had been achieved or could be achieved in the future. The current position was also looked at from the point of view of the opponent (the eventual loser), i.e. defensively from the winner's perspective.

Each column was given a score for both players leading to 14 attributes. Each attribute is an assessment of the worth of playing that column next. These scores were based on the identification of patterns as follows (in order of merit, i.e. strength to the player concerned where a low score is high strength):

1 = makes four in a row
2 = makes more than one three in a row (i.e. may be augmented in more than one way subsequently to make four in a row)
3 = makes three in a row (not blockable)
4 = makes three in a row (blockable)
5 = possible four in a row (sometime)
6 = never makes four in a row
7 = full (the column is already full)

The attribute values are identified by examining all of the squares effected in the four directions around the position to be played. Those four directions are the horizontal, the vertical, the left diagonal and the right diagonal. There are potentially seven squares affected in each direction, with less for those board positions near the edges. From the examination of the four directions within a column four preliminary scores will be returned. The most prominent of these (with respect to the order of merit indicated above) will be the score for that attribute.

The scoring is illustrated using the game state shown in figure 6.

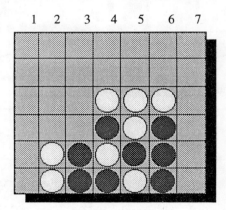

Figure 6 : An example board position - red to play.

Here red (the eventual winner) is about to play. The final scores, i.e. values for all 14 attributes are shown in table 2.

Table 2 : Summary of evaluation of position in figure 6

Column	Winner's score	Loser's score
1	5	4
2	3	4
3	4	2
4	5	5
5	5	2
6	5	5
7	5	5

From this position, the eventual winner, red, played column 2. The example obtained thus associates the 14 attribute values in table 2 with the class 2.

Again this approach produced 531 examples from the recorded games.

4 Rule induction and evaluating learning

Induced rule sets from the 531 examples were obtained for both sets of attributes using ID3 together with Niblett-Bratko pruning. It is common practice in assessing algorithms such as ID3 to induce from about 2/3 of the example set and retain the remaining 1/3 as a test set from which the classification rate can be estimated. Accordingly, new inductions were made, for each definition, from 350 randomly selected examples with the remaining 181 reserved for testing. The numbers of rules obtained are shown in are shown in table 3.

Table 3 : Statistics for pruned rule sets for attribute definitions

No. exs used in induction	No. pruned rules	
	Ist defn	2nd defn
531	133	361
350	105	199

When the rules sets induced from 350 examples were used to classify the remaining 181 examples, the percentage of correct predictions was 16.0% for the first definition and 27.6% for the second. The rate for the first definition is not much above that for predictions made at random which would yield a success rate of 1/7 (14.3%). In many domains there will be one best class associated with a condition but in the present application there may several sensible moves with perhaps one or two which are definitely bad. Thus a low rate of correct classification is not entirely due to noise in the data or lack of informativeness in the attributes.

Clearly, though, the attempt by the first set of attributes to evaluate a current position purely in terms of what happens within individual columns is inadequate. Consider, for instance, the game position shown in figure 5 above. The eventual winner (red) did not play column six because of the pattern of strong and weak columns across the board. But rather looking at the board a potential pin can be identified on column one. Given the current board position, yellow cannot play column one because this would enable red to consequently play column one and win. Furthermore, yellow cannot play column four. Doing so enables red to play column four and force yellow to block the win on column one. However, yellow playing column one is a losing move. The move played by red was column six. Doing so at this stage also stops yellow from being able to play column five. If yellow did follow that move with column five this allows red to also play column five, producing a possible win on column four which yellow must block. However, as has just been shown, yellow playing column four is a losing move.

The induced rule sets can be incorporated in a game playing program. All that is required in addition to the rules is a mechanism for choosing an alternative move when the one recommended by the rules is not available because that column is already full. In this situation a column was chosen at random from those not yet full.

To gather further evidence of the relative merits of the first and second rule sets they were used in play against a random strategy, i.e. one which chooses a move at random from the available columns. The choice of which player (random or induced) made the first move in any game was also made at random. Over 100 games played against each rule set, the first set won 48 while the second won 85. This confirms the lack of informativeness in the first attribute set and suggests that the second is to some degree informative.

Finally each of the authors played the second rule set. As one of us is an expert he had no difficult in consistently winning. The other, though, is of only average ability and is regularly beaten by the program!

5 Conclusion

Use of a large attribute set of 14 attributes each having 7 values and in which the opponents perspective is taken into account (thus doubling the number of attributes) has produced some useful game playing knowledge without any overt attempt to develop a playing strategy. This was achieved in spite of the fact that the number of examples used was very small in relation to the size of the induction problem. It is still not clear how good this set of attributes is.

Further work will involve collecting more examples and comparing the pruning effects of different pruning algorithms, including the improved Niblett Bratko version (Cestnik and Bratko, 1991). Algorithms other than ID3 - CN2 for example - could also be employed. The domain is of a useful size and could prove to be of benefit for investigation of constructive induction techniques.

References

Anantharaman, T., Campbell, M., & Hsu, F. (1990). Singular extensions: Adding selectivity to brute force searching. *Artificial Intelligence,* **43,** 99-110.

Bratko, I. (1990). *Prolog programming for artificial intelligence.* Wokingham: Addison-Wesley.

Cestnik, B. & Bratko, I. (1991). On estimating probabilities in tree pruning. In Y. Kodratoff (Ed.), *Machine learning-EWSL-91* (pp. 138-150). Berlin: Springer-Verlag.

Clark, P. & Boswell, R. (1991). Rule induction with CN2: Some recent improvements. In Y. Kodratoff (Ed.), *Machine learning-EWSL-91* (pp. 151-163). Berlin: Springer-Verlag.

Epstein, S. L. (1990). Learning plans for competitive domains. In B. W. Porter & R. J. Mooney (Eds.), *Proceedings of the seventh international conference on machine learning* (pp. 190-197). San Mateo: Morgan Kaufmann.

Matheus, C. J. (1991). The need for constructive induction. In L. A. Birnbaum, & G. C. Collins (Eds.), *Proceedings of the eighth international workshop in machine learning* (pp. 173-177). San Mateo: Morgan Kaufmann.

Michalski, R. S. (1983). A theory and methodology of inductive learning. *Artificial Intelligence,* **20,** 111-161.

Mingers, J. (1989). An empirical comparison of pruning methods for decision tree induction. *Machine Learning,* **4,** 227-243.

Quinlan, J.R. (1986). Induction of decision trees. *Machine Learning,* **1,** 81-106.

Machine learning under felicity conditions: exploiting pedagogical behavior

John D. Lewis and Bruce A. MacDonald

Knowledge Sciences Inst., Computer Science Dept., The University of Calgary, Calgary, Canada

Abstract

In task instruction a knowledgeable and well-intentioned teacher will provide focusing information, highlight decision points, indicate structure, and so forth. Recognition of these communicative acts is critical for practical learning. This paper describes an extension to explanation-based learning that augments domain knowledge with knowledge of constraints on the structure of instructional discourse. VanLehn's felicity conditions have formalized some of the constraints guiding a teacher during the presentation of examples. In this paper we weaken his condition that each lesson be an optimal set of examples, instead assuming that the teacher provides an explanation. Instruction is viewed as planned explanation, and plan recognition is applied to the problem at both domain and discourse levels, and extended to allow the learner to have incomplete knowledge. The model includes (a) a domain level plan recognizer and (b) a discourse level plan recognizer that cues (c) a third level of plan structure rewriting rules. Details are given of an example in which a robot apprentice is instructed in the building of arches. Tractable learning and instructability are our goals.

1 Introduction

An important distinction in learning from observation is whether the actor is aware of being observed — *intended recognition* — or not — *keyhole recognition* — (Cohen, Perrault & Allen, 1981). There is a marked difference in the way the observed events are understood. In keyhole recognition the steps taken by the actor can be assumed to be part of the actor's planning process (whether fortuitous or not), and so the set of events can be considered an example for that particular environment and goal. Intended recognition is not so simple; the actor could be manufacturing behavior to bias the observer's interpretation. In the instructional setting the actor behaves to simplify recognition of the underlying plan — an instance of *cooperative observation*. A variety of didactic tools are employed to help the learner; teachers combine examples and explanatory information in a complex discourse (Coulthard 1970). Verbal and nonverbal cues are used to focus the learner, to highlight decision points, indicate structure, and so on. The observed steps should no longer be regarded as an example, rather as an explanation.

Machine learning models generally ignore intended recognition. It is typically assumed that the instructor will provide optimal or near optimal examples. This eliminates an important source of information for the student — a teacher's communicative acts are critical for practical learning. Theoretical results show the impact of a knowledgeable and well-intentioned teacher on learning from examples (Gold 1967; Valiant 1984; Rivest & Sloan 1988). One result shows that when few of the many attributes are relevant, polynomial learnability can be achieved in only two ways: (a) at the expense of reliability (Littlestone 1988), or (b) by having a teacher point out the important attributes (Valiant 1984). Studies with human learners suggest the same limitation. Failing to adequately tie lessons together causes students to misunderstand relationships and learn incorrect procedures (VanLehn 1990).

Further practical impetus for exploiting prior knowledge of pedagogical behavior comes from the instructable systems arena. Insisting that the instructor supply only optimal examples is an unnatural constraint — a teacher is not merely a bag of examples. Attempting to remove this constraint by extracting and discarding communicative acts (Segre 1988) is guaranteed only limited success as suggested by the theoretical results above. A few examples with an accompanying discourse can replace many examples without. In addition this approach assumes that the explanation provided by the instructor contains a solution; this may not be the case. For example, the desired goal may be achievable by repeated application of some simpler operation — the teacher may just show the first few repetitions and expect the learner to extrapolate. Learning systems that cannot understand the role of the discourse cues may misunderstand the entire explanation, and so induce a suboptimal or incorrect procedure.

1.1 Felicity Conditions

Regarding the observed actions as an explanation rather than an example overcomes these difficulties, but complicates interpretation. Examples allow the application of explanation-based or similarity-based learning techniques, but understanding an explanation is akin to natural language understanding — the teacher's actions are a discourse in which the teacher attempts to communicate the form of a procedure to the student. The difficulty is that the procedure can be a complex structure while the action sequence is linear. The saving grace is that the teacher is cooperative. In *cooperative conversation* the speaker's meaning depends on recognition of the goals behind their utterances, and they behave to facilitate this recognition (Grice 1969). It has been proposed that a set of constraints, called *felicity conditions* are (sometimes unknowingly) adhered to by speaker's in cooperative conversation, and that these tacit conventions are exploited by the listener. There is a direct parallel between cooperative conversation and cooperative observation. In cooperative observation the speaker is replaced by a visual actor, as in charades, who similarly attempts to make obvious his or her underlying intentions. Felicity conditions may also be exploited by students in cooperative observation to allow the construction of an appropriate structure from the linear sequence of observed actions.

VanLehn tested this hypothesis in a generative model of human skill acquisition (VanLehn 1983; VanLehn 1987; VanLehn 1990). His model accepts a lesson sequence and evolves procedures that are consistent with each subsequent lesson. Three felicity constraints are used, all of which are compatible with the format of the texts: (1) each lesson requires the addition of at most a single disjunction to the evolving procedure, (2) all objects referenced by examples must be visible (relationships between objects must be shown explicitly), and (3) the lessons should exemplify the procedure. The procedures generated by the model were used to solve tests on the material as the lessons progressed, and the results predicted by the procedures were compared to those of human learners. The success of the theory in duplicating the mistakes that human learners made suggests that people do exploit felicity conditions.

Although his work was intended as a generative model of human skill acquisition, the practical implications for machine learning are significant. Exploiting knowledge of felicity conditions to constrain the learning algorithm allows arbitrary concepts to be learned, but reduces the combinatorics enough to make learning practical (Rivest & Sloan 1988). The one–disjunct–per–lesson constraint addresses the unlimited disjunction problem which is usually handled by placing restrictions on the representation of procedures. Similarly, the show–work principle does not restrict the types of relations that can be shown, between observed events and the environment, but it limits the search depth for these relations.

The third felicity condition however, places an untenable restriction on teaching: each lesson must consist of a set of optimal examples (with intermediate steps shown) and a representative set of exercises. More complex explanatory information is disallowed. In multi-digit subtraction lessons the data suggested that some students failed to recognize the connection between a lesson showing digit regrouping and the following lessons that required borrowing (VanLehn 1990). This relationship could be elucidated by walking through an example to the point where borrowing is required (and hence an impasse is reached) to motivate the regroup lesson — Goldstein's "setup step" (Ernst & Goldstein 1982). Indeed some teacher's do this; unfortunately there is no data to verify the conjecture that there is a correlation between this style of teaching and students who avoided the bugs associated with that misunderstanding (VanLehn 1991).

This paper concentrates on how this felicity constraint can be relaxed. If the teacher provides appropriate explanatory information in a lesson, then the examples need not be optimal, in fact there need not be examples at all. Prior knowledge of discourse structure can be used to distinguish and understand this explanatory information, and thus to constrain the learning algorithm.

1.2 Outline of the paper

The next section discusses instruction as discourse. Section 3 is an overview of our model. Section 4 gives an example from a robot apprentice implementation.[1]

2 Instructional discourse

Discourse analysis has shown that the type of discourse defines the structure, rather than the content. Studies of classroom transcripts have noted a hierarchical structure in instructional discourse (Sinclair & Coulthard 1975). This structure provides

---

[1]The robot–apprentice implementation is discussed in more detail in (Lewis & MacDonald 1992).

restrictions on what kinds of acts can follow what others. Each of the top levels gives a decomposition in terms of the schemata at the level below, and the identity of the primitives is determined from the discourse context — both the environment and the acts that precede and follow.

Examining the content of the acts can provide even further restrictions on interpretation. Theories of cooperative (Pollack 1986) and rational behavior (Cohen & Levesque 1990) are useful in this endeavor. Cooperative behavior has previously been used to explain helpful responses in question–answering (e.g., if one can recognize the questioner's intentions from their query, and can foresee obstacles, then cooperative behavior demands that additional information be provided rather than just answering the query directly). In the instructional setting cooperation demands that the teacher provide clarification for obscure or ambiguous actions. Rational behavior is the notion that an agent will not perform actions that achieve conditions that already hold. This is useful in the instructional setting where the instructor's actions effect both the domain and the mental state of the learner. Although the domain conditions already hold, the instructor may carry out such actions as a means of identifying relevant attributes and indicating the structure of the procedure. Another behavior constraint on instructors is that the actor will not execute unnecessary actions. This means that each of the observed actions in an example is either necessary to accomplish the task or necessary to communicate it efficiently.

Our model of learning from instruction is based on these ideas: the interpretation of observed events is restricted by both the syntactic structure of the discourse and its content.

3 The model

Accounting for discourse strategies in instruction is seen as a form of knowledge-based generalization. Our model incorporates discourse knowledge into the explanation–based learning framework. This provides several advantages: (1) both discourse and domain knowledge can be used to interpret and generalize the teacher's explanation, (2) the original framework insisted that the final state satisfy the goal; incorporating discourse knowledge allows the learner to interpret the teach sequence even though the goal may not be satisfied by the final state, and (3) the process becomes naturally incremental.

The job of the learner is to evolve the schemata necessary to understand the instructor's actions, and to generalize them as suggested by the presentation style and domain knowledge. These schemata evolve in the discourse context as a structure capturing the relation of the operators. The structure is a sequence of AND/OR graphs where each graph expresses the hierarchical relation of the operators it contains, and top-level items in the sequence are related temporally. The discourse context is composed of the original knowledge (including a goal expression), the operators recognized to the current point of processing, and the perceived structure of the teach sequence.

The model has two parts: a plan recognizer (of which there are two instances, one using domain knowledge, the other using discourse knowledge), and a plan rewriter. The plan recognizers read schemata from their respective sources and pattern match against the evolving structure. The body of a schema lists the operators that make

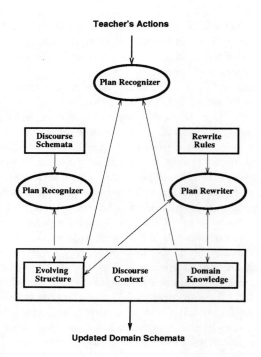

Figure 1: The model

up the abstraction. If a schema matches then it is added to the structure as the parent of the operators in its body. The plan rewriter is similar, but can potentially alter both the domain knowledge and the evolving structure. A rewrite rule is a pair: an abstract schema and a function. If the schema for a rewrite rule can be recognized, then the function is instantiated with the same variable substitutions as the operator and applied to the discourse context. The model is shown in figure 1.

The instructor's actions are fed one at a time to a plan recognition module that uses only domain knowledge to add the action to the discourse context and group it with previous actions. The largest groups (i.e. the most abstract schema in the domain hierarchy) are preferred. When no further domain grouping can be performed, control switches to a second plan recognition module that uses dialogue knowledge to identify a single communicative act. If a discourse schema is recognized then a rewrite rule is sought, and the discourse context is modified accordingly. These rewrite rules use the structure provided by the discourse schema to bias the search for relations that the actions might be intended to communicate. If novel compositions of domain schemata result from rewriting then they are added into the domain hierarchy, and can be used to process the rest of the teach sequence. After altering the discourse context, in whatever ways dialogue knowledge suggests, control is returned to the first plan recognition module which again tries to group according to domain knowledge. This continues until neither domain nor discourse schemata can be recognized; then another of the teacher's actions is added, and the process resumed. When the teach sequence is exhausted the discourse context should contain all the schemata necessary to achieve the goal. The main algorithm is

```
LEARN( Teach-Sequence, Discourse-Context, Mode, Modified )
  If the Discourse-Context was not Modified and Mode = domain
  Then
    If the Teach-Sequence is empty
    Then DONE
    Else
    LEARN( Teach-sequence tail,
           Discourse-Context + primitive action matching Teach-Sequence head,
           Mode remains domain,
           Modified is set to True )
  Else
    Let Updated-Discourse-Context be RECOGNIZE( Mode, Discourse-Context )
    If a schema was recognized
    Then
       If Mode = discourse
         Then set Updated-Discourse-Context to
                                           REWRITE( Updated-Discourse-Context )

       LEARN( Teach-sequence,
              Updated-Discourse-Context
              Mode remains domain,
              Modified is set to True )
    Else
    LEARN( Teach-sequence,
           Discourse-Context,
           Mode inverted,
           Modified is set to False )

RECOGNIZE( Mode, Discourse-Context )
  For each schema in the Mode hierarchy
    Match its body against all ordered subsequences of the root nodes in
                                           the Discourse-Context structure.
    If a match is found
    Then
        Add the schema to the Discourse-Context. (In the structure it becomes a root
                   node, and the root nodes that matched its body become its children.)
        Instantiate the new operator, backtracking when necessary.
        Return the updated Discourse-Context, and save the state for back-tracking.
    Else Return Discourse-Context.

REWRITE( Discourse-Context )
  For each schema in the rewrite rules
    Match its body against all ordered subsequences of the root nodes in
                                           the Discourse-Context structure.
    If a match is found
    Then
        Add the schema to the Discourse-Context. (In the structure it becomes a root
                   node, and the nodes that matched its body become its children.)
        Instantiate the new operator and function, backtracking when necessary.
        Apply the rewriting function to the Discource-Context.
        Return the updated Discourse-Context and save the state for back-tracking.
    Else Return Discourse-Context.
```

Figure 2: The LEARN, RECOGNIZE, and REWRITE algorithms.

given at the top of figure 2, with the supporting recognition and rewrite algorithms below it.

3.1 Representation

Both domain and discourse knowledge exist as schema hierarchies. The schema representation is an offspring of the original STRIPS operators (Fikes 1972). Our operators include:

- The name: a pattern consisting of a key and a set of arguments to match.
- Constraints, which express applicability conditions.
- Preconditions, which express variable conditions that must be true before the operator can be applied.
- A body, which restricts the search for a series of operators that achieve the postconditions.
- Postconditions, which express conditions that must hold after the operator has been applied in order for its application to be considered successful. This ensures that there were no conflicts at lower levels.
- Effects, which assert changes to the world.

At the bottom of the hierarchy, primitive operators have empty bodies and empty postconditions; the effects stipulate how the environment has changed, as shown in figure 3(a) by a primitive operator that moves the robot apprentice.

```
((name (goto $to))
 (constraints (
      (not equal $from $to)))
 (preconditions (
      (armempty)
      (at robot $from)))
 (body ())
 (postconditions ())
 (effects (
      (not at robot $from)
      (at robot $to))))
```

Figure 3(a): A primitive domain operator

Abstract operators are similar to primitive operators; they have the same fields, but their body and postconditions fields will no longer be empty. The postconditions field specifies a set of goals to be achieved, and the body contains names for other operators that will be used to solve the goals. Operator names are not unique so the sequence of names in the bodies does not specify a decomposition in the normal sense, but does serve to restrict the search space. Furthermore, the application of any of the combinations of the operators matching the names does not ensure success. The preconditions must hold before the body can be executed; and the constraints must remain true throughout the life of the operator. The effects field contains a set of additions and deletions to the state, that are made if the operator is successful. These effects will augment those asserted by the operators in the body; they may add information that the operators in the body were not concerned with.

```
((name (stack $obj1 $obj2))
 (constraints (
     (Block $obj1)
     (Block $obj2)
     (tower $obj2 $h2)
     (equal $h1 (+ $h2 1))
     ))
 (preconditions (
     (armempty)
     ))
 (body (AND
     (pickup $obj1 _)
     (putdown $obj1 _)
     ))
 (postconditions (
     (on $obj1 $obj2)
     ))
 (effects (
     (on $obj1 $obj2)
     (tower $obj1 $h1)
     )))
```

Figure 3(b): An abstract operator for stacking blocks.

The example abstract operator shown in figure 3(b) is taken from the robot apprentice implementation; it stacks one block on another. It has as one of its effects (tower $obj2 $h1). This asserts new information as the two operators in its body know nothing of towers. The other effect is redundant in the sense that the putdown operator will have asserted it as well. It is reasserted because the effects field is examined during problem solving; a goal in search of an operator will search the effects field of all operators to find those that may be useful. The abstract operator in 3(c) shows two things: recursion, and the use of an AND/OR graph in the body. It expresses that a tower of height two can be built by moving the base block, and stacking a block on top; or that a tower of height n is an extension of a tower of height $n-1$.

```
((name (build-tower $obj1 $h1))
 (constraints (
                (Block $obj1)
                (Block $obj2)
                ))
 (preconditions (
                (armempty)
                (clear $obj2)
                ))
 (body (AND
        (OR (move $obj2 _ _)
            (build-tower $obj2 (- $h1 1)))
        (OR (stack $obj1 $obj2))
        ))
 (postconditions (set
                  (on $obj1 $obj2)
                  (tower $obj1 $h1)
                  ))
 (effects ()))
```

Figure 3(c): An abstract operator for building a tower of a given height.

3.2 Discourse knowledge

Discourse operators are responsible for recognizing communicative acts, but only mark the structure. The discourse markings fire rewrite rules, and focus the analysis. Felicity constraints and discourse cues are encoded in discourse schemata. The discourse schema corresponding to the theory of rational behavior is shown in figure 4(a). Discourse knowledge also encodes patterns that are expected to be recognized and cued on to construct the procedure. Figure 4(b) shows an example discourse schema from the robot apprentice, which recognizes when goal conditions are achieved.

```
((name (redo $postconds))
 (constraints
  (operator-postconds $opid $postconds)
  (not equal () $postconds)
  (not equal (redo $postconds) $name))
 (preconditions $postconds)
 (body
  ((id $opid)
   ($name ($start-time $stop-time))))
 (postconditions $postconds)
 (effects ()))
```

Figure 4(a): Example discourse schema

```
((name (show-example (goal-pattern)))
 (constraints (
   (not member (goal-pattern) $body)))
 (preconditions ())
 (body $body)
 (postconditions (goal-state))
 (effects ()))
```

Figure 4(b): Example rewrite schema

Once a discourse schema is recognized rewrite rules are sought to update the discourse context. Figure 4(c) shows the schema portion of the rewrite rule corresponding to the discourse schema of figure 4(b), which generates a schema from the example and adds it to the domain knowledge. The discourse schema identifies the steps in the example; the rewrite schema passes this information to a function (not shown here) which manipulates the discourse context — generalizing the operators and adding them to the domain knowledge. Rewrite rules are separated from discourse schema in this way because discourse cues may be ambiguous — a single cue may suggest multiple possibilities for rewriting. Examples can be related to each other to suggest recursive or iterative procedures.

```
((name (generalize-and-add-to-domain-knowledge))
 (constraints
   (append $head
           ((id $opid)
            (value
              ((show-example (goal-pattern))
               ($t1 $t2)))
            (AND | $steps))
           | $tail)
           (structure)
   ))
 (preconditions ())
 (body
   (AND (OR ((id $opid)
             (value
               ((show-example (goal-pattern))
                ($t1 $t2)))
             (AND | _)))
   ))
 (postconditions ())
 (effects ()))
```

Figure 4(c): Example rewrite schema.

4 A robot apprentice

In constructing an arch the columns must be of equal height and must be separated by an amount equal to the length of the lintel. Assume the arch is to rest on towers of height two, and that one such tower already exists. The teach sequence begins with the instructor picking up the top block of the existing tower and putting it down in its original position. This violates the rational behavior principle allowing the student to identify the actions as communicative. The steps in figure 5 describe pictorially how the model analyzes the teach sequence for these first few steps. It shows the recognition of the domain schemata for stacking the block back in its original position, recognizing this as a communicative act, and applying the rewrite rule that discards the components and maintains the information that the tower is necessary.

The remainder of the teach sequence involves moving the lintel to directly beside the location where the columns will stand. This is done to show the distance relationship between the base of the two towers (the show–work felicity condition). The instructor may then build the second column and place the lintel on top — an arch. When the goal is achieved analysis identifies the construction of the second column and the placement of the lintel as the body of the task. The fortuitous existence of the tower (previously marked by irrational behavior) is added to the operator's preconditions. Cooperative behavior demands that all actions taken in a plan are necessary (Pollack 1986), and therefore the movement of the lintel must be communicative. This initiates a search (limited by the show–work principle) for an inference to explain the existence of the actions in the teach sequence, and leads to the inclusion of the distance condition in the constraints. Finally the new schema is generalized and added to the domain knowledge.

5 Concluding remarks

Felicity conditions and discourse cues offer biases and constraints necessary for practical learning. Felicity conditions directly constrain the interpretation of the instructor's actions, as well as making the identification of communicative actions possible. This addresses instructability (MacDonald 1991) — our system can be programmed via explanatory demonstrations. As well as improved instruction complexity, computational benefits are to be expected. As indicated by Valiant's (1984) results, the ability to recognize significant actions in a task sequence removes the need for an exhaustive search to find relevant features.

Future work must analyze the computational penalty of recognizing communicative acts, as this must not outweigh the benefits.

Acknowledgements

This work is supported by the Natural Sciences and Engineering Research Council of Canada.

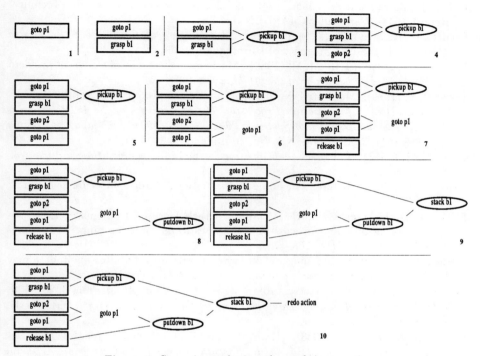

Figure 5: Steps in analyzing the arch instruction.

Boxes indicate primitive domain operators, ovals abstract domain operators, and unframed text represent discourse operators. In step 1 the discourse structure is empty, so a teacher's action is added. In step 2 no domain or discourse abstractions can be recognized, so another teacher's action is added. Step 3 is the recognition of the domain pickup abstraction. In the next two steps no domain or discourse abstractions can be recognized, so teacher's actions are added. In step 6 the discourse complex abstraction is recognized, and the two adjacent gotos are rewritten as one. Step 7 adds a teacher's action. In step 8 the domain putdown abstraction is recognized, and in step 9 the domain stack abstraction. Step 10 recognizes the discourse redo schema, which is interpreted as a communicative act.

References

Cohen, P. & Levesque, H. (1990). Rational interaction as the basis for communication. In P. Cohen, J. Morgan, & M. Pollack, (Eds.), *Intentions in Communication* (pp. 221–256). Cambridge, MA: MIT Press.

Cohen, P., Perrault, R., & Allen, J. (1981). *Beyond question–answering.* (Technical Report 4644.) Cambridge, MA: BBN Inc.

Coulthard, R. M. (1970). *An Empirical Investigation of Restricted & Elaborated Codes.* PhD thesis, University of Birmingham.

Ernst, G. W. & Goldstein, M. M. (1982). Mechanical discovery of classes of problem solving strategies. *Journal of the ACM*, **29**, 1–23.

Fikes, R. E. (1972). Monitored execution of robot plans produced by STRIPS. In C.V. Freiman, (Eds.), *Information Processing 71: Proceedings of the IFIP Congress 71, Vol. 1, Foundations and Systems, 23–28 August 1971, Ljubljana, Yugoslavia* (pp. 189-194). Amsterdam, ND: North–Holland.

Gold, E. M. (1967). Language identification in the limit. *Information and Control*, **10**, 447–474.

Grice, H. P. (1969). Utterer's meaning and intentions. *Philisophical Review*, **68**(2), 147–177.

Lewis, J. D. & MacDonald, B. A. (1992). Task learning from instruction: An application of discourse processing techniques to machine learning. *Proceedings of Applications of AI X*, 168–179.

Littlestone, N. (1988). Learning quickly when irrelevant attributes abound: a new linear–threshold algorithm. *Machine Learning*, **2**(4), 285–318.

MacDonald, B. A. (1991). Instructable systems. *Knowledge Acquisition*, **3**, 381–420.

Pollack, M. (1986). *Inferring Domain Plans in Question Answering.* PhD thesis, University of Pennsylvania.

Rivest, R. & Sloan, R. (1988). Learning complicated concepts reliably and usefully. *AAAI-88, Proceedings of the Seventh National Conference on Artificial Intelligence, St. Paul Minnesota.* Volume 2, 635–640.

Segre, A. M. (1988). *Explanation–based learning of generalized robot assembly plans.* Amsterdam, ND: Kluwer Academic Publishers. [Also Ph.D. thesis, Department of Electrical and Computer Engineering, University of Illinois at Urbana–Champaign, 1988].

Sinclair, J. M. & Coulthard, R. M. (1975). *Towards an Analysis of Discourse: The English Used by Teachers and Pupils.* Oxford: Oxford University Press.

Valiant, L. G. (1984). A theory of the learnable. *Communications of the ACM*, **27**(11), 1134–1142.

VanLehn, K. (1983). *Felicity conditions for human skill acquisition: validating an AI–based theory.* (Interim Report P8300048). Palo Alto, CA: XEROX Palo Alto Research Centers.

VanLehn, K. (1987). Learning one subprocedure per lesson. *Artificial Intelligence*, **31**(1), 1–40.

VanLehn, K. (1990). *Mind Bugs: The Origins of Procedural Misconceptions.* Cambridge, MA: MIT Press.

VanLehn, K. (1991). Personal communication, October.

A Blackboard Based, Recursive, Case-Based Reasoning System for Software Development

Barry Smyth[1] and Pádraig Cunningham

Hitachi Dublin Laboratory, Trinity College

Dublin, Ireland.

Abstract

This paper describes Déjà Vu, a Case-Based Reasoning system for software design, which uses a recursive problem solving technique, mirroring the human design process of successive refinement, to automatically generate solution code for a given target specification. This recursive approach is reflected in the structure of the case-base in that solutions have an explicit partonomic structure with each solution represented by a top level case and several lower level cases representing solution components at greater levels of detail. Currently the system has two such levels; complex top-level cases are composed of low-level detailed cases. The problem solving mechanism proceeds by using top-level cases to guide the retrieval and adaptation of low-level cases thereby generating the solution code in a modular fashion. The control issues introduced by this multi-stage process are addressed by organising the system as a blackboard system with dedicated control agents guiding the reasoning process.

1. Introduction

Case-based reasoning (CBR) is motivated in large part by the intuition that much of human expertise is based on the retrieval and modification of previously solved problems stored in memory. Automatic software design is a promising application area for CBR in that human expert competence in software development is often based on the retrieval and adaptation of solutions to previously solved problems.

---

[1]E-mail: bsmyth@vax1.tcd.ie;
 Tel: 01-6798911; Fax: 01-6798926

Automatic program generation has been the goal of much research efforts but has met with limited success. Attempts to automate software development have the advantage that requirement specifications can be formalised, giving rise to optimism about the prospects for automating the program generation process. The task of generating the code for a given specification can viewed as a theorem proving process by viewing the specification as a theorem to prove. The resulting proof steps are effectively the required solution. The main difficulty with such a perspective is the huge search problem associated with the theorem proving process. The approach advocated in this paper addresses this search problem by recasting the software development task as a CBR task. The idea being that adapting a retrieved solution which is similar to the target problem is expected to be easier than generating the target code from the original specification.

The advantage of formalization in software specification for automated software development is offset somewhat by the problem of solution brittleness. A program must be exactly correct to be of any use, there is very little room (if any) for errors. This is in contrast to many traditional CBR application domains where a family of solutions are acceptable. Our experience is that the problem of brittleness is exacerbated if the solution is to be in a low-level language. The more remote the solution is from the specification, the more difficult it is to automate the adaptation process.

Déjà Vu is a case-based reasoning system for software design working with a very high level target language and consequently the adaptation process is more tractable (Smyth & Cunningham, 1992). The system is novel to the extent that the generation of a new solution is a recursive operation involving several iterations of the CBR process each concentrating on different aspects of the target problem. This recursive process reflects the human expert's process of design by successive refinement. However the increased complexity of a problem solving episode introduces the problem of control into the system.

This paper describes the case-base structure used in Déjà Vu and how the complexity of reasoning with this case-base is addressed by organising the overall process as a blackboard system that orchestrates the composition of complete solutions.

In the next section the problem domain and target language will be discussed before examining the details of the system. The hierarchical structure of the case-base and how it relates to Déjà Vu's reasoning process is described in section 3 with the blackboard control architecture presented in the fourth section. Section 5 illustrates the problem solving process using a detailed example.

2. The Problem Domain

The aim of Déjà Vu is to generate plant control software for controlling autonomous vehicles in a steel mill environment involved in the loading and unloading of metal coils during the milling process, a typical example is shown in Figure 1.

Very briefly, spools are objects which hold coils of steel. Tension Reels are used to hold spools of steel during the milling process. In this case we are dealing with an empty spool. Skids are a repository for spools or coils. Buggies are vehicles which can carry spools or coils and have a lifting device which allows their load to be raised, lowered or tilted. Both the lifting device and buggy itself have various operational parameters (e.g., speed, capacity).

The domain in which Déjà Vu is expected to operate has the advantage that the programming language is very high level with operations represented as nodes on a solution graph (see Figure 2 for an example). This graph can be compiled by a code generation system to produce machine executable code (Sakurai, Shibagaki, Shinbori, and Itoh, 1990; Ono, Tanimoto, Matsudaira, and Takeuchi, 1988)

The example in Figure 2 is a program which unloads a spool from a tension reel to a loading bay (skid) using a two speed buggy. The program appears quite complex but can be broken down into a number of steps. Basically the buggy is moved to the tension reel and aligned with the spool. The spool is released onto the buggy which can then transport its load to the skid before releasing it onto the skid platform. An important point to note about this domain is that vehicle control is achieved by sensor activity. For example, in order to move the buggy to the tension reel it is first set in motion, then at some later time the sensor indicating arrival at the tension reel is activated causing the buggy to stop. This action is captured in the code section called Forward*1 (Figure 2).

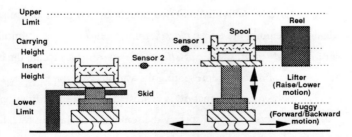

Figure 1. Déjà Vu produces code for controlling the movement of a buggy and lifter for handling reels of material in a steel mill

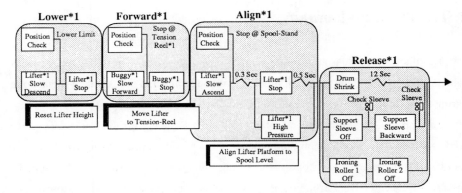

Figure 2. An example of a portion of a solution for unloading a spool from a tension reel.. The solution (which corresponds to the first four nodes of figure 3) is segmented into its various components, each titled by their detailed case (DC) name.

3. Case-Base Structure

It has been argued that the structure of memory should mirror its intended function (Hammond, 1989). In Déjà Vu the case-base can be viewed as having two main functions. It is primarily a memory of past experiences organised to improve the efficiency of the retrieval process. Secondly, the composite structure of cases reflects the recursive nature of Déjà Vu's problem solving technique. In the following sections we will examine the case-base structure in the context of these two requirements.

3.1 Taxonomic Case-Base Structure for Retrieval

The problem of retrieval in CBR involves retrieving cases from the case-base that best match the target specification. This matching is done on the basis of case features and, in Déjà Vu, the features have been divided into two categories in order to optimise retrieval. The more fundamental features are called *classification features* and in addition each case will have a set of *internal features*. The *classification features* (CF) locate the case within a *taxonomy* of classification nodes representing the various categories of cases stored in the case-base[2]. Each classification feature is represented by a classification node and defines a set of *internal features* which specify the semantics of the classification feature

---

[2] This taxonomy has some of the characteristics of a shared feature hierarchy with the important distinction that it is a graph rather than a tree (nodes can have more than one superclass). This means that attributes are not strictly ordered by importance as is the case with shared feature networks.

in terms of concepts known to the system. Actual cases are stored in case nodes which are the terminal nodes of the case-base and each case node inherits its internal features from its ancestral classification nodes. Figure 3 illustrates a portion of a sample case-base. The main advantage of such a taxonomic organisation is the increased efficiency of the retrieval process.

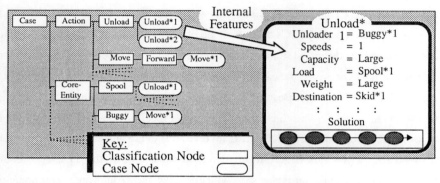

Figure 3. A simplified example of a portion of the case-base. The insert shows the internal features of the top-level case (TLC), Unload*1.

The retrieval process has a two stages. A given target specification is formulated as a retrieval frame which specifies both the classification features and internal features of the target problem. The first step uses the target classification features to indicate a number of possible routes through the case-base which lead to suitable clusters of base cases. The second stage matches the internal features of the target with those of each candidate base case from the previous step and results in the retrieval of a single *best* case.

The determination of a *best* case is based on a similarity metric which estimates the *adaptation distance* (AD) between the target frame and the base case (i.e. a measure of the work needed to adapt the base case to the target case).

3.2 Partonomic Structure for Solution Formation

The case-base also has an explicit partonomic structure with each solution represented by a top level case and several lower level cases represented at greater levels of detail. (see also (Thompson & Langley, 1989) for a similar approach to case-base organisation). Currently there are two such levels, top-level cases (TLC) and detailed cases (DC).

TLCs are represented as the composition of a number of DCs. It is worth pointing out that the architecture is by no means restricted to this two level scheme and will support intermediate levels if necessary.

Figure 4 is an example of a TLC composed of a number of DCs. The TLC corresponds to the solution example in figure 2. It is important to note that TLCs do not reference their DCs directly. Instead *link descriptions* are used that define the parameters of the required DC. This level of abstraction in the partonomic structure allows for solutions built from adapted TLCs to be made up of a different set of DCs to the original solution. This will be discussed in more detail in section 5.1 but suffice it to say that such an approach allows for greater flexibility during a problem solving episode by providing access to all the DCs in the case-base so that one may be chosen which best fits the current situation. In brief, when presented with a target specification, Déjà Vu retrieves a *best* case TLC. The adaptation of this TLC involves retrieving and adapting DCs which correspond the TLCs link descriptions. Thus the structure of the TLC acts as an decomposition/integration template for connecting the individual code components generated during the various CBR cycles.

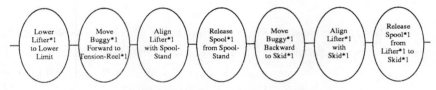

Figure 4. An example TLC called Unload*1.

This partonomic structure builds considerable flexibility into the case-base and it has the added attraction that it supports problem solving by successive refinement. However it introduces a considerable problem of control into the reasoning process. In Déjà Vu this control problem is addressed by organising the recursive CBR process as a blackboard system.

4 Blackboard Based Control

The hierarchical case-base structure described in the previous section reflects one of the central tenets of our research. In a problem solving process, when a human expert draws on experience stored as episodes in memory several episodes may be retrieved to help with different aspects of the problem in hand. A comprehensive approach to CBR must allow for the retrieval of several cases that contribute to the formation of the final solution. This requirement for recursive CBR introduces a fundamental problem of control in the design of the CBR system. In knowledge based systems development the issue of control of interaction between different knowledge sources strongly suggest the adoption of a blackboard architecture and this is the approach taken in Déjà Vu. Before describing how the interaction of the different problem solving stages in Déjà Vu is coordinated through a blackboard we will introduce blackboard based problem solving in general terms.

4.1 A Perspective on Blackboard Systems

The conventional view of blackboard systems reflects the origins of the blackboard idea as a communication and control methodology in the two seminal blackboard systems Hearsay II and Hasp (see (Erman, Hayes-Roth, Lesser, and Reddy, 1980; Nii, 1986). These systems perform what are essentially analytical tasks of interpretation or classification. Hearsay II is a speech recognition system taking the basic speech signal as input. Hasp interprets continuous sonar signals collected by hydrophone arrays monitoring areas of the ocean, the objective being to identify the vessel that is the source of the signal. The common feature is that the analysis can be partitioned into separate knowledge agents (e.g., phoneme-hypothesizer or word-candidate-generator in Hearsay II) and, in the blackboard implementation, these agents cooperate in building a solution by communicating through a blackboard.

So this perspective of blackboards is of the blackboard as a global data structure through which heterogeneous knowledge agents (KAs) communicate in forming a solution. The communication is based on KAs posting intermediate hypotheses on the blackboard where they can be picked up and developed by the other KAs. This allows for opportunistic scheduling of the agents as they respond 'eagerly' to the hypotheses posted on the blackboard (Nii, 1986). It should be clear that this perspective on what blackboards do reflects the programming-language implementation of the idea on an abstract machine. Before proceeding with our realisation of the blackboard idea in Déjà Vu we would like to consider a broader inference based perspective.

It has been argued (Clancy, 1990, 1992) that the blackboard architecture is an embodiment of his System Model Operator perspective on expert systems development. According to this argument expert systems are programs that construct a situation-specific **model** of some **system** so that it can be diagnosed, assembled, etc. These models represent processes and structures as relational networks. A situation specific model (SSM) is a model of the system, specific to the current context, that is generated from a general model of the problem domain (see Figure 5). Control knowledge for constructing such a model can be described as **operators** that construct a graph linking processes and structures causally, temporally, spatially, by subtype etc.

The blackboard architecture conforms to this perspective very well, with the KAs considered as operators constructing the situation specific model of the system on the blackboard.[3] This view of blackboard systems emphasises the reasoning process and de-emphasises the programming implementation. One implication of this perspective is that all expert systems incorporate, either explicitly or implicitly, models of aspects of the problem domain that are non numeric in character. This elevates qualitative reasoning to a position of considerable importance as it implies that qualitative modelling is fundamental to all expert systems development.

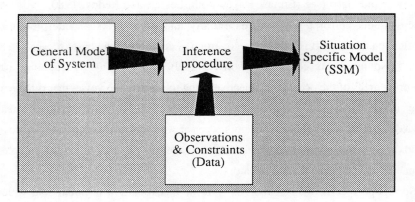

Figure 5 Clancey's System Model Operator perspective recasts experts systems as the process of building an situation specific model reflecting the current task.

The main contribution that this broader perspective offers to our research on Déjà Vu is the emphasis on considering the blackboard as a situation specific model of the current task rather than as a shared data structure where hypotheses are posted by the knowledge agents.

---

[3] Clancey is so keen on this perspective that he advocates that all expert systems can fruitfully be considered as relational networks for modelling processes; even MYCIN, a system that would conventionally be considered a disease classification hierarchy rather than a model-based process [Clancey '92]. Indeed the context tree in MYCIN can usefully be considered a blackboard.

4.2 Déjà Vu's Blackboard Architecture

Déjà Vu's blackboard structure can been seen in Figure 6 in terms of its *knowledge agents* and blackboard areas or *panels*. In the context of Clancey's perspective mentioned above, the focus of the situation specific model is in the Solution Panel with threads of the inference associated with the SSM coming from the other panels, particularly the Adaptation Panel and the Specification Panels.

Before giving an example of Déjà Vu at work we will briefly describe the panels and agents of the blackboard:-

- **Target Specification Panel:** Target problem descriptions are supplied to Déjà Vu in a very high-level format (e.g., "Remove Spool*1 from Tension-Reel*1 using Buggy*5). These target descriptions are deposited in this panel where they are made available to the analysis agent. The panel is also used during a problem solving episode to hold the high-level DC link descriptions which invoke further CBR cycles.

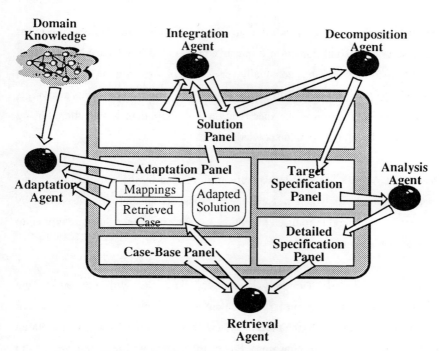

Figure 6. Déjà Vu's Blackboard Architecture.

- **Detailed Specification Panel:** This panel is used to contain a detailed frame representation of the specification that is currently the focus of the retrieval and adaptation process.

- **Adaptation Panel:** The result of a retrieval process is posted to the *retrieved case* sub-panel of this panel and the mappings established are written to the *mapping* sub-panel. The *adapted solution* sub-panel will contain the adapted version of the retrieved solution.

- **Case-Base Panel:** This panel provides the retrieval agent with access to the case-base.

- **Solution Panel:** As SSM construction proceeds this panel contains the resulting solution code.

The following are the five knowledge agents currently in the system:

- **Analysis Agent:** This agent's task is to take the high-level target description (or link description) and transform it into a more detailed frame based description. This transformation task is potentially very complex involving the gathering of all knowledge pertinent to the current target problem. The product of this analysis is a frame representation of the specification with extra details instantiated using information from the domain knowledge base.

- **Retrieval Agent:** The detailed specification is used by the retrieval agent as a retrieval probe with which to search the case-base for an appropriate base case by the retrieval method introduced in section 3.1. The retrieval results are written to the adaptation panel.

- **Adaptation Agent:** Typically a retrieved base case will not satisfy the requirements of the current target problem and so some form of adaptation or repair must be done. In the current system this adaptation proceeds in both an automatic and interactive fashion. Firstly the system will use the mappings to perform a rudimentary adaptation of the retrieved solution so that it conforms, at an entity level, to the target situation. Further automatic adaptation methods will utilise the domain knowledge captured in the domain model to perform a more detailed adaptation of the retrieved solution. It would in fact be more accurate to describe this agent as a set of agents (or *agency*) since adaptation will be the result of many adaptation processes at work, each specialising in a specific type of adaptation. It may happen that the errors in the solution are beyond the capability of Déjà Vu's adaptation methods and so the user may intervene at this stage to aid the production of the required piece of code. Very briefly, during the interactive adaptation process the user is guided by the mappings set up during retrieval and the domain knowledge structures, and in this way Déjà Vu acts as a programmer's "assistant" rather than an "expert".

- **Decomposition Agent:** This agent simply takes each link description structure of a TLC and writes them in turn to the target specification panel thus invoking further CBR cycles.

- **Integration Agent:** This agent is responsible for integrating the components of the solution produced by the adaptation agent into the evolving solution.

5 A Problem Solving Episode

The task of generating the solution for a given target specification proceeds in a decomposition - generation - integration fashion where the generation component takes the form of a number of CBR retrieval-adaptation cycles. The target specification when expanded into a detailed specification causes the retrieval of a TLC. This TLC acts as a decomposition/integration template indicating the low-level components of the TLC as a set of link descriptions. Each link description is sent to the target specification panel by the decomposition agent and causes further CBR retrieval-adaptation cycles resulting in the generation of the appropriate code segments which may then be integrated into the target TLC by the integration agent.

5.1 Déjà Vu in Action: An Example

As an example of Déjà Vu's problem solving mechanisms at work consider the target problem of *"Remove Spool*1 from Tension-Reel*1 using Buggy*2 and deposit at Skid*1"* . We present this example in three stages indicating the state of the system during the first, second and k'th. cycles. At each stage we illustrate the state of the system using a diagramatical representation of the blackboard.

Cycle 1: Figure 7 shows the result of the first cycle. The cycle starts with the initial target specification in the target specification panel. This is transformed through the action of the analysis agent into the frame structure shown in the detailed specification panel.

The activation of the retrieval agent causes the TLC called Unload*1 to be retrieved and placed in the retrieved case sub-panel of the adaptation panel (this is the TLC shown earlier in Figure 4). This retrieved case is similar to the target except that instead it uses Buggy*1, a one speed buggy, rather than Buggy*2, a 2 speed buggy. This discrepancy is captured in the mappings set up during the retrieval and the adaptation agents use this knowledge to change the link descriptions of Unload*1 appropriately (as can be seen in the adapted solution sub-panel).

The adapted TLC (henceforth called the target TLC) is picked up by the integration agent and written to the solution panel as the top-level solution. This the first integration stage is very simple as other solution components have not yet been generated. However, future integration stages must consider the effects that the current solution component will have on those already generated.

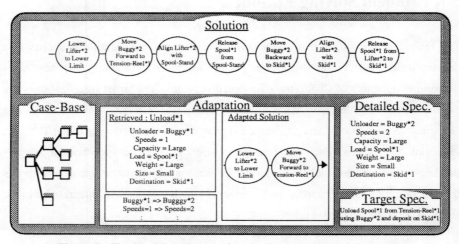

Figure 7. The blackboard state after the first problem solving stage.

Having established a TLC to use as a decomposition/integration template, Déjà Vu proceeds by generating the code for each component of the target TLC (refer to Figure 8).

Cycle 2: In this cycle the decomposition agent has written the first link description to the target specification panel and in a similar fashion to the previous stage this causes the retrieval and adaptation of an appropriate base case from the case-base. This base case is a DC and thus the integration agent must integrate the adapted solution into the target TLC at the appropriate place.

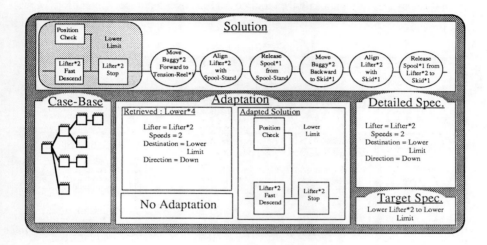

Figure 8. The blackboard state after the second problem solving stage.

A point of interest here is that originally the DC Lower*1 was linked to Unload*1 (see figure 3). However the retrieval component of this cycle returned the DC Lower*4 since it fitted the target situation better than Lower*1. Clearly the use of link descriptions adds greater flexibility to the system, relieving the adaptation problem somewhat. This can be seen if we consider what would have happened had a direct linking mechanism been employed to specify the component DCs of TLCs. In such a situation the target TLC would have directly referenced the DC Lower*1 which would have been retrieved. Déjà Vu would then have had to adapt this DC to allow for two speed action. This adaptation is not necessary however, since the DC Lower*4 exists and requires no changes.

Cycle k: Figure 9 represents the k'th cycle where an appropriate DC is retrieved as in the first cycle. However a direct match was not achieved and thus DC adaptation must be done. As in the case of adapting the initial TLC a rudimentary direct mapping form of adaptation is done to rectify the discrepancies between the retrieved DC and the target specification, namely the destination of the movement action is changed from Spool-

Stand to Skid*1. The adapted solution is then integrated into the solution. We have not elaborated on the problem of integration in this example; the solution components used did not result in the occurrence of any conflicts with other components. However, as we mentioned earlier, integration must ensure that such conflicts, should they arise, are catered for appropriately.

In this way Déjà Vu builds up the required solution over a number of iterations of the CBR process.

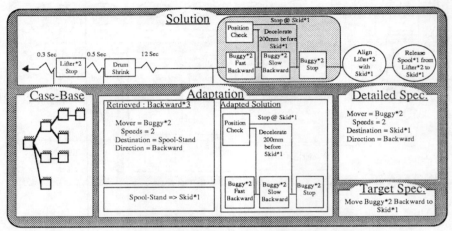

Figure 9. The blackboard state after the kth problem solving stage.

It is worth noting here again that Déjà Vu's problem structure is not restricted to only two levels of detail as indicated in this example. In general, a problem may have any number of levels of detail, and in response Déjà Vu problem solving method will reflect this.

6 Conclusion

This paper described Déjà Vu, a case-based reasoning system for software design which uses multiple CBR cycles to generate a target solution. We consider this recursive CBR process to be similar to the human expert's process of design by successive refinement.

The increased complexity of the recursive process introduces a control problem. The blackboard solution to this problem shows considerable promise in handling this added complexity. The successive refinement process requires that the case-base contain cases at different levels of abstraction (currently we are dealing with just two levels, top-level and detail-level, although the architecture will support more if needed). Consequently top-level cases are organised as a partonomic hierarchy of detailed cases. As a whole the case-base is also structured as a taxonomy that facilitates the retrieval process.

The example problem solving episode was fairly simple in nature involving adaptation due to a capability change of a domain entity, namely the speed of the buggy-lifter vehicle. This required the alteration of individual solution units without the need to consider the overall program structure. A related problem concerns the integration of detail cases into the target top level case. In more complex situations the incorporation of a given detail case into an evolving solution may introduce goal interaction problems which must be attended to. Currently such *difficult* adaptation/repair tasks are left up to the user through the interactive adaptation facility.

Current research efforts are focused on the blackboard architecture controlling the reasoning process and on the CBR retrieval component. Long term research will concentrate on introducing automatic techniques for overcoming the more *difficult* adaptation/repair problems such as those mentioned above.

References

Clancey W. J. (1992). Model construction operators. *Artificial Intelligence*, **53**(1), 1-115

Clancey W. J. (1990). Implications of the system-model-operator metaphor for knowledge acquisition. In H. Motoda, R. Mizoguchi, J. Boose, B. Gaines (Eds.), *Proceedings of the First Japanese Knowledge Acquisition for Knowledge-Based Systems Workshop*, (pp. 65-80).

Erman L.D., Hayes-Roth F., Lesser V.R., & Reddy D.R. (1980). The Hearsay-II speech-understanding system: Integrating knowledge to resolve uncertainty. *Computing Surveys*, **12** (2), 213-253.

Hammond K. (1989). *Case-Based Planning*. Academic Press.

Nii H. P. (1986). Blackboard Systems: The Blackboard Model of Problem Solving and the Evolution of Blackboard Architectures. *AI Magazine,* **7** (2), 39-53.

Ono Y., Tanimoto I., Matsudaira T., & Takeuchi Y. (1988). Artificial Intelligence Based Programmable Controller Software Designing. *International Workshop on Artificial Intelligence for Industrial Applications*, (pp. 85-90).

Sakurai T., Shibagaki T., Shinbori T., & Itoh S (1990). An Automatic Programming System based on Modular Integrated-Concept Architecture (MICA). *Proceedings of International Conference on Industrial Electronics, Control and Instrumentation* (pp. 1303-1308). Monterey, California, U.S.A..

Smyth B., & Cunningham P. (1992). Déjà Vu: A Hierarchical Case-Based Reasoning System for Software Design. *Proceedings of the Tenth European Conference on Artificial Intelligence.* Vienna, Austria.

Thompson K., & Langley P. (1989). Organization and Retrieval of Composite Concepts. *Proceedings of the Case-Based Reasoning Workshop* (pp. 329-333). Pensacola Beach, Florida, U.S.A..

Session 5: Natural Language 2

The Construction and Use of Scope Neutral Discourse Entities.

Norman Creaney[1]
Faculty of Informatics, The University of Ulster at Coleraine,
Coleraine, N. Ireland.

Abstract

This paper is concerned with the quantifier scoping problem in English. In particular an approach to scoping ambiguities is described which is distinguished by being both scope neutral and incremental. A level of representation is described which is capable of encoding all the possible quantifier scoping arrangements in a given sentence. A set of constraints are presented which account for the way in which scoping options are sensitive to the logical structure of English and an algorithm for propagating these constraints is described. Of particular concern is the interaction between incremental scope interpretation and discourse processing. The notion of a scope neutral discourse entity is developed as a level of representation suitable for mediating this interaction.

1 Introduction

This paper is concerned with the quantifier scoping problem in natural language. In particular, a method of handling scoping ambiguities is described which is both incremental and scope neutral. That is to say, our method makes the least possible commitment to quantifier scoping choices but is able to reflect the incremental disambiguation which occurs as a discourse is processed. The classic example used to illustrate quantifier scoping ambiguity is:

(1)   *"Every man loves some woman."*

which has two readings illustrated by, for example, the following small models:

(2)   a   love(fred, wilma)
              love(barney, wilma)
              ie. the same woman for each man

---

[1]tel: (0265) 44141, ext. 4502, fax: (0265) 40916, email: cbga23@uk.ac.ulster.ujvax

b love(fred, wilma)
   love(barney, betty)
   ie. a different woman for each man

These two readings are often distinguished at a logical form (LF) level of representation as:

(3) a $\exists Y: \forall X: man(X) \Rightarrow woman(Y)\ \&\ love(X,Y)$

    b $\forall X: \exists Y: man(X) \Rightarrow woman(Y)\ \&\ love(X,Y)$

respectively. In (3a) *"some woman"* is said to outscope or have scope over *"Every man"* while in (3b) *"Every man"* is said to outscope *"some woman"*.

A common approach to resolving scoping ambiguities is to generate all the possible readings and use a combination of, semantic information, contextual information and heuristics to select the preferred or most likely one. For example, Grosz et al. (1987) use the notion of a *scope critic* to implement linguistic preferences. Scope critics assign scores to different scoping arrangements according to such things as, the particular determiners used and the order in which the quantifiers appear in the surface form. The scoping with the highest score is then taken to be the preferred one. Hobbs and Shieber (1987) present an algorithm for generating all the structurally possible scopings of a sentence.

An alternative approach, which relies on the notion of *scope neutral form* is argued for in Hobbs (1983). This is a representation with the following properties.

- It is neutral with respect to any scoping ambiguities.

- It is capable of expressing the narrowing of the space of possibilities that takes place as an expression is incrementally disambiguated.

2   Why Scope Neutrality ?

The arguments in favour of scope neutral representations (some of which have been made by Hobbs 1983) are summarised below.

2.1 Computational expense

Disambiguating, or generating and selecting between competing scopings can be computationally expensive, especially if, as is often the case, it requires the use of inference. Of course any scope neutral representation is likely to carry its own computational overhead and this must also be taken into account.

2.2 Discourse sufficiency

Some information can be obtained even from a totally ambiguous (with respect to quantifier scoping) sentence and it may be the case that this limited amount of information is sufficient for the given sentence to serve its purpose in the discourse. For example, the sentence:

(4) *"Every representative of a company saw most samples."*

has five different relative scopings but even in this totally ambiguous form it is clear that there were some representatives and some samples and that some seeing went on between the representatives and the samples. In answer to the question, *"Did any representatives see any of the samples ?"* this information might be entirely adequate. If the unscoped form gives all the information needed then why go to the trouble of scoping ? Furthermore, even if more information is required from a sentence than is available from the totally unscoped form it does not follow that a precise scoping needs to be computed for every quantifier in the sentence. For example, if (4) is followed by the sentence:

(5) *"Representatives of that company are always interested in our samples."*

then a referent is required for *"that company"*. This can be derived from (4) by forcing *"a company"* to have scope over *"Every representative"* and as a side effect (4) is partially disambiguated. However nothing need be said about the relative scoping of *"a company"* and *"most samples"*, or *"Every representative"* and *"most samples"*. In this case the partially ambiguous (4) is sufficient to allow the discourse to proceed.

2.3 Timing

Arbitrary[2] amounts of material may appear between an ambiguous sentence (like (4)) and a further one (like (5)) which (partially) disambiguates it. Consider, for example, *"Every representative of a company saw most samples. Some saw all of them. Representatives of that company are always interested in our samples."*.

Since the following both hold:

(5) could easily be replaced by another sentence (eg. *"Most of their companies placed an order the next day."*) which reverses the scoping of *"Every representative"* and *"a company"*.

---

[2]The amount of intervening material is obviously not without limit however the point is that the limit is not based on any simple measure (eg a fixed number of sentences). Rather, it is more likely to be based on the discourse structure.

The intervening material "*Some saw all of them*" could easily be extended (eg. "*Some saw all of them. In fact most saw all*").

there can be no point at which a once and for all disambiguation of (4) is possible.

3   Scope Neutrality and Constraint Propagation

The representation used to achieve scope neutrality is a constraint network and the mechanism used to refine it is constraint propagation. Constraint propagation is an inherently incremental process whereby given:

- A set of variables, V_i.

- A set of domains, D_i; where D_i is the set of possible values which V_i may take.

- A set of constraints which must hold between the variables.

all those values from each D_i, which cannot (on the basis of the constraints) form part of a solution of the constraint satisfaction problem, are deleted from that D_i. A solution being a unique assignment of a value from each D_i to the respective V_i such that all the constraints are satisfied. Constraint satisfaction problems are often described in terms of a constraint network where the variables (V_is) are seen as nodes[3] and the constraints as links. Many problems in AI have been successfully characterised as constraint satisfaction problems, eg. vision: Waltz (1975), reference resolution: Mellish (1985), Haddock (1988,1989).

For the purposes of constraint propagation there is one node for each pair of variables in the logical form. The constraint network may be seen as a table where the variables from the LF are tabulated both horizontally and vertically, and cells within the table are nodes. Following this the constraint network is referred to as a *table of nodes* (TON). For example if a LF has variables V1, V2 and V3, then the following TON is constructed.

	V1	V2	V3
V1		<V1,V2>	<V1,V3>
V2			<V2,V3>
V3			

---

[3]We use the term "node" rather than "variable" to avoid confusion with the variables which appear in logical forms.

Since any node <X,Y> contains all the scoping information for <Y,X>, and since the scoping of <X,X> has no meaning, only cells above the diagonal are considered.

Each node <X,Y> may be ordered, X before Y or Y before X depending on whether X has scope over Y or Y has scope over X, respectively. This much is straightforward, however, X & Y may also be unordered or parallel. Consider, for example, the following sentence.

(6) "A man with three wives has two problems."

If "A man" outscopes both "three wives" and "two problems" (ie. there is only one man) then no question about the relative scoping of "three wives" and "two problems" even arises. They are considered to be parallel. For each pair of variables, X & Y, in the LF then, there are three possibilities:

   ord(X,Y)   X outscopes Y
   par(X,Y)   X and Y are parallel
   rev(X,Y)   Y outscopes X

Equivalently, for each node, <X,Y>, in the TON there are three possibilities:

   <X,Y> is ord
   <X,Y> is par
   <X,Y> is rev

Each node in the TON has a domain of possible values which encodes the extent of knowledge about the relative scoping of the given variables. Since there are generally three possibilities for each pair of variables, each domain initially has three values. As disambiguating information becomes available the TON is refined by deleting values from domains. For example:

   domain(<X,Y>) = { ord, par, rev }
      =>   nothing is known about the relative scoping of X & Y
   domain(<X,Y>) = { ord, rev }
      =>   either X outscopes Y or Y outscopes X
           (but they are not parallel)
   domain(<X,Y>) = { ord }
      =>   X outscopes Y

Representing scoping information in this way allows either; no commitment about scoping to be made, one particular scoping to be committed to, or any intermediate stance to be taken. This is the basis of our scope neutral representation.

As has been observed by Hobbs and Shieber (1987) not all combinatorially possible scoping arrangements of a sentence are always possible. In particular, the logical structure of the sentence directly constrains the number of possible readings. Consider the following sentence.

(7)   "A representative of most companies brought a sample"

If "*A representative*" outscopes both "*most companies*" and "*a sample*" (ie. there is only one representative) then no question about the relative scoping of "*most companies*" and "*a sample*" even arises (ie. they are parallel). This is a consequence of the fact that "*most companies*" is part of the restriction of "*A representative*" while "*a sample*" is not, which is a property of the logical structure of (7). Note that once the possibility that "*most companies*" outscopes "*A representative*" is allowed, the relative scoping of "*most companies*" and "*a sample*" again becomes a question. This illustrates how constraints which arise from logical structure interact with scoping commitments about other quantifiers in the sentence, and underlines the complexity of the problem.

4   Making the Constraints Explicit

The constraints make use of the *sib* (sibling) relationship which holds between any two variables which participate directly in some relationship in a sentence. For example, the sentence (4) has the LF.

(8) $\forall$ R (representative(R) & $\exists$ C company(C) representative_of(C,R))
       =>MOST[4] S sample(S) & saw(R,S)

sib holds of R and C because they both participate in the relationship *representative_of*. On the other hand *sib* does not hold of C and S because they participate in no such direct relationship in (8). Use is also made of the *res* relationship which holds of any two variables, X & Y, where Y is in the restriction of X. For example res(R,C) holds in (8).

The first constraint states that, in the case of two siblings, one must outscope the other. That is to say, they cannot be parallel.

(9)   sib(X, Y) => not par(X, Y)

The second constraint enforces the transitivity of the *ord* relationship.

(10)   ord(X, Y) & ord (Y, Z) => ord(X, Z)

The remaining constraints enforce the interaction between logical structure and scoping which are discussed in Hobbs and Shieber (1987).

---

[4]The quantifier MOST is meant to gloss over the semantics of the English "*most*".

(11) res(X, Y) & sib(X, Z) =>
  ((ord(Y, X) => ~par(Y, Z))   &
      (ord(Z, X) => ~par(Y, Z))   &
      (ord(Y, Z) => ord(Y, X))   &
      (ord(Z, Y) => ord(Z, X))   )

(12) res(X1, Y) & res(X2, Z) & sib(X1, X2) =>
  ((ord(Y, X2) => ~par(Y, Z))   &
      (ord(Z, X1) => ~par(Y, Z))   &
      (ord(Y, Z) => ord(Y, X2))   &
      (ord(Z, Y) => ord(Z, X1))   )

(13) res(X, Y) & res(Y, Z) =>
  ((ord(Y, X) => (ord(Z, X) <=> ord(Z, Y)))   &
      (ord(Y, X) => (par(Z, X) <=> ord(Y, Z)))   &
      (ord(Y, X) => ~ord(X, Z))   )

Creaney and McTear (1991) describe an earlier implementation of these ideas which uses a different but related set of constraints. While the earlier constraints give identical results in the examples we considered they were later discovered to be inadequate. The above constraints are a correction of our earlier ones.

The algorithm used to propagate the constraints is a Waltz (1975) type one. When a change is made to the domain of some node, that node becomes awake. While there remain wakened nodes, the constraints are applied to each of them in turn in an attempt to find values from the domains of other nodes which no longer participate in any solution of the constraint problem. If any such values are found they are deleted and those nodes becomes wakened. The process continues until there are no remaining wakened nodes. The TON encodes the most general scoping arrangement of a sentence which is consistent with both its logical structure and any explicit scoping commitments which have been made. Explicit scoping commitments are added to the TON as a consequence of discourse and semantic operations (eg. anaphor resloution, inference) and the process of constraint propagation then makes further scoping commitments. Constraint propagation makes *only* those scoping commitments which are necessary to maintain consistency with logical structure.

5 Scope Neutral Discourse Entities

The underlying representation of the discourse, or discourse model (DM), contains a set of discourse entities (DEs). However, because of scope neutrality, an enriched notion of discourse entity is required. According to Webber's (1978) rules for DE construction, an existential quantifier within the scope of a universal evokes a set type discourse entity, so that, in the case of (1) and assuming wide scope for "*Every man*", the following DEs are evoked.

(14) e1.1 = { X: man(X) }

  e1.2 = { X: woman(X) & ∃ Y ε e1.1: love(X,Y) }

where e1.1 is assumed to exist already in the DM and does not concern us here. In order to accommodate scope neutrality a variation in the form of e1.2 is adopted as follows.

(15) e1.2' = range(λ(Y ε e1.1) ι(X: woman) love(X,Y))

The function λ(Y ε e1.1) ι(X: woman) love(X,Y) may be thought of as a Skolem function from members of e1.1 to women, and e1.2' is then the range of this function and is exactly e1.2.

It should be clear that any variables (in general there may be zero or more) bound by a lambda operator in a DE correspond to variables which distribute over the DE in logical form. If this set of variables is called LAM, then LAM encodes the relevant scoping information and is derivable from the TON. However, in general, the TON does not give a definitive LAM. Rather it gives two sets. A set of variables which *necessarily* distribute over the DE (NECLAM), and a set which may *possibly* distribute over it (POSLAM). If the assumption of wide scope for "*Every man*" is dropped, then NECLAM = {} and POSLAM ={Y}. NECLAM and POSLAM are used to give lower and upper bounds on the DE and are necessary to accommodate scope neutrality at the level of discourse entities.

$$\text{range}(\ \lambda()\ \iota(X: woman)\ \forall(Y\ \varepsilon\ e1.1)\ love(X,Y)\ )$$
$$\leq e1.2' \leq$$
$$\text{range}(\ \lambda(Y\ \varepsilon\ e1.1)\ \iota(X: woman)\ love(X,Y)\ )$$

According to this scheme a scope neutral discourse entity (SNDE) consists of five parts, as follows.

NECLAM   The set of variables which necessarily distribute over the DE on LF.
POSLAM   The set of variables which possibly distribute over the DE in LF.
HEAD   The head of the DE containing its variable and type.
RES   The restriction of the DE containing any predications on it.
BODY   The body of the DE containing any predications on it.

The complete scope neutral version of DE e1.2' in (15), assuming no scoping commitments, is:

(16) e1.2" = NECLAM = λ()

```
POSLAM   =   λ(Y ε e1.1)
HEAD     =   ι(X: woman)
RES      =   true
BODY     =   love(X, Y)
```

In fact, it is more convenient to drop all variables from NECLAM, POSLAM, NECDIS and POSDIS since there are occasions when it is difficult to use them consistently while the SNDE still contains ambiguities. Following this e2.1″ becomes:

(17) e1.2‴ =
```
NECLAM   =   { }
POSLAM   =   { e1.1 }
HEAD     =   ι(X: woman)
RES      =   true
BODY     =   love(X, e1.1)
```

It should be noted that SNDEs are not stored but they are derived when needed, from the TON, which is the sole repository of scoping information. As a discourse proceeds, so the TON is incrementally refined. Consequently the SNDEs derivable from it are a function of the stage at which they are derived. This is a reflection of the incremental refining of scoping information which occurs as a discourse is processed.

6   Interaction with Discourse Processing

The following example illustrates the way in which SNDEs change as a discourse is processed. Recall sentence (4), *"Every representative of a company saw most samples."* and its LF:

(18)   ∀ R (representative(R) & ∃ C company(C) representative_of(C,R))
           =>MOST S sample(S) & saw(R,S)

Assuming no scoping commitments the following initial TON is constructed.

	R	C	S
R		{ord,par,rev}	{ord,par,rev}
C			{ord,par,rev}
S			

According to constraint C1, <R,C> cannot be parallel because R & C are siblings. This also applies to <R,S>, giving the following TON.

	R	C	S
R		{ ord, rev }	{ ord, rev }
C			{ord,par,rev}
S			

The following SNDEs can be derived from this TON.

(18) e2.1 = { X: representative(X) }

    e2.2 =
- NECLAM = { }
- POSLAM = { e2.1, e2.3 }
- HEAD = ι(X: company)
- RES = true
- BODY = representative_of(X, e2.1)

    e2.3 =
- NECLAM = { }
- POSLAM = { e2.1 }
- HEAD = ι(X: most'(sample))
- RES = true
- BODY = saw(e2.1, X)

where: most'(sample) is a predicate which is true of any set of samples, Ss, such that, Ss contains most of the set { X: sample(X) }
{ X: sample(X) } and { X: representative(X) } are assumed to exist already in the DM.

Note that, e2.2 does not appear in POSLAM of e2.3. This is because e2.2 is evoked by the noun phrase, "*a company*" which does not have a distributive quantifier. The procedure for deriving SNDEs from the TON is sensitive to this fact and constructs POSLAM accordingly.

Suppose now that the semantics of the relation *representative_of* requires that a company may have only one representative. In particular then, it cannot be the case that "*a company*" has scope over "*Every representative*". This commitment is applied directly to the TON to give the following.

	R	C	S
R		{ ord }	{ ord, rev }
C			{ord,par,rev}
S			

Constraint propagation then takes place but makes no further changes, and the derived SNDEs are now as follows.

(19) e2.2' =
- NECLAM = { e2.1 }
- POSLAM = { e2.3' }
- HEAD = ι(X: company)
- RES = true
- BODY = representative_of(X, e2.1)

    e2.3' = NECLAM = { }

```
POSLAM  =  { e2.1 }
HEAD    =  ι(X: most'(sample))
RES     =  true
BODY    =  saw(e2.1,X)
```

Suppose now that the following sentence is processed.

(20) *"Each representative saw those samples which his own company distributes."*

This suggests that it is a different set of samples for each representative. In particular (20) evokes the following SNDE.

```
(21) e3.2 =  NECLAM  =  { e2.1 }
             POSLAM  =  { }
             HEAD    =  ι(X: set(sample))
             RES     =  true
             BODY    =  distributes(company_of(e2.1), X) &
                        saw(e2.1, X)
```

Since e3.2 is consistent with e2.3' they are taken to be co-referential. Since the scoping information associated with e3.2 is less ambiguous (in fact it is not ambiguous at all) than that associated with e2.3', e2.3' is updated with the less ambiguous information. That is to say, *rev* is deleted from the domain of <R,S>, to give the following TON.

	R	C	S
R		{ ord }	{ ord }
C			{ord,par,rev}
S			

At this point constraint propagation takes place which results in further deletions from the TON to give.

	R	C	S
R		{ ord }	{ ord }
C			{ par }
S			

The following SNDEs can now be derived.

```
(22) e2.2" =  NECLAM  =  { e2.1 }
              POSLAM  =  { }
              HEAD    =  ι(X: company)
              RES     =  true
              BODY    =  representative_of(X, e2.1)

     e2.3" =  NECLAM  =  { e2.1 }
```

                    POSLAM   =   { }
                    HEAD   =   ι(X: most'(sample))
                    RES   =   true
                    BODY   =   saw(e2.1, X)

These SNDEs now contain no scoping ambiguities and may be rewritten as the following conventional DEs. At this stage appropriate variables are re-inserted.

(23)   e2.2''' =   range(λ(R ε e2.1)
                            ι(X: company) representative_of(X, R))

       e2.3''' =   range(λ(R ε e2.1) ι(X: most'(sample)) saw(R, X))

7 Conclusion

An approach to scoping ambiguities in English has been described which is distinguished by being both scope neutral and incremental. The problem has been characterised as one of constraint propagation and a scheme for encoding all the relevant scoping information in a table of nodes has been described. A set of constraints which account for the way in which scoping options are sensitive to the logical structure of English has also been presented. Of particularly concern has been the interaction between incremental scope interpretation and discourse processing, and a method for accomplishing such interaction based on scope neutral discourse entities has been described.

It is concluded that constraint propagation is an appropriate mechanism to implement scope neutral form. The table of nodes encodes all relevant scoping information, while constraint propagation narrows the space of possible interpretations in a natural way as disambiguating information becomes available. One disadvantage of the present approach is the underlying assumption that all scoping commitments made are certain. Once a commitment is applied to the TON there is no facility to undo it later. This assumption is reasonable when scoping commitments arise from discourse or semantic considerations like those described above. However, it is difficult to see how linguistic preferences like those described in Grosz et. al. (1987) could be accommodated since they are by definition not completely reliable.

Research in this area continues and of particular interest is the interaction between the present model of scope interpretation and bound variable anaphora.

References.

Creaney, N., McTear, M.F. (1991). The Incremental Disambiguation of Quantifier Scoping Ambiguities. *Proceedings of the 11th Conf. on Expert Systems & their Applications*, Avignon.

Freuder, E.C. (1978). Synthesizing Constraint Expressions, *Commmunications of the ACM*, **21**, 958-965.

Grosz, B.J., Appelt, D.E., Martin, P.A., Pereira, F.C.N. (1987). TEAM: An Experiment in the Design of Transportable Natural-Language Interfaces. *Artificial Intelligence*, **32**, 173-243.

Haddock, N.J. (1988). Incremental Semantics and Interactive Syntactic Processing, PhD. Thesis, University of Edinburgh.

Haddock, N.J. (1989). Computational Models of Incremental Semantic Interpretation. *Language and Cognitive Processes*, **4** (3/4), 337-368.

Hobbs, J.R. (1983). An Improper Treatment of Quantification in Ordinary English, *Proceedings of the 21st Annual ACL*, 57-63.

Hobbs, J.R., Shieber, S.M. (1987). An Algorithm for Generating Quantifier Scopings. *Computational Linguistics*, **3** (1), 47-63.

Mackworth, A.K. (1977). Consistency in Networks of Relations. *Artificial Intelligence*, **8**, 99-118.

Mellish, C.S. (1985). *Computer Interpretation of Natural Language Descriptions*. Chichester: Ellis Horwood.

Smith, B.M. (1990). An Application of Constraint Satisfaction in Constructing an Anaesthetists' Weekly Rota. *9th Alvey Planning SIG Workshop*, Nottingham University.

Tsang, P.K. (1987). The Consistent Labeling Problem in Temporal Reasoning. *Proceedings of AAAI-87*, 251-255.

Waltz, D. (1975). Understanding Line Drawings of Scenes with Shadows. In P.H. Winston, (Ed.). *The Psychology of Computer Vision*. NY, McGraw-Hill.

Webber, B.L. (1978). A Formal Approach to Discourse Anaphora. (BBN report no. 3761). Cambridge, MA: Bolt Beranek & Newman Inc.

Constructing Distributed Semantic Lexical Representations using a Machine Readable Dictionary

Richard F.E. Sutcliffe[1]

University of Limerick

Limerick, Ireland

Abstract

We present an experiment in which distributed semantic representations are automatically constructed for 4263 nouns taken from the Merriam Webster Compact Electronic Dictionary. The algorithm for defining a concept involves extracting taxonomic links from dictionary definitions and using these to infer weighted features to add to that concept's representation. Initial analysis of the results is promising: word representations tend to cluster in a semantically intuitive fashion. The approach can be generalised to the rest of the dictionary by improving the parser.

1 Introduction

In order to construct systems which can process natural language in a sophisticated fashion it is highly desirable to be able to represent linguistic meanings in a computationally tractable fashion. One approach to the problem of capturing meanings at the lexical level is to use a form of distributed representation where each word meaning is converted into a point in an n-dimensional space (Sutcliffe, 1992a). Such representations can capture a wide variety of word meanings within the same formalism. In addition they can be used within distributed representations for capturing higher level information such as that expressed by sentences (Sutcliffe, 1991a). Moreover, they can be scaled to suit a particular tradeoff of specificity and memory usage (Sutcliffe, 1991b). Finally, distributed representations can be processed conveniently by vector processing methods or connectionist algorithms and can be used either as part of a symbolic system (Sutcliffe, 1992b) or within a connectionist architecture (Sutcliffe, 1988).

Apart from the issue of meaning representation itself, there is also the problem of scale. To construct a useful natural language processing (NLP) system we need

---

[1] I am most grateful to Tom Brehony, Lorraine Bennett, Colman Collins, Denis Hickey, Ken Litkowski, Tony Molloy and Oliver Murphy for their help with this research. The assistance of Eileen Madden and Gemma Ryan is also gratefully acknowledged. The author's address is Department of Computer Science and Information Systems, University of Limerick, Ireland. Tel. +353 61 333644 Ext 5006 (Direct) Ext 5084 (Messages) email sutcliffer@ul.ie .

to be able to recognise the meanings of a reasonable proportion of the words which a native speaker understands. This might be somewhere between 100,000 and 500,000 words. The sheer number of words to be considered presents a formidable problem to NLP researchers. However, in recent years considerable interest has been shown in the possibilities opened up by the use of large scale linguistic resources such as machine readable dictionaries (MRDs) (Amsler, 1984, Evens, 1989). [1] If some of the semantic information within an MRD can be extracted automatically, this provides an objective and replicable means of constructing a useful lexicon.

Distributed representations are particularly amenable to the use of automated techniques. In this paper we present one possible method by which such representations can be constructed for nouns. The method involves extracting features from the dictionary definition of a word and then using the implicit taxonomic information in the dictionary to inherit other features extracted from the definitions of semantically related words. The result is a bundle of weighted features which can then be converted into an n-dimensional vector.

The work described here is closely related to a number of other projects. Extracting taxonomic information from MRDs has been the focus of much dictionary work. For example Amsler constructed a taxonomy from the Merriam Webster Pocket Dictionary as early as 1980 and there has been much work in this area since. In particular, Vossen has constructed semantic taxonomies from parses of dictionary definitions in various MRDs including the Longmans Dictionary of Contemporary English as part of the ACQUILEX project (Vossen, 1990, 1991a, 1991b). Guthrie et al. have also constructed a taxonomy from the LDOCE, using the parser developed by Slator (Guthrie et al. 1990). Nutter and Fox have also worked on extraction of semantic information from MRDs. In their case the objective is to construct a more general semantic network particularly for use in Information Retrieval applications (Nutter and Fox, 1990, Fox and Nutter, 1991). Wilks et al. also construct distributed representations for words using the LDOCE (Wilks et al. 1990). However their method involves using all words in dictionary definitions as features, and does not involve taxonomic analysis. Finally, Sharkey, Day and Sharkey have shown how connectionist nets can learn to perform sophisticated word sense disambiguation when distributed lexical representations based on features are used.

The remainder of this paper is structured as follows. First, we motivate the work by briefly describing the kind of distributed lexical representations we are working with. Next, we describe the extraction algorithm which we have used. Thirdly we describe the results of the work so far. Finally we draw conclusions from our results and point to possible future directions for the research.

2 Distributed Semantic Representations

In this section we describe how concepts, such as are conjured up by nouns or noun phrases, can be expressed using distributed representations. We use a set of *semantic features* for this purpose. Each concept is encoded by assigning to each feature a numerical *centrality* which indicates how strongly that feature applies to the concept. Suppose for example that we wish to encode the concept 'person' using

---

[1] In fact there are now some dictionaries such as WordNet (Beckwith et al., In Press) which have been created especially for machine use rather than being derived from printed versions.

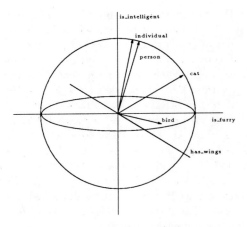

Figure 1. Amorphous concepts mapped onto a sphere

a set of just three features. We might come up with the following:

person
is_furry   1
has_wings   0
is_intelligent 10

This is interesting because other concepts can also be encoded in terms of the same three features. Here are two examples:

cat
is_furry   10
has_wings   0
is_intelligent 5

bird
is_furry   4
has_wings   10
is_intelligent 3

Representations constructed in this way have several important characteristics. First of all, meaning comparison can be accomplished quite easily by considering each concept as a vector in an n-dimensional space, one dimension for each feature. Once each vector is *normalised* (scaled so that its length is one) we can compute the closeness of meaning of a pair of concepts by using one of the standard vector comparison methods such as the dot product. The diagram shows three three example concepts plotted in a three dimensional space so that the similarity of concepts can be judged visually.

The second characteristic of these representations is that any number of concepts can be captured using the same set of features. This effectively allows us to work with concepts which are related to each other but which differ in subtle ways. At the same time we can handle nebulous words such as 'charity' or 'life' alongside words describing specific objects in the world. These two characteristics make feature-based distributed representations very suitable candidates for use in capturing linguistic meanings.

Naturally it is necessary to use feature sets containing more than just three elements for this representation scheme to work effectively. In earlier systems we used a set of 166 features to represent nouns (Sutcliffe, 1988, 1991a). In practice the representations described here have been found to work well. However, in all cases each word was encoded by hand. The purpose of the present work, therefore, was to investigate whether the same kind of representations could be extracted automatically from an MRD. We turn to the details of how this was done in the next section.

3 Extracting Feature Representations

The dictionary used for this work is the Merriam Webster Compact Electronic Dictionary (CED) which is for American English and contains 79,515 entry points. The method adopted is, however, equally applicable to other dictionaries with simple definitions, such as the LDOCE or Oxford Advanced Learners Dictionary. Each word entry in the CED consists of the word itself (in a particular inflection), an entry type tag (regular or idiom starter), a unique identifier and a list of word sense entries. Each word sense entry includes a syntactic category (selected from a set of twelve), a definition, a list of <feature,value> pairs and a usage note. Not all fields are filled for all words. In general, word definitions are extremely short and simple, which makes them potentially easy to parse. Indeed, it was found that the definitions for 4263 nouns in the dictionary could be parsed using a simple noun phrase grammar and it was these words which were used for the present study.

The objective of this experiment was to construct for each of the set of nouns a distributed representation consisting of a list of <Feature,Centrality> pairs. In the case of nouns a feature is taken to be either an adjective or a noun modifier. The essential idea of the algorithm is to extract all features from the definition of a particular word and then to extract further features by chaining through the taxonomic links implicit in the dictionary. For example, consider the definition for *tightwad* in CED:

tightwad: "stingy person"

From this we conclude that **stingy** is a feature of *tightwad*. Next we consult the definition for *person*:

person: "human being"

We now conclude that **human** is also a feature of *tightwad*. Next we inspect the definition for *being*:

being: "living thing"

We add **living** and consult the definition for *thing*. This is not present because its definition is not a noun phrase, and hence analysis stops here. This process yields the following feature set for *tightwad* : { stingy, human, living }. In fact we associate a numerical centrality with each feature and these decrease by a fixed increment as we progress up the taxonomic tree. In the present experiment we start with a weight of 1.0 at the initial definition and subtract 0.1 with each further definition consulted. The resulting definition for *tightwad* is thus {<stingy,1.0>,

Freq Threshold	No. of Features
1	2013 (ie ALL)
2	1005
3	670
4	492
5	392
10	176
15	120
20	90
25	67
30	55
40	33
50	23
60	17
80	9
100	7

Table 1. Frequency of features above certain thresholds. For example seven features have a frequency of 100 or more.

<human,0.9>, <living,0.8>}.

Two further problems need to be considered, both of which are shown by the following example:

design : "decorative pattern"
design : "preliminary sketch"
pattern : "artistic design"
sketch : "rough drawing"

The first problem is that we have a loop in the taxonomy: *design* is defined as "decorative pattern", while *pattern* is defined as "artistic design". This will cause the extraction algorithm to loop with the features **decorative** and **artistic** being added with progressively smaller centralities. We control this by stopping all searches once the centrality reaches zero, and then only using the highest centrality found for each feature.

The second problem is that many words have various different senses not all of which are appropriate to the concept being defined. Suppose we are defining *design* in the first sense, "decorative pattern". Because *pattern* is defined as "artistic design" we are back to the definition of *design* again, and must decide which of the definitions to use, "decorative pattern" or "preliminary sketch". At present we simply ignore this problem. When we come to a word with several meanings, the taxonomy from each such sense is pursued as far as possible, subject to the centralities not falling below zero.

Once each word has been encoded as a set of <feature,value> pairs the final step in building the lexicon is to construct a list of all the features which have been used in at least one definition and to define each word meaning as a normalised vector over the n-dimensional space so defined, where n is the number of features.

4 Assessment of Results

Having constructed an initial lexicon in the fashion described in the previous section, we then carried out an initial test on the effectiveness of the semantic representations we had generated. The test chosen was to construct a hierarchical cluster analysis of a set of word representations using the Ward method. For practical reasons it was not possible to cluster the whole lexicon. As a first attempt at determining the efficacy of the extraction process, therefore, 63 words were selected to be used in the cluster experiment. These words represent various kinds of vehicle, person and dog and were chosen because one would expect them to cluster naturally into three groups. Three analyses were carried out using the same words, but with different numbers of features. The frequency of occurrence of each feature is defined to be the number of word representations in which it is to be found. Thus if a feature is only used in one word its frequency is one. In each clustering experiment we chose a different frequency threshold, eliminated all occurrences within word representations of features with a lower frequency and renormalised the representations before carrying out the clustering. The results are shown in the attached figures. As can be seen, in all cases we see a good clustering of semantically related concepts. For example consider the results using a threshold of four. We have a cluster of female persons from *girl* down to *woman*, a cluster of vehicles from *surrey* down to *sulky* and a cluster of dogs from *bulldog* down to *wolf*. This indicates that our semantic representation has captured at least some of the uniformities which exist over female persons, vehicles, and dogs respectively. The other outputs show clustering using thresholds of two and ten. Once again, in the former we have clusters such as *chaise* to *surrey* and *terrier* to *wolf* while in the latter we can see *slattern* to *woman* and *dog* to *wolf*. However, the clustering appears more fine grained with the lower frequency thresholds.

5 Summary and Conclusions

In summary we have presented here a simple algorithm for extracting distributed semantic representations from a set of noun definitions in the Merriam-Webster CED. The algorithm involves extracting adjectives and noun modifiers from a word's definition and using them as features. We then use the implicit taxonomic information within the dictionary to inherit further features from higher up in the taxonomy. The further up the hierarchy a feature is found, the lower its weight, and thus the lower its contribution to the meaning of the word being encoded.

Initial results seem encouraging, especially considering that the problem has been simplified in several critical ways: First, We are only extracting information from nouns defined using simple noun phrases. This means that we are only extracting a small part of the taxonomic information which is relevant to each word's meaning. The extraction algorithm can be generalised to handle more complicated definitions provided they can be parsed, although it will be necessary to make decisions about the semantic connotations of different constructions. For example there a several different ways in which features could be extracted from a definition such as that for *buoy* — "a floating marker anchored in water".

Second, we are not restricting taxonomic expansion of polysemous words to salient word senses but are expanding the taxonomic paths emanating from all word senses. This means that our word representation may be acquiring irrelevant features when inappropriate word senses are being expanded. Thirdly, there are many synonyms within the feature set itself which it was planned to unify using a thesaurus. For example if one word has the feature **large** while another has the feature **big** we may wish to combine the features by substituting all occurrences of **big** in word representations by **large** or vice versa. It is expected that a considerable improvement in performance could be achieved by addressing these three limitations even partially. This should be possible as there has already been considerable work in this general area. For example both Vossen (1991a) and Guthrie et al. (1990) have constructed exhaustive taxonomies for the LDOCE and can therefore parse the majority of definitions accurately, while Lesk (1986) and Ide and Veronis (1990) have both addressed the issue of contextual word sense disambiguation. Addressing these three shortcomings would seem to be the most promising next steps to take.

6 References

Amsler, R. A. (1984). Machine-Readable Dictionaries. *Annual Review of Information Science and Technology (ARIST)*, **19**, 161-209.

Beckwith, R., Fellbaum, C., Gross, D., & Miller, G. A. (In Press). WordNet: A Lexical Database Organized on Psycholinguistic Principles. In U. Zernik (Ed.), *Using On-line Resources to Build a Lexicon*. Hillsdale, NJ: Erlbaum.

Evens, M. (1989). Computer-Readable Dictionaries. *Annual Review of Information Science and Technology (ARIST)*, **24**, 85-117.

Fox, E. A., Nutter, T. J., & Evens, M. W. (1991). A Lexicon Server using Lexical Relations for Information Retrieval. In C.-M. Guo (Ed.), *Machine-Tractable Dictionaries*. Norwood, NJ: Ablex.

Guthrie, L, Slator, B. M., Wilks, Y., & Bruce, R. (1990). Is there Content in Empty Heads? *Proceedings of the 13th International Conference on Computational Linguistics (COLING-90), Helsinki, Finland, Vol. III*, 138-143.

Ide, N. M., & Veronis, J. (1990). Very Large Neural Networks for Word Sense Disambiguation. *Proceedings of the European Conference on Artificial Intelligence, ECAI'90, Stockholm, August 1990, 366-368*.

Lesk, M. (1986). Automated Word Sense Disambiguation using Machine-Readable Dictionaries: How to Tell a Pine Cone from an Ice Cream Cone. *Proceedings of the 1986 SIGDOC Conference*.

Nutter, J. T., Fox, E. A., & Evens, M. W (1990). Building a Lexicon from Machine Readable Dictionaries for Improved Information Retrieval. *Literary and Linguistic Computing*, **5**(2), 129-138.

Sharkey, N. E., Day, P. A., & Sharkey, A. J. C. (1991). A Connectionist Machine Tractable Dictionary: The Very Idea. In C.-M. Guo (Ed.), *Machine Readable Dictionaries*. Norwood, NJ: Ablex.

Sutcliffe, R. F. E. (1988). *A Parallel Distributed Processing Approach to the Representation of Knowledge for Natural Language Understanding*. Unpublished doctoral thesis, University of Essex, UK.

Sutcliffe, R. F. E. (1991a). Distributed Representations in a Text Based Information Retrieval System: A New Way of Using the Vector Space Model. In A.

Bookstein, Y. Chiaramella, G. Salton & V. V. Raghavan (Eds.), *Proceedings of the Fourteenth Annual International ACM/SIGIR Conference on Research and Development in Information Retrieval, Chicago, Il., October 13-16, 1991* (pp. 123-132). New York, NY: ACM Press.

Sutcliffe, R. F. E. (1991b). Distributed Subsymbolic Representations for Natural Language: How many Features Do You Need? In M. F. McTear & N. Creaney (Eds.), *Proceedings of the 3rd Irish Conference on Artificial Intelligence and Cognitive Science, 20-21 September 1990, University of Ulster at Jordanstown, Northern Ireland.* Berlin, Germany, Heidelberg, Germany, New York, NY: Springer-Verlag.

Sutcliffe, R. F. E. (1992a). Representing Meaning using Microfeatures. In R. Reilly and N. E. Sharkey (Eds.), *Connectionist Approaches to Natural Language Processing* (pp. 49-73). Englewood Cliffs, NJ: Lawrence Erlbaum Associates.

Sutcliffe, R. F. E. (1992b). PELICAN: A Prototype Information Retrieval System using Distributed Propositional Representations. To Appear in H. Sorensen (Ed.), *Proceedings of AICS-91 - The Fourth Irish Conference on Artificial Intelligence and Cognitive Science, University College Cork, 19-20 September 1991.* London, UK, Berlin, FRG, Heidelberg, FGR, New York, NY: Springer-Verlag. Also available as Technical Report UL-CSIS-91-9, Department of Computer Science and Information Systems, University of Limerick, April 1991.

Vossen, P. (1990). The End of the Chain: Where Does Decomposition of Lexical Knowledge Lead us Eventually? (Technical Report Esprit BRA-3030 ACQUILEX WP 010.) Amsterdam, The Netherlands: University of Amsterdam, English Department.

Vossen, P. (1991a). Comparing Noun-Taxonomies Cross-Linguistically (Technical Report Esprit BRA-3030 ACQUILEX WP 014). Amsterdam, The Netherlands: University of Amsterdam, English Department.

Vossen, P. (1991b). Converting Data from a Lexical Database to a Knowledge Base (Technical Report Esprit BRA-3030 ACQUILEX WP 027). Amsterdam, The Netherlands: University of Amsterdam, English Department.

Wilks, Y., Fass, D., Guo, C.-M., Macdonald, J., Plate, T., & Slator, B. (1990). Providing Machine Tractable Dictionary Tools. *Machine Translation*, **5**, 99-154.

Dendrogram using Ward Method

```
                                  Rescaled Distance Cluster Combine

           C A S E           0         5        10        15        20        25
      Label              Seq +---------+---------+---------+---------+---------+

      tot                 55  -+------------+
      tyke                56  -+           |
      pug                 38  -+           +------------+
      terrier             54  -+           |            |
      Pekinese             2  -+-+         |            |
      lapdog              25  -+ +---------+            |
      Chihuahua            1  ---+                      |
      dog                 18  -+                        |
      pooch               37  -+                        +-----------------------+
      bulldog              8  -+                        |                       |
      chow                15  -+-+                      |                       |
      cur                 17  -+ |                      |                       |
      canine              10  -+ +-+                    |                       |
      puppy               41  ---+ +---------------+    |                       |
      malamute            26  ---+ |               |    |                       |
      bitch                4  -----+               +-----+                      |
      collie              16  -+---+               |                            |
      mastiff             30  -+   +---------------+                            |
      wolf                62  -----+                                            |
      chaise              13  -+                                                |
      surrey              53  -+                                                |
      caisson              9  -+---------------+                                |
      carriage            11  -+               |                                |
      cart                12  -+               +-----------------------+        |
      wagon               57  -+------+        |                       |        |
      wagonette           58  -+      +---------+                      |        |
      buggy                7  -+-+    |                                |        |
      phaeton             36  -+ +----+                                |        |
      sulky               52  ---+                                     |        |
      sage                44  -+                                       |        |
      senor               45  -+                                       |        |
      milksop             31  -+                                       |        |
      pundit              40  -+                                       |        |
      rake                42  -+                                       +--------+
      youth               65  -+                                       |
      signor              46  -+                                       |
      patriarch           35  -+                                       |
      bachelor             3  -+--------------+                        |
      bruiser              6  -+              +------------------+     |
```

Figure 2a. A dendrogram showing the representations of 63 words selected from the lexicon, clustered using the Ward method. In this case the frequency threshold was four, meaning that features which occurred in less than four of the 4263 words were eliminated from the representations before performing the analysis on the re-normalised patterns.

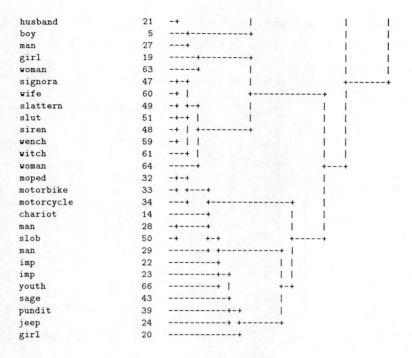

Figure 2b. The second half of Figure 2a.

```
Dendrogram using Ward Method

                                    Rescaled Distance Cluster Combine

          C A S E          0         5        10        15        20        25
     Label          Seq    +---------+---------+---------+---------+---------+

     tyke            57   -+
     youth           67   -+
     boy              5   -+
     sage            44   -+
     tot             56   -+
     girl            21   -+
     imp             24   -+
     cart            13   -+
     chaise          14   -+
     caisson          9   -+-----+
     carriage        12   -+     +---+
     buggy            7   -+     |   +-+
     imp             23   -------+   | |
     phaeton         37   -+---------+ +-+
     surrey          54   -+           | +-+
     pundit          40   -------------+ | +-+
     jeep            25   ---------------+ | +---+
     canine          10   -----------------+ |   |
     wagon           58   -+-----------------+   |
     wagonette       59   -+                     +------------+
     moped           33   -+                     |            |
     motorbike       34   -+-----+               |            |
     motorcycle      35   -+     +---------------+            |
     chariot         15   ---+---+                            |
     sulky           53   ---+                                |
     man             29   -+-+                                |
     slob            51   -+ +-----+                          |
     man             30   ---+     |                          |
     milksop         32   -+       |                          |
     rake            43   -+       |                          +-------------+
     sage            45   -+       |                          |             |
     youth           66   -+       +--------------------+     |             |
     bachelor         3   -+       |                    |     |             |
     patriarch       36   -+       |                    |     |             |
     pundit          41   -+       |                    |     |             |
     husband         22   -+       |                    |     |             |
     signor          47   -+-----+ |                    |     |             |
     senor           46   -+     +-+                    |     |             |
     bruiser          6   -+     |                      |     |             |
```

Figure 3a. A dendrogram showing the representations of 63 words selected from the lexicon, clustered using the Ward method. In this case the frequency threshold was two, meaning that features which occurred in only one of the 4263 words were eliminated from the representations before performing the analysis on the re-normalised patterns.

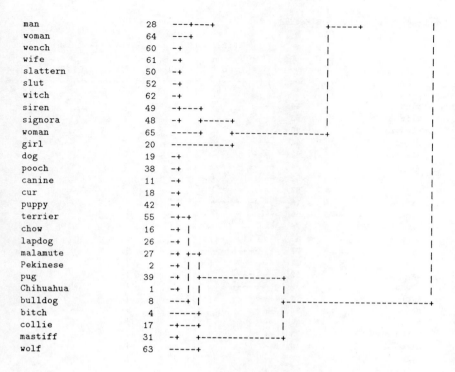

Figure 3b. The second half of Figure 3a.

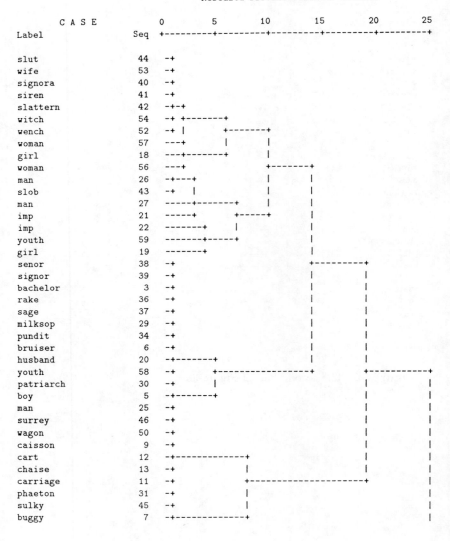

Figure 4a. A dendrogram showing the representations of 63 words selected from the lexicon, clustered using the Ward method. In this case the frequency threshold was ten, meaning that features which occurred in less than ten of the 4263 words were eliminated from the representations before performing the analysis on the re-normalised patterns.

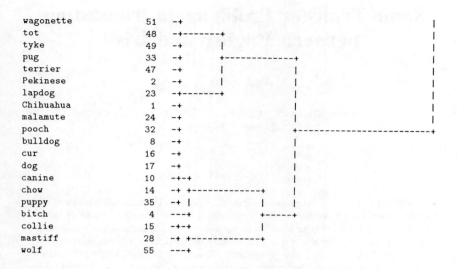

Figure 4b. The second half of Figure 4a.

Some Transfer Problems in translating between English and Irish

Andrew Way[1]

School of Computer Applications, Dublin City University
Dublin 9, Ireland[2]

Abstract

There are two native languages in Ireland, Irish and English. All parliamentary proceedings have to appear in both languages. Owing to the sheer volume of work, and the fact that there is an insufficient number of translators capable of performing the task, there is currently a five year backlog in translation. It is apparent that the translators' workload could be significantly alleviated if an English-Irish MT system were to be developed. This paper outlines some of the problems involved, and describes future intentions.

1 Introduction

Ireland has two native languages, English and Irish. Consequently all proceedings in the Dáil (The Irish Parliament) have to be transcribed in both languages. With the amount of work which this entails, and owing to the paucity of qualified Irish speakers, there is currently a backlog of some five years translation, despite the fine work done by those translators currently involved in this work.

Consequently, it was decided to investigate the viability of alleviating this backlog by computational methods. More specifically, this entails the development of an English-Irish machine translation (MT) system capable of translating specifically in this particular area. The advantages of translating in one specific subject field are well known (e.g. Kittredge & Lehrberger, 1982), and will not be repeated here.

It is intended that a pilot study be undertaken involving (i) an investigation of the sublanguage; (ii) the writing of toy grammars for English and Irish capable of parsing in this field; (iii) an implementation of an MT system in the ET-6 prototype[3] when this is eventually released.

With this in mind, this paper is an attempt to illustrate some of the problems which will have to be addressed if the current goals are to be realised. The examples are discussed in the framework of the Eurotra MT system, a multi-level, multilingual transfer-based system, whose architecture is widely familiar, and a more detailed description of which is beyond the scope of this paper (see Arnold, 1986;

---

[1]Many thanks to colleagues on the Eurotra project at DCU for their help with the Irish translations and structural representations.

[2]Email: **waya@dcu.ie**, Fax: **+353-1-7045442**

[3]The follow-up to the Eurotra software (Schütz, 1991).

Raw, van Eynde, ten Hacken, Hoekstra & Vandecapelle, 1989).

2 The work so far

As stated above, it is intended that the MT system will initially be tailored specifically to translate a particular sublanguage, namely that of Irish parliamentary proceedings. Consequently an initial study of this sublanguage was necessary, and as an example text we selected "The Influence of Computerisation on the Irish Language" (1988), drawn up by the Second Joint Committee on the Irish Language.

This text was selected for several reasons: firstly, and most importantly, it was available in both languages; secondly, it seemed to be a typical text in this field, containing the sorts of sentences likely to be encountered by the system (i.e. "official speak": no idiomatic phrases, intricate syntactic patterns, well-formed sentences and so forth[4]); thirdly, the report was written by TD's (members of the Dáil), i.e. people whose texts the system would be confronted with.

From this text we selected a sample corpus consisting of thirteen English sentences, which apart from minor amendments (e.g. altering "a lot of" to "many" for the sake of simplicity, or deleting the additional sentence "when research plans are being drawn up" from the end of sentence 1 to prevent over long processing time by the software) were drawn in entirety from the original. These sentences were chosen because they contained a wide range of syntactic phenomena, including Negation, Modality, Comparison, Tense, Quantification, Relative Clauses and Passive. The sentences are as follows:

> It is often possible to include the needs of the Irish language. The Committee proposes projects in the area of Information Technology. Computers offer many advantages when used with large bodies of data. They do not become tired. They do not make mistakes. They are much faster than humans. The new technology can be employed in many areas.
> One must ensure that the basic data is readily available. The Committee recommends that a central database should be established. The proposals will cost money. The amounts involved are relatively small. Much change can take place in a very short period of time. The Irish language groups should form a network of communication.

As can be seen, despite the fact that these sentences were drawn from different parts of the document, they form a (more or less) coherent text. This is important because the system will be confronted with texts rather than individual sentences, so it seemed sensible to try and replicate as closely as possible the conditions under which the system will be expected to perform.   The Irish equivalents were then extracted from the same bilingual source, and the two corpora compared for translational problems.

---

[4]The sublanguage will be commented upon in greater detail at a later stage.

2.1 Transfer Problems

A framework was needed in which to describe the set of translational problems occurring in these small texts. As the Eurotra framework was both widely known and typical of a transfer-based system, the following structural representations are set out in the form of **IS** (Interface Structure) trees in that formalism. IS is the highest level of representation in the Eurotra formalism, and the representations are in the form of dependency structures. These are structures where primary importance is attached to the verb, which appears as the **governor** of the sentence with its **arguments** (compulsory constituents) and **modifiers** (optional constituents).

With this in mind, it seems that Irish would be a good candidate for a Eurotra language. As it is a **verb-initial** language (VSO), the surface order of sentence constituents replicates even at this level[5] the order of components at IS. For an SVO language like English, various rules have to be written in order to transform the surface order into that required at IS; it seems, at this juncture at least, that this will not be necessary for Irish.

Transfer problems are seen where source and target sentences have different IS architectures, for in these instances rules will have to be written to manipulate one structure in order to attain the other. Typical examples are those where a verb has a different **argument structure** in one language to that of another (and where these are translations of each other, of course), which can be divided, among others, into cases of **head-switching** (e.g. the "like-graag" case), **relation-changing** (e.g. the "like-plaire" case) and **shoehorning** (e.g. the "viser-aim at" case; Sadler, Crookston & Way, 1989; Crookston, Simcoe-Shelton & Way, 1990; Brockmann, 1991).

2.2 Corpus Analysis

<u>Sentence 1</u>: It is often possible to include the needs of the Irish language.

The IS structure of the English sentence is as follows:

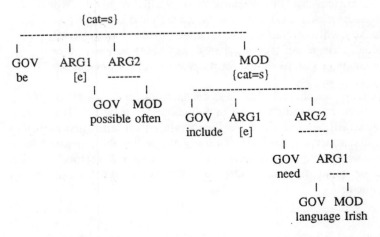

We note that when this is compared with its Irish translation, namely "Is minic

---

[5]In the Eurotra system this would be **ECS** (Eurotra Constituent Structure).

is féidir freastal a dhéanamh ar chuid de riachtanais na Gaeilge" (literally: be frequent be possible service to make on some of_the requirements of Irish), there are three points of divergence: firstly, the translation of the MOD of the ARG2 of the main sentence (namely "often") is a sentence in Irish rather than an adverb, and the ARG2 (i.e. the object) of the subordinate sentence is complicated by the presence of a QUANT (representing "chuid"). The major difference, however, would appear to be the representation of the modifying sentence, which in Irish needs to be considered as a **Support Verb** construction.

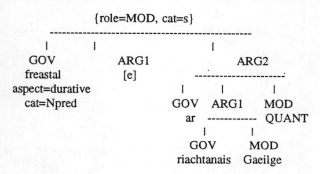

This is where we have the **frozen** expression consisting of "déan + freastal", where the verb is **semantically empty** and the noun is **predicative**, i.e. it has the same argument structure as the verbal form from which it is derived (hence the alternative "deverbal" noun). The predicative noun "freastal" therefore becomes the GOV of the sentence, the only instance where a category other than a verb is permissible in this position. A further point to note here is that the verb to be selected is reduced to a problem of lexical lookup, where differences in meaning such as "give/maintain/lose hope" (for example) are captured by aspectual differences.

Finally at this juncture some comment seems warranted regarding the status of the given Irish translations. As already stated these were extracted from the report drawn up by The Second Joint Committee on the Irish Language (1988), but according to informants the translation of this sentence (among others) is neither wholly accurate nor the best available. Better translations (i.e. representing more accurately the content of the source sentences) would be:

Is minic is féidir riachtanais na Gaeilge a áireamh
Is minic is féidir riachtanais na Gaeilge a chur san áireamh

where the first is stylistically imperfect, although the latter seems to be the best available, accurate both semantically and stylistically.

Sentence 2: The Committee proposes projects in the area of IT.

The translation of this sentence is "Molann an coiste tionscnaimh a bhaineann le cúrsaí ionformeolais", literally "recommends the Committee projects which apply to matters of_IT", where we note the presence of a subordinate sentence in Irish where there is a PP in English, again resulting in complex transfer, owing to the different IS structures.

Sentence 3: Computers offer many advantages when used with large bodies of data.

This translates as "Bíonn buntáistí móra ag ríomhairí nuair a bhíonn siad ag plé le bailiúcháin móra eolais" (lit: be advantages large to computers when be they dealing with bodies large of_knowledge). The English IS representation is:

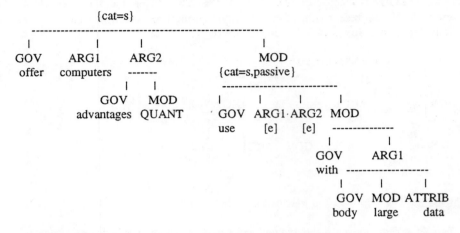

The Irish sentence is rather difficult to analyse. This is because "tá" (meaning "be", from which "bíonn" is derived) and "ag" together mean "have" **in this instance**! We can compare this example with the following:

(1) Bíonn Seán ag an bhfleá => Seán is at the party.
(2) Bíonn bia ag an bhfleá => There is food at the party.
　　　　　　　　　　　(? The party has food)
(3) Bíonn críoch ag an bhfleá => The party has an end.

The first two examples illustrate that the attribute [+human] is irrelevant when it comes to deciphering the translation of this construct, while example (3) contrasts with (2) in that an inanimate object occurs in each yet the translation differs.

If we can delete "ag" from the analysis then this aids our analysis somewhat, and indeed we **have** to, as we cannot have a PP as ARG1 of a sentence; as can be seen by the above gloss, we need "computers" as ARG1 rather than "advantages" (i.e. "many advantages have to computers ..."). If we are permitted this freedom, our analysis becomes much simpler (although at a lower level we would need to state in a rule that "tá" plus "ag" means "have"), as does translation, as this part of the Irish sentence now receives the same representation as the English. The subordinate sentence, however, still contains some differences, as can be seen by examining this section of the Irish sentence:

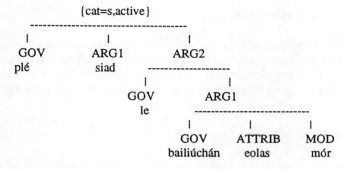

We note that the Irish sentence is an active sentence, as opposed to the passive English equivalent (although the possibility exists to have a passive sentence in Irish using "agus iad ag plé" - literally "and them dealing" - yet the representation of this construction is far from clear). Consequently, in this instance there is an overt NP as ARG1, and the subordinate sentence's MOD in English becomes the ARG2 of the corresponding Irish sentence.

<u>Sentence 4</u>: They do not become tired.

Even a simple sentence like this proves to be problematic when translating into Irish. The English representation is:

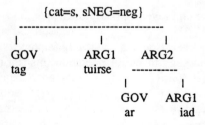

whilst the Irish translation "Ní thagann tuirse orthu" is represented thus:

```
         {cat=s, sNEG=neg}
       ---------------------------
  |            |           |
 GOV         ARG1        ARG2
 tag         tuirse    -----------
                         |       |
                        GOV    ARG1
                         ar     iad
```

which literally means "NEG tiredness comes on them". Here we see that the ARGS not only change position, but also category,

i.e. ARG1:NP(PRON)they --> ARG2:PP(ar iad)
     ARG2:AP(ADJ)tired --> ARG1:NP(N)tuirse

The translation "Ní éiríonn siad tuirseach" ("NEG become they tired") is also permissible here, although less acceptable than the chosen translation.

Sentence 5: They do not make mistakes.

This translates as "Ní dhéanann siad botúin", which is the first instance of simple transfer. One potential problem, however, is the fact that the negation process in Irish includes the well known phenomenon of **lenition**, which adds an "h" to the noun "déan". This should, however, be solved in morphological analysis at lower levels.

Sentence 6: They are faster than humans.

The Irish translation, "Tá siad níos sciopa ná daoine" (literally "Be they more fast than people"), receives the same architectural representation as its English counterpart, thus resulting in simple transfer. The IS tree is:

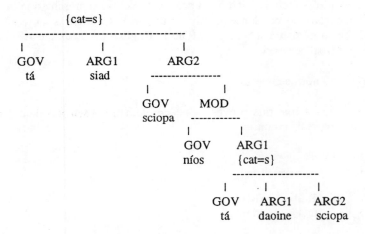

i.e. (in English) "they are more fast than people are fast".

Sentence 7: The new technology can be employed in many areas.

This translates as "Is feídir an nuatheicneolaíocht a chur i bhfeidhm i go leor áiteanna" (lit: be possible the new_technology to put in operation in a_lot places). The IS representation of the English is as follows:

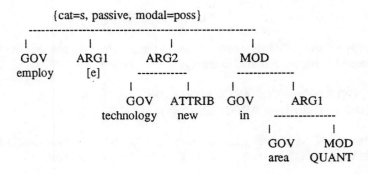

We note that the modal "can" is featurised at this level. The Irish translation maintains this sentence en bloc, except that the modality feature in English is realised by an overt sentence "Is féidir" (as we saw in sentence 1), under which the above representation (with Irish lexemes transposed, of course) is attached as a subordinate sentence. Naturally, this is yet another instance of complex transfer.

<u>Sentence 8</u>: One must ensure that the basic data is readily available.

This translates as "Caithfear a dhéanamh cinnte de go mbíonn an buneolas ar fáil go réidh" (lit: must to make certain of_it that be the basic_data be available readily). The English representation is as follows:

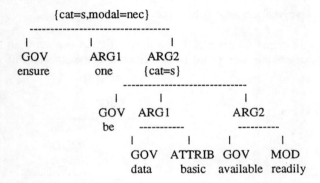

There are many changes in the Irish IS, which can be categorised as follows:

     S1:ARG1:NP(PRON)one --> S1:ARG1:NP(Empty)
     S1:ARG2:S --> S1:MOD:S
     S2:ARG1:NP(N,ATTRIB)basic data --> S2:ARG1:NP(N)buneolas
     S2:ARG2:AP --> S2:ARG2:S

The IS representation of the Irish sentence is the following:

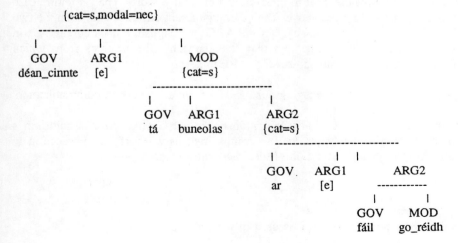

Sentence 9: The Committee recommends that a central DB should be established.

The Irish translation reads "Molann an coiste go gcuirfí teasclann ar bun" (lit: recommends the Committee that would_be_put database on base), which differs in structure in the subordinate sentence, owing to the choice of a multi-word translation of "establish" in this instance[6].

Sentence 10: The proposals will cost money.

The Irish translation "Cosnóidh na moltaí airgead" presents no problems, hence simple transfer.

Sentence 11: The amounts involved are relatively small.

This translates as "Tá na suimeanna i gceist sách réasúnta" (lit: be the sums in question fairly reasonable), where the only problem is a change in the structure of the ARG1 node, as in:

i.e.   S1:ARG1:MOD:S --> S1:ARG1:MOD:PP.

Sentence 12: Much change can take place in a very short period of time.

The Irish translation of this sentence is "Is féidir go leor a bhaint amach in am fíorghairid" (lit: be possible a lot to take out in time really short). The "is féidir" case has already been described above. The other significant difference between the two IS's is that the NP "much change" is the ARG1 in English, but its counterpart in Irish is ARG2, owing to the fact that "bain amach" (the lexemes from which "bhaint" is derived) has no subject (i.e. ARG1) here.

Sentence 13: The Irish language groups should form a network of communication.

The Irish translation here is "B'fhearr do na grúpaí Gaeilge gréasán caidrimh a bhunú" (lit: would_be_better for the groups Irish network of_communication to establish). The IS representation of the Irish sentence is:

---

[6]The one word translation "bun" is used in sentence 13.

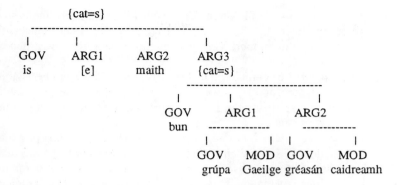

We see that the string "b'fhearr" is composed of the lexemes "is" and "fearr" (the comparative of "maith", meaning "good"), and as with similar structures the English IS becomes a subordinate sentence in Irish, and this is almost identical in both languages ("Gaeilge" means "Irish", whereas the English sentence contains "the Irish language". Obviously we could delete "language" in the English sentence without losing any expressiveness, resulting in an identical IS for this substructure).

3 Observations and Conclusions

One of the first things to note is the influence of the sublanguage on translation. As is typical in a report of this kind, there are several examples of impersonal constructions ("One must ...", "It is possible ..."), as the authors are making general recommendations. We note that this often involved the English main clauses being relegated to subordinate sentences in Irish, with impersonal sentences built on top. Although this resulted in complex transfer, we should not feel dismayed: as we expect this kind of construction to proliferate in the texts in this field, the rules we will have to write will be used a great deal. We can surmise that they are not being written for a few isolated instances.

One potential problem, however, would be in calculating the **closure** of the sublanguage; indeed, in the worst possible case it might turn out not to be a sublanguage after all, given that parliamentary reports necessarily deal with a wide spectrum of subjects. If, as suspected, this does cause problems, then we shall have to restrict our area of study to a particular subset of reports.

We discovered many transfer problems in the selected sentences. As noted earlier, the sentences are complex[7], so perhaps such results were to be expected. This would probably be the case in translating between **any** two pairs of languages, even two of the same type (e.g. Germanic languages -- Dutch and German, or Romance languages - Spanish and French).

As for the **types** of complex transfer encountered, these seem to be the standard ones of category changing, role switching, and compounds in one language being NP's in another. The only cases particular to the pair of languages dealt with here are those containing modals, which necessitate sentences being built on top of the

---

[7]Although as noted in the discussion of Sentence 1, the translations chosen are not necessarily the most appropriate.

corresponding English sentences (in which the modals are featurised), the case of "tá" plus "ag" meaning either "be" or "have" depending on the circumstances, the translation of "often", and the Support Verb construction where there was none in English.

Despite the large number of complex transfer cases, our initial hypothesis about Irish being a good candidate for MT proved to be well-founded. The fact that the surface word-order **already** reflects the IS ordering would definitely result in less processing of the input sentences.

This leads us on to our plans for future work. Obviously any claims presented here need to be substantiated by implementing the system. We await the arrival of ET-6 with relish. Assuming the above corpora can be dealt with, and a running prototype can be developed, we intend to expand the number and type of sentences, along with the development of a larger lexicon. This, we hope, will result in a large-scale system which may attract research funding either at governmental level or, if Irish becomes a recognised EC language (as we hear is a possibility, along with Luxemburgish), from the Commission as well. If the intended aim of alleviating the workload of Irish translators is achieved, we would then hope to apply the system to texts of different sublanguages, in order to test its portability and extensibility.

References

Arnold, D. (1986). Eurotra: a European Perspective on MT. *Proceedings of the IEEE*, **74**(7), 979-992.

Brockmann, D. (1991). *Transfer Problems*. (Internal Memo ETIM 026) Colchester, UK: University of Essex, Department of Language and Linguistics.

Crookston, I., Simcoe-Shelton, J., & Way, A. (1990). *Eurotra Problem Office Interlevel Syntax Research Pool Area B: Interlevel Processing Final Research Report*; Luxembourg, Luxembourg: DG-XIII, CEC (to appear as: *Eurotra Studies Volume 4*; Luxembourg, Luxembourg: CEC, forthcoming).

Kittredge, R., & Lehrberger, J. (Eds). (1982). *Sublanguage: Studies of Language in Restricted Semantic Domains*. Berlin: Walter de Gruyter.

Raw, A., van Eynde, F., ten Hacken, P., Hoekstra, H., & Vandecapelle, B. (1989). *An Introduction to the Eurotra Machine Translation System*.(Working Papers in NLP) Leuven, Belgium: Eurotra Leuven, Katholieke Universiteit.

Sadler, L., Crookston, I., & Way, A. (1989). *Co-description, Projection and "Difficult" Translation*. (Working Papers in Language Processing No 9) Colchester, UK: University of Essex, Department of Language and Linguistics.

Schütz, J. (Ed). (1991). *ET6/2 - Software Environment Study Final Report*. Luxembourg, Luxembourg: Commission of the European Communities.

The Second Joint Committee on the Irish Language. (1988). *The Fifth report from the Joint Committee: The Influence of Computerisation on the Irish Language*. Dublin, Ireland: Rialtas na hÉireann.

Session 6: Novel Aspects of AI/CS

Why Go for Virtual Reality if you can have Virtual Magic? A Study of Different Approaches to Manoeuvring an Object on Screen

Roddy Cowie and Fiona Walsh
School of Psychology, Queen's University
Belfast, Northern Ireland

Abstract

We describe a system for manoeuvring a three-dimensional object on screen which combines analogue elements with icons. We call it ICAS (Icon cum Analogue System). An experiment examines the benefits of each component of ICAS. Psychological considerations led us to use multiple measures in the experiment, and to use both regular and irregular objects. Results show that two of our controls are useful at least in certain circumstances, whereas the third is never beneficial and sometimes impairs performance. The results are not compatible with design principles that are widely accepted, or with extreme commitment to the ideal of virtual reality. We argue that this illustrates the need for interface design to exploit the duality of its medium - as a screen in the real world and a depiction of things in a virtual world - as older graphic media have done. That involves accepting that virtual realities can and should be magical, and that in interface design, understanding the mind is as fundamental as understanding the world.

1 Introduction

This paper deals with a system for manoeuvring an object on screen which we have developed over a number of years. The system was developed for straightforward reasons, and it is not particularly spectacular. However it takes on an added interest because of the high profile that has been given to alternative ways of doing similar things. Those alternatives have been given the title virtual reality (Rhingold, 1991). It was not our intention, but we find that our system sums up some of the reservations we have with extreme versions of the virtual reality ideal.

The term virtual reality is has a range of meanings: for example, "desktop VR" is far removed from the archetypal helmet-and-glove systems. However it seems fair to say that the core of its meaning is allowing users to interact with objects in a virtual world in such a way that the interaction resembles interacting with objects in the real world as closely as possible. In that sense, virtual reality is the logical extension of ideas which have been current in interface design for some time. Consider, for instance, the widely accepted criteria for a control system that have emerged as a result of work by Britton, Chen and others (Chen, Mountford & Sellen, 1988).

A) There should be a high level of kinesthetic correspondence (i.e. the direct and natural relationship between movement of muscles and joints, and the corresponding rotations about the axes).

B) Complex rotations should be possible i.e. simultaneous rotation in XYZ or at least coupled rotations in XY, XZ, YZ, rather than merely single-axis control.

C) The user should feel that he/she is directly manipulating the object itself.

These clearly emphasise analogy between real-world control and control on screen, and their logical conclusion is virtual reality.

Certainly progress towards virtual reality can have uses, but it is important not to be swept away by the idea. After all, humans are notoriously discontent with reality as it is. If we are aiming to give people the best possible control over a virtual world, the best route may include giving them powers that are not available in real reality. The only way to know is to explore alternatives and test them systematically. This paper does that in a small way.

The paper describes two innovations. The first is a new set of controls which let an observer manipulate objects on screen. It includes some features which are akin to virtual reality techniques and some which are quite different. The second is a new set of tasks for testing control systems. These will be introduced in turn. Then we describe an experiment which measures the benefit of different types of control as measured by different types of task. Finally we come back to the larger theme of design philosophies for interactive interfaces between people and virtual worlds.

2 The Icon cum Analog system

The control system described in this paper evolved, and it is useful to describe it in historical terms. The system is implemented on Apple MacIntosh computers in a dialect of BASIC.

At an earlier conference we described a prototype system based purely on virtual reality type principles (Cowie, Bradley & Livingstone 1991). Observers looked at a three-dimensional object on screen and controlled their viewpoint using a mouse. We imagined that the object was encased in a glass sphere, and the mouse was coupled to a camera which moved round the sphere. Moving the mouse changed the camera position, and so it changed the view of the object that the observer received.

This system belongs to the class we will call virtual sphere systems. The general class is well known (e.g. Chen, Mountford & Sellen, 1988). However, ours has two features which are worth making clear.

The first feature involves alternatives which always arise with a virtual sphere system. One is to couple the mouse and the camera directly. In that case the mouse represents the observer's position, and points on the near half of the object move left across the screen as the mouse moves right. The other option is to couple them inversely. In that case the mouse represents a towrope attached to the object, and points on the near half of the object move right across the screen as the mouse moves right. The ambiguity between these two is a serious problem for virtual reality systems (Cobb, personal communication). We settled for the second alternative where the object moves rather than the observer, because users found it much easier to understand.

The second feature of our virtual sphere system is more unusual. Only part of the sphere is accessible at any given time. More precisely, mouse movement is only effective while the mouse pointer remains within the area of the screen: that corresponds to keeping movement within a roughly rectangular window on the surface of the virtual sphere. The window is always nearly parallel to the screen. The mouse can be disconnected from the object by clicking once. Moving the mouse while it is disconnected pulls the window across the virtual sphere. The next click reconnects the mouse. At that point, movement is bounded by a new window. The virtue of that arrangement is that the mouse is never on parts of the sphere which are at a steep angle to the screen surface: that is important because allowing the mouse to move into those areas leads to highly counterintuitive coupling between the mouse and the object.

Initially we used the control system that we have outlined as a stand-alone system. Later it was incorporated into a larger system. In that context, we call it basic control.

A large number of students used our pure basic control system, and many of them found great difficulty with it. As a result, we developed a second system which we call an icon bar. The icon bar represents a standard type of Macintosh design. There is a bar at the right of the screen. It is divided into eight squares, each of which contains an icon. Each icon shows a simple type of movement. The top one shows anticlockwise rotation in the plane of the screen. The one immediately below it shows exactly the opposite motion, clockwise rotation in the plane of the screen. Below that is a pair of icons showing rotation about a vertical axis, one in each direction, and below that is a pair of icons showing rotation about a horizontal axis in the plane of the screen. The last pair of icons show approach and recession: using the first makes perspective stronger, using the second reduces perspective. All of the icons work in the same way. First the mouse has to be disconnected from basic control and the pointer has to be located on an icon. When that is done, pressing down on the mouse button generates the kind of movement that the icon shows for as long as the button stays down. Clicking reconnects the mouse to basic control.

The dual system using system of basic control and icons was used extensively and appeared to be much better received than basic control alone. However for this study we added a third element. It is a more conventional type of virtual sphere control. It consists of a circle in the bottom right hand corner of the screen. Once again, it is accessed by disconnecting the mouse and pressing down on the button within the circle. The circle is a projection of the virtual sphere which encloses the object. When the button is pressed down, the mouse pointer 'sticks' to the point on the sphere which corresponds to the point in the circle which it is touching. Moving the mouse pulls the sphere with it, and the effect is to rotate the object inside the sphere. To make it obvious what is happening, axes are drawn within the circle and they rotate with the mouse: for instance, if the mouse pointer picks up the end of one of the axes, it stays attached to it so long as the button is pressed down. In this study, the axes were not related to the natural axes of the object: they were simply a device for showing how the sphere was moving.

The system that we have outlined contains a mixture of iconic and analogue elements, and so we call it the Icon cum Analogue system of control, or ICAS control for short. Figure 1 illustrates the screen as it appears with all controls present.

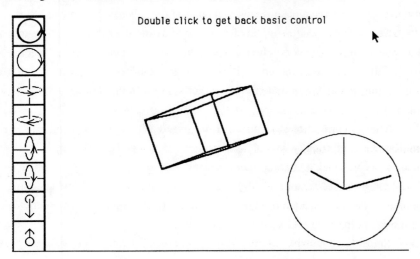

Figure 1: ICAS control screen (with all controls present).

3 The theory of vision and the design and test of control systems

The design of ICAS control was based loosely on an understanding of the way human vision recovers information about tridimensional shape from two-dimensional images. Two points were critical.

The first is that human vision prefers not to use truly general methods. It trades on regularities in the world if it can. We have made the case for that view in a number of papers, for instance, Cowie & Clements (1989); Cowie, Bradley & Livingstone (1991).

Often the objects we look at exhibit a mass of regularities - right angles, parallels, symmetries, and so on. In those cases, it may not be important to supply additional regularities. However, ICAS control does offer regularities to trade on even when the object is irregular. The icons provide a form of movement which is highly regular - long unbroken rotations. Mathematically, recovering structure from pure rotational movement is very much simpler than recovering it from arbitrary motion. Psychologically, people seem to manage pure rotation better than arbitrary movement when they have no choice in the matter (Green, 1961), and it is one of the main patterns which people try to create when they can use basic control to choose their own viewpoints (Cowie, Bradley & Livingstone, 1991). Second, the axes in the virtual sphere are at right angles to each other. The projective geometry of rectangular corners is particularly simple (Attneave & Frost, 1969), and so the axes in the virtual sphere provide a simple way of monitoring the observer's position and motion by vision. In short, ICAS control provides regularities which are potentially useful, but not necessarily available in the real world.

The second basic point behind ICAS is that vision is purposive, particularly when people are able to manoeuvre for a good vantage point. Manoeuvres that help to answer one question may be irrelevant to another, and vice versa. Hence it is sensible to provide alternative forms of control and let people choose among them rather than looking for a single optimum which is probably illusory.

The ideas which lay behind the design of ICAS control were also incorporated into the tests we developed. Again, two main points need to be made.

First, we considered that a key variable was the regularity of the objects being viewed. That follows from our theory that the regularities ICAS provides may not be important when the object to be viewed is regular. Hence we need to look at both regular and irregular objects to evaluate the system properly.

Second, we considered that different tasks had to be used. It follows from our theory that an option which is not useful for one task may still be very useful for another, and so a test that considers only one task has the potential to be quite misleading.

4 Method

Our experiment used four variants of the ICAS control which has been described. All of them used basic control. However the either or both of the icon and the virtual sphere could be absent, giving rise to the four variants.

The task we used had two components. Subjects were asked to judge the angle between two target edges and their relative lengths. The edges were embedded in skeleton parallelepipeds: one was red and one was green to distinguish them from the other lines. The objects varied in three ways. First, the lengths of the target edges were set by a random process for each trial. Second, the angles between the target lines varied between 90° and 150° in 15° steps. Third, the line which joined the target lines to form a vertex could so in two ways. It could be at 90° to both, so that the whole object formed a prism. Alternatively it could meet one at 50° and the other at 70°, so that the whole object showed no particular regularity. An additional measure was also taken: after each trial, the program recorded how long the subject had spent using each of the controls that was available.

There were twelve subjects. Each one completed a session consisting of five blocks. The first was a preliminary practice with all the controls available. Each of the others involved one of the four combinations of controls. In each block, a subject saw ten objects - five prisms, one with each of the five possible angles between the target lines, and five corresponding irregular objects. On each trial, subjects viewed the object and manipulated it for as long as they wanted, and then wrote down their judgements of the angle between the target lines and their judgements of their relative lengths.

5 Results

Separate analyses of variance were carried out for the angle and line judgements. Results can be presented by considering the main variables in turn.

We begin by considering the factors which are secondary from our point of view. Object type had a highly significant effect on judgement of line lengths, and a significant effect on the judgement of angle size. In both cases the prisms were more accurately described. The size of the angle between the target lines also affected accuracy at judging angles size - estimates were best when the angles were nearest 90°. These results reinforce the point made above that human vision does trade on regularities in the objects it is looking at, and tends to suffer when they are not available.

Our main concern, though, is with the controls. Two factors involve the controls. One is the presence or absence of the virtual sphere display, the other is the presence or absence of the icon bar.

There is only one significant effect involving the presence or absence of the virtual sphere. It is an interaction between the presence of the sphere display and object type ($F(1,11) = 11.00$; $p < .01$). Figure 2 shows the effect. If the sphere is not present, subjects gain a considerable benefit when the object is regular. That benefit almost disappears when the sphere is present. The corresponding effect for judgements of length was not significant ($F(1,11) = 2.67$; $p > .1$), but the pattern is similar (Figure 3). The sphere gives some advantage when the objects are irregular. Much more surprisingly, and more clearly, it is counterproductive when the objects are regular. These effects were not expected, but they makes sense in retrospect: we will come back to the reason in the discussion.

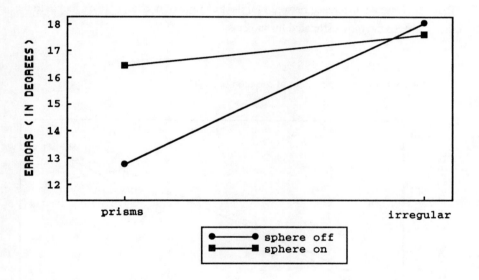

Figure 2 Absolute difference between (a) true angle between target lines and (b) subjects' estimates of the angle.

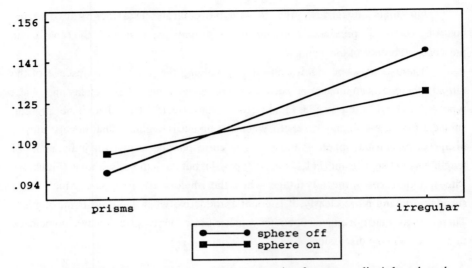

Figure 3. Absolute difference between (a) the true ratio of one target line's length to the other's and (b) the ratio estimated by subjects.

We now consider the effect of the icon bar.

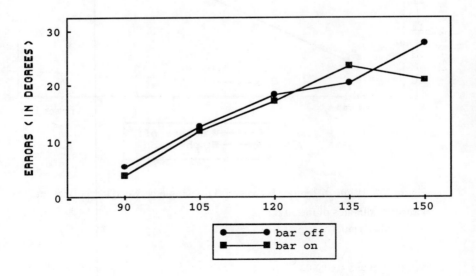

Figure 4 Absolute difference between (a) true angle between target lines and (b) subjects' estimates of the angle.

There is no overall effect of the icon bar on angle judgements. However, there is a significant interaction between the size of the target angle and the presence of the icon bar ($F(4,44) = 3.0489$; $p < .05$). Figure 4 shows the pattern. The visual impression is that the bar benefits performance at the extreme target angle, where judgements are worst without it. Analysis of simple effects confirms that that is the source of the interaction. Again, this makes some sense, and we will discuss it later.

With length judgements, there is a highly significant main effect showing that errors are lower with the icon bar on ($F(1,11) = 10.891$; $p < .01$), and a significant interaction between the presence of the bar and object type ($F(1,11) = 7.369$; $p < .05$). Figure 5 shows both effects. It is clear that the main benefit of the bar is with the irregular objects. Simple effects analysis shows that the advantage it gives with the prisms is not significant.

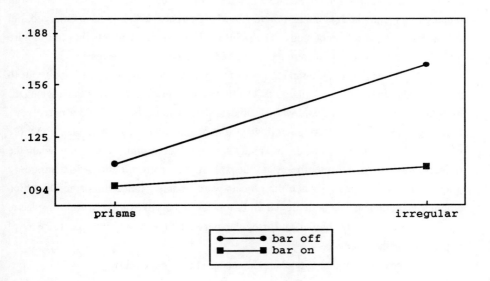

Figure 5 Absolute difference between (a) the true ratio of one target line's length to the other's and (b) the ratio estimated by subjects.

We turn finally to the use made of the three types of control in the case where all three were available. Two effects emerged. First, the nearer the target angle was to 90°, the lower the total subjects spent on the task. Second, subjects had clear preferences among the controls. Mean time spent using the Sphere control was 5.7 seconds; mean time spent using the icon bar was 20.9 seconds; and mean time spent using basic control was 32.1 seconds. Analysis of variance confirms that the means are significantly different ($F(2,22) = 5.70$; $p < .05$): only the two extremes are different on a post hoc test.

6 Discussion

The discussion divides into three parts. The first two look at relatively specific implications of the study, in terms of evaluation techniques and in terms of the techniques embodied in ICAS control. The third looks at the general issues that the study raises for the design of interactive systems.

In terms of evaluating techniques, the experiment vindicates our emphasis on using alternate measures and a range of objects.

As regards measures, the advantage of the icon bar was only clear from the length judgement task and the negative impact of the virtual sphere was only clear from the angle judgements. It is interesting to ask why that pattern should have emerged, but that is matter for a different study. The point we want to make here is that our picture would have been seriously incomplete if we had not used both measures.

A similar point can be made about using a range of objects. It is important to recognise that two principles emerged. The first is unsurprising. It is that the advantages of some controls may only show with objects that pose problems for the perceiver - typically objects which are irregular in ways that people tend not to anticipate. The more surprising finding is that some controls may have limitations which only show when the objects are regular. Remember that the sphere control's potential for impeding judgements only appeared when the objects were prisms. Hence the principle which emerges is that one needs both regular and irregular objects to evaluate a control system. Our approach of using regular and irregular parallelepipeds may be a useful way of implementing the principle.

We now look at the particular elements used in ICAS control.

Basic control is the mode subjects used most, and in that sense it appears to be a success. We do not have direct evidence on the way it affects accuracy because there was no condition where it was absent. That is clearly a target for future research. There is also room for comparisons between the particular form of control we used and relatives which use different ways of coupling mouse movement to object movement. It is important to register that basic control meets two of the traditional criteria that were set out at the start: it provides a degree of kinesthetic correspondence and a sense that the observer is manipulating the object. To that extent, the experiment vindicates the view that these criteria are significant. What the experiment calls into question is the suggestion that they are the only significant criteria. Even basic control violates the other criterion, in that using a limited window effectively restricts observers to rotation about two axes in the short term.

The icon bar is the most distinctive feature of the design. It comes out well, in the sense that it receives a substantial amount of use and also provides a very clear enhancement of accuracy for one task (judging length). It meets only one of the traditional criteria, which is that the observer has a sense of manipulating the object directly.

The sphere display is the only display which meets all three traditional criteria, and it is both the least used of the control devices and the only one which actually makes performance worse. This underlines the limitations of the traditional criteria. However, it is important to register that modified versions of the sphere display could well prove much more useful.

The key feature of the sphere is probably that it showed three orthogonal edges - they are effectively a set of axes for the virtual space. The image of those axes provided a degree of feedback on the effect of movement. However it was not simply related to the orientation of the object because the axes in the sphere were not aligned with the natural axes of the object. The sphere probably impaired performance by distracting people from a slightly harder, but more effective strategy - that is, identifying the natural axes of the object and using their movement for feedback. That had a negative effect with the prisms because they did have edges which corresponded to two of the object's natural axes, so that people could do quite well at tracking the natural axes of the object if they were not distracted from it. With the irregular objects, the natural axes were difficult to track, and the axes provided by the sphere were is anything a slight asset.

The implication of that reading is that the sphere display could be useful if it provided axes which aligned with the object's natural axes. That is an obvious target for study. However, it underlines the point that if the sphere display is useful, it will be for reasons that are very loosely related to the traditional criteria set out at the start of the paper, and a great deal to do with providing thoroughly unrealistic forms of enhanced feedback.

We now turn to the general principles which the study highlights.

Our general view is that trying to duplicate reality is a dubious goal for interface design. Our experiment illustrates the point that thoroughly unrealistic control systems can be very effective. This reflects a recurring theme in the history of depiction. It has been possible to create *trompe l'oeil* displays which genuinely confuse the viewer since the seventeenth century, but they have never been more than a diversion (Gombrich, 1977). High art has always played on the duality between the depicted scene and the picture surface with its own carefully constructed pattern, and in the late nineteenth century Impressionism and its successors forcibly reasserted the primacy of the picture surface. Engineering and architectural drawing have been even further from realism with their adoption of highly conventional systems of parallel projections (Dubery & Willats, 1972). There seem to be two factors at work which are potentially relevant to interface design.

The first factor is that people seem to have limited tolerance for techniques which blur the line between reality and illusion. This is speculative, but it is based on inference from observations that are reasonably clear. It is reasonably clear that there has been limited acceptance for techniques that could generate highly realistic experiences, such as viewing through peepholes for traditional trompe l'oeil and viewing through special lenses for 3-d films. There are inconveniences associated with these techniques, but they are not particularly severe: one has to suspect that a significant obstacle to acceptance is discomfort with depictions that encroach too much on reality. It is also reasonably clear that people find it unsettling to be exposed to compelling illusions, such as the traditional geometric illusions or the illusory movement and dazzle created by op artists such as Bridget Riley. Informal experience with virtual reality systems suggests that related problems do arise there: many people find immersion in a virtual world intriguing as a novelty, but disquieting in the longer term.

The second factor is that human ingenuity can capitalise on the fact that a display is not real. Engineering and architectural drawing provides good examples. They are constructed so that the picture surface can give simple access to key measurements, and so that several views can be seen at once at once in a way that is impossible in reality. They are also shot through with conventional symbols for components and relationships. It seems to us that these are not crude limitations of an old fashioned medium: they illustrate the way mature use of a medium can capitalise on its dual reality, and are an example worth taking to heart.

That general view leads us to propose two principles for interactive interface design which we believe are a valuable counterweight to the intuitions that make virtual reality such a compelling idea.

The first principle is that virtual realities can and should be magical. What we mean by magic is that they give users powers - of action and of perception - that they cannot have in real reality. The birthright of virtual worlds is their ability to give people powers which are magical in that sense, and it seems perverse to disown that birthright.

Introducing the notion of magic points to a rich source of ideas about controlling virtual worlds. Myths, fairy tales and fantasy novels reflect long, disciplined and imaginative thinking about powers which the human mind seems to feel it could use if the world would only allow it - and about the pitfalls associated with superficially attractive gifts. As a result, aiming to make the user a wizard in a virtual world is a powerful heuristic. Thinking about it leads easily to principles which are manifestly relevant to interface design. For instance, a key issue is to make sure that the user can control the power. That hinges on five main requirements. Commands should be easy to give. They should invoke actions which are intuitive. The consequences of giving the commands should be easy to see. Not too much should be expected of the user's memory. And there should be ways of reverting to a secure state if things start to go catastrophically wrong. We have no doubt that these kinds of intuition are widely shared, but it seems to us that they are worth articulating.

The second principle is that in interface design, understanding the mind is as fundamental as understanding the world. We have argued in the context of vision research that computational approaches tend to be attracted towards solutions based on knowledge about the external world, and resist coming to terms with the awkward facts of human performance (Cowie & Clements, 1989). Something similar may contribute to the appeal of the virtual reality ideal: it suggests that the place to look for solutions is the orderly, consistent world of physics rather than the quirks of the human mind.

We have indicated at various points some quirks of mind which we believe are important. One is that certain kinds of operation feel natural even though they are rarely possible for a human to execute in the natural world. ICAS control uses an example, which is spinning an object at will. We have suggested that magical tales are an important source of ideas about operations like this. The idea of ICAS control is strongly influenced by another set of psychological concerns, that is concern with the mechanisms of vision. So long as screen displays remain the main way observers receive information from the computer, it is essential to consider what the visual system is likely to make of a display and not to assume uncritically that it will reconstruct the object specification that was used to generate the display. That point is related to a third, which is understanding the frames of reference in which people locate themselves, and their actions, and the structures they perceive. The problem of meshing different frames of reference showed in the malinteraction between the structure displayed by visible spheres and the structure inherent in prisms, it is central to the issue of how we choose between observer and object movement, and it is central to the distinction between accepting displays as a kind of reality and treating them as a depiction with no location the actual world.

We have no wish to claim that these ideas are original. However, they seem to be worth restating in the context of contemporary fascination with virtual reality. And we believe that ICAS control, although it is less glamorous than alternatives, reflects an approach which is both effective and reasonable enough to be worth developing.

References

Attneave, F. & Frost, D. (1969) The determination of perceived tridimensional orientation by minimum criteria. *Perception & Psychophysics* **6**, 391-396.

Chen, M., Mountford, S.J. & Sellen, A. (1988). A Study in Interactive 3-D Rotation Using 2-D Control Devices. Proceedings of SIGGRAPH '88. In *Computer Graphics,* **22**, 4, 121-129.

Cowie, R. & Clements, D. (1989). The logical basis of Visual Perception. Computation and Empirical Evidence. *Irish Journal of Psychology,* **10**, 2, 232-246.

Cowie, R., Bradley, D. & Livingstone, M. (1991). Using observer-controlled movement and expectations of regularity to recover tridimensional structure. In A. Smeaton (Ed), *Artificial Intelligence and Cognitive Science '90.* Berlin: Springer Verlag. pp.178-192.

Cobb, L. Personal communication.

Dubery, F. and Willats, J. (1972) *Drawing Systems.* London: Studio Vista.

Gombrich, E. (1977) *Art and Illusion* (5th edn) London: Phaidon Press.

Green, B.F. (1961) Figure coherence in the kinetic depth effect. *Journal of Experimental Psychology* **62**, 272-282.

Rhingold, H. (1991) *Virtual Reality.* London: Secker & Warburg.

A Context-free Approach to the Semantic Discrimination of Image Edges

Ezio Catanzariti

Dipartimento di Scienze Fisiche, Università di Napoli,

Napoli, Italy

Abstract

It is generally believed in the image analysis field of research that the geometrical characteristics of intensity edges can be related to the different physical processes that gave rise to them. As well, it is usually accepted that this *intrinsic nature* of image edges can be retrieved by some kind of context-free process acting locally on image points. However, the many attempts at classifying edges on the basis of their local properties have so far failed to do so. We have here attempted to succeed at the task of classifying different types of intensity edges, namely, steps, ramps and roofs, by using Gabor Elementary Functions as local visual filters and the Maximum Likelihood scheme of classification. Our intention was to develop a computational tool for looking into the relationship between image edges and their 3D counterparts. Results are presented, visually evaluated and discussed, for the case of simple images. These results show that our algorithm successfully performs the assignement given to it, that is, it correctly discriminates most candidate image edges on the basis of their shape. However, the physical interpretation corresponding to each of the three edge classes thus found is not unique. According to our analysis, the given edge semantics appear to have an inherent ambiguity which cannot be solved by a context-free approach only.

1 Introduction

1.1 Background

This paper introduces an edge classification scheme that identifies edges in a digital image and uniquely assigns each of them to one of a number of predefined edge categories such as step, ramp or roof. The purpose of this section is to introduce the problem of discriminating different types of image edges as well as to review briefly the two main approaches which have been given so far to its solution.

Many theoretical studies in early image formation and description (Horn, 1975) (Marr and Hildreth, 1980) claim that the geometrical characteristics of intensity edges in images can be related to the different physical processes that give rise to them. If this is so, the recovery of the shape of image edges constitutes by itself an immediate matter of image interpretation. This is the reason we need to discriminate between differently shaped image edges.

This issue has been widely addressed from a practical, as well as from a theoretical point of view, in the edge detection literature.

In an early study on detecting boundaries in images of polyhedral objects (Herskovitz and Binford, 1970), Herskovitz and Binford noticed that discontinuities in image brightness can be classified, according to their one dimensional shape characteristics, that is, the shape of the intensity profile taken in a direction orthogonal to the edge in the image, in three clearly defined classes, namely, *step* edges, *roof* shaped edges and *edge-effects* (peaks). By carefully studiyng the behaviour of the image intensity function, they also observed that the prevalence of one type of edge over the other depends on such physical factors as the distribution of the light sources in the scene or the properties of the imaged surfaces (texture, translucency, etc.). The interesting point about these findings is that the different geometric shapes so found, which are purely image categories, are in this way described in terms of purely scene characteristics, such us *occluding edge* and *illumination edge*.

In line with these results, Herskowitz and Binford made the assumption that the process of edge detection must be performed differently according to each specific edge type situation. The outcome of the edge detection process is therefore sought by them as a set of edge points which are marked with information about the *type* of the edge discontinuity to which they belong.

The same problem is tackled by Canny (Canny, 1986). He also suggests the use of specific operators for the independent detection of steps, roofs and symmetric ridge intensity edge types. However, in this solution, as in the previous one, the problem of the possible combination of the different edge detector outputs is not solved. Canny admits that, since different operators are likely to respond to the same intensity edge, there is no obvious reason for preferring one edge type over another. Along the same methodological line can be situated some more recent work on edge detection where the problem of detecting image edges whose shape fall into a specified category is addressed, as for example in (Cheng and Don, 1991) and (Petrou and Kittler, 1991).

An edge-type descriptor is also included in the complete edge description in Marr's primal sketch model of the image. In (Marr, 1976) he gives an extension of the Herskovitz and Binford catalog derived, this time, rather than from image observations, from observations of how specific peak patterns group together in the one dimensional convolutions of image with edge-shaped (first derivative) and bar-shaped (second derivative) masks. Some other edge categories, such as *extended-edges* and *shading-edges*, are added to the previously seen edge types. His approach to the detection and discrimination of the different edge types is rather different from the one of Herskovitz and Binford: mainly, it is a problem of searching for the same specific pattern of extrema which is characteristic for that edge type in the output profile of the mentioned edge detectors. A fairly similar approach was proposed by Shirai in (Shirai, 1975).

The main shortcoming of Marr's and Shirai's solution is that the amount of data one has to deal with is, because of the noise, totally unmaneageable. In fact, in a real word situation not only the points of the image corresponding to physical edges, but almost every point in the image, give rise to a non zero convolution value. One consequence, for example, is that many extrema in the filter response will not generally correspond to edges in the input signal. The way Marr tries to overcome this problem is by suggesting the use of masks of different sizes: the way the response of a set of masks evolves when the mask's size changes should provide enough information for the relevant peaks to be selected among the multitude of unimportant ones. Once this is accomplished, the task of parsing local configurations of peaks in order to obtain a specific edge type becomes in principle possible. While the idea of using different mask sizes in order to obtain power spectra of the different edge types has certainly inspired a great deal of work in computer vision, in order for this specific approach to become feasible, the problem of matching the displaced peaks, and so the patterns, obtained by the masks of different size has first to be solved, and an efficient solution to this problem has not, so far, been found.

Despite some more recent results (Lee, 1989) (Perona and Malik, 1990) on the same subject, no much progress has been made so far in the task of taking into account the varied or composite nature of image edges when performing edge detection. In all cases, the final result of the edge detection process consists in a partition of the image into edge and no-edge points, where some attributes, such as sharpness and contrast may be associated with the edge points. The information about the *intrinsic nature* of the edge contained in its geometric characteristics, is missed, and its recovery is perhaps left to higher levels of the image interpretation process acting on different representations of the image.

In this section we have tried to suggest that the problem of the discrimination of the different types of edges, as generally approached in the context of edge detection, that is, as a computational process based on local cues only, has turned out to be a very difficult task to deal with. One of the reasons might be that the recognition of image edge types involves more global, context-sensitive computational routines. On the other hand, the human visual system seems to perform very well at the early perception of different kind of edges; we do not seem to have much trouble in perceiving blurred changes of luminance or blob-shaped intense variations of luminance by themselves, although it is not clear if, and to which extent, our recognition of them as distinctive local features is due to high level processing of contextual information. As with almost every image understanding task, there is here a chicken and egg cycle one has certainly to deal with, but which goes beyond the scope of the present work. In this work we have approached the same problem strictly maintaining the assumption that the recognition of the different types of image edges is an early image processing task, that is, one which, in principle, may be accomplished by a processing routine acting context-free on image data. This hypothesis constitutes the starting point of our algorithm for edge recognition. For this task, we need powerful local descriptors for image edges, and a global mechanism for combining the local descriptions. The method we propose employs the special class of analytic functions known as Gabor elementary functions (or Gabor EF's), for retrieving the structural properties of image edges as encoded in their spectral and spatial characteristics.

1.2 Gabor elementary functions

The recent popularity of the Gabor transform in the fields of Image Analysis and low-level Computer Vision is due to the fact that theoretical evidence has recently been found (Marcelja, 1980) that the representation of the image in the visual cortex must include both the spatial and the spatial frequency variables in its description.

A generalization of the Gabor EF's for analysis of two dimensional signals has been given by Daugman (Daugman, 1985). According to this scheme, an image can be represented as a linear combination of Gabors's EF with impulse response h:

$$h(x,y) = g(x,y) \times exp[2\pi i(u_0 x + v_0 y)]$$

where:

$$g(x,y) = exp\{-[(x/\sigma_x)^2 + (y/\sigma_y)^2]/2\}$$

The 2D Gabor EF is therefore the product of a two-dimensional sinusoidal lattice with radial frequency $\sqrt{u_0^2 + v_0^2}$ and orientation $\arctan(v_0/u_0)$ modulated by an elliptical Gaussian, centered at the origin and scaled by widths σ_x and σ_y. In their two-dimensional formulation, such filters simultaneously capture all the fundamental properties of linear neural receptive fields in the visual cortex: spatial localization, spatial frequency selectivity and orientation selectivity (Daugman, 1985). What make Gabor EF's particularly attractive to computer vision research is the mathematical property of being maximally localized both in space and spatial frequency (Daugman, 1985). Because of these properties, the Gabor representation has established itself as a successful framework for a computational theory of visual perception. And indeed, this schema has been used in many different computer vision tasks, including image texture segmentation (Bovik et al., 1990), motion (Fleet and Jepson, 1989) and image compression and reconstruction (Daugman, 1988).

We will use the Gabor representation for the task of discriminating different types of image edges.

2 The method

If a set of Gabor EF's with varying spatial extent, tuned to a number of spatial frequencies, are used as visual filters, an image will be represented by a set of independent 2D feature-maps. Each map is the output of the convolution of the impulse response of a specific Gabor operator with the image. As in the multiresolution approach, the results of the analysis so obtained have to be compared accross the different spatial and spatial frequency channels. We have already noted that the solution given by Marr to this problem is not a viable one. On the other hand, there is still no information available on how this operation is performed in the visual system. The way we suggest to confront this is from the standpoint of Bayesian decision theory. Specifically, the approach we propose is the supervised version of Bayes criterion known as maximum likelihood classification. With this approach, for each image edge of interest, the probability of that edge belonging to each edge-type class is computed on the basis of its spectral characteristics. If the classification decision must be made context-free, the edge may be assigned to the class that gives the highest probability of membership.

A classification scheme based on a set of image measurements taken by a bank of Gabor feature sensors was advocated by Watson (Watson, 1987) as a computational model for vision. A similar idea was implemented in (Porat and Zeevi, 1989) and (Bonifacio et al., 1992) for the task of image texture discrimination.

The Gabor scheme fits nicely inside the classification paradigm in that, being a type of joint spatial / spatial-frequency representation, is able to improve pattern separability relative to pure space or spatial-frequency representations. Besides, Bayes decision theory is general and capable of representing the many dimensions of our domain in a way that present multiresolution schemes fail to do. Specifically, it gives us a method for combining all the information extracted at different spatial scales and frequency bandwidths. Furthermore, the assumptions required, in absence of specific image information, are usually reasonable.

3 Results

The version of 2D Gabor EF's we used for convolution with our image data is defined as:

$$h(x, y) = g(x', y') \times exp[2\pi i(u_0 x + v_0 y)]$$

where $(x', y') = (x \cos\theta_0 + y \sin\theta_0, -x \sin\theta_0 + y \cos\theta_0)$ are rotated coordinates. In this equation, $g(x, y)$ (as defined in sect.1.2) is a two dimensional Gaussian function with major axis oriented at an angle θ_0 from the horizontal axis, that is, parallel to the direction of the modulating waveform (fig.1)

As far as the values of the parameters defining the specific form of Gabor EF's is concerned, one observes that, in order to capture at the best the frequency characteristics of image edges, filters must be chosen in such a way as to produce a uniform covering of the spatial frequency plane, possibly with little overlapping at the extremes of their bandwidths. Moreover, to cover also the possible range of spatial scales of the digital edges, filters with different spatial extent must be employed. Therefore, we generated a variety of filters of increasing spatial support, specifically, having $\sigma_x = 2, 4, 8$[1], and, for each Gaussian window thus obtained, we considered a fixed number of values for the central radial frequency ω_0. We also restricted to a fixed value the ratio of the

---

[1] Once the value of σ is set, the value for the radial frequency bandwidth $\Delta\nu = \frac{\alpha}{2\pi\sigma}$ also remain set, where α is a factor of proportionality.

filter's linear dimensions. Our Gabor EF's were elliptical with an aspect ratio $\sigma_x/\sigma_y = 2$. The set of admissible frequencies was therefore derived from the ratio $r = 1/(\omega_0 \times \sigma_x)$, for $r = 1, 2, 3$ and 4. As for the orientation θ, reasons of ease of implementation suggested a fixed number of orientation values be chosen for the whole range of frequencies. Thus we considered Gabor filters at an angle $\Delta\theta = 30°$ apart from each other[2]. Since the resolution in orientation of a Gabor EF depends on the radial frequency ω_0 (Catanzariti, 1992), this choice does not provide us with a uniform covering of the 360° angle at all frequencies. It should be pointed out that other quite different choises for previous parameters are possible, and sometimes more appropriate (Bonifacio et al., 1992), depending on the specific application. However, we believe that the specific spatial frequency characteristics of Gabor filters are not at all critical to the performance of our method.

A further measure toward reducing the dimensionality of the feature space was obtained by first compairing local orientations from the same frequency channel. In fact, orientation does not affect membership of an edge class. Therefore, although every candidate edge point was convolved with Gabor EF's oriented at all directions, only the value of the convolution output corresponding to the direction of maximum response was retained.

Finally, although other solutions are possible (Bonifacio et al., 1992), in actual computations we treated each of the symmetrical and antisymmetrical parts of a Gabor EF as independently responding to the input signal.

The first image we used to test our method (fig.2a), represents a scene from the *block*'s world, namely, a wedge, illuminated from the side, resting on a flat surface. This image was purposely chosen simple so that the relation between image edges and their physical causes be as close to a one-to-one relationship as possible.

In order for our method to be applied, the classifier had first to be trained. According to the edge theory previously mentioned, we expected three main types of edges to stand out in this image, namely, *steps*, due to the discontinuities between surfaces and between surface orientations, *ramps*, originated by discontinuities in illumination, and *roofs*, generated by the self obscuring concave edge between the wedge surface and the background surface. This hypothesis was verified by visually examining a number of intensity profiles taken in a direction perpendicular to the image edges. This scrutiny revealed that, contrary to the previously suggested semantics for edges, there was no simple relationship between image edges and scene edges.

For example, while it is true that the intensity profiles across the blurred part of the cast shadow edge show the characteristic pattern of the ramp (fig.3a), on the contrary, the one collected along the sides of the shadow, in the proximity of the wedge surface, displays a step-like shape (fig.3b). Similarly, although the obscuring edge (roof) shows a consistent shape along most of the concave border (fig.3c), a certain amount of light reflected from the support surface on the side of the light source, makes one of the slopes of the roof rise quite abruptly, in a step-like fashion, on the side of the wedge face (fig.3d). Finally, many of the image edges corresponding to the lighting transition between the illuminated and the self-shadowed face of the wedge show a composite pattern recalling Marr's extended-edge (fig.4b). As well, the transition between

---

[2]Only six convolutions were actually required, in fact, θ and $\theta + \pi$ differ only in sign.

illuminated and self-shadowed face of the wedge, in the case when the former one is not directly illuminated, is not as abrupt as one would expect from a luminance profile belonging to the step-edge category (fig.4c).

Therefore, when collecting the training set, one has to make a decision on where to hold on to the ideal set of pattern classes defined *a priori* on the basis of the geometrical shape of their elements, namely, step, ramp and roof, or rather to perform the collection on the basis of the final result one seeks from the edge recognition process, in which case the pattern classes will be convex-edge, shadow-edge and concave-edge. In both cases, the very same process of collecting the samples for each class will be one of image interpretation. Although this fact is partly an inevitable result of some inherent ambiguity of the problem under study, it is also a consequence of the approach we have taken. In fact, according to Bayes decision theory, choices are made on the basis of a certain amount of *a priori* knowledge of the domain under study. Probably a clustering procedure should be used in case one wanted to pursue the first of the suggested possible paths. However, this has not be done. We think that, in principle, both choices are equally acceptable. In fact, the ultimate goal of edge detection is one of image interpretation, that is, all image edges have to be eventually (correctly) identified by a unique label specifying the physical edge which gave rise to them. In our case, for example, this implies that all shadow edges be eventually marked with the same label. The approach we are presenting here is only part of the complete edge recognition process. The edge label resulting from the application of our algorithm is meant to represent the fact that the likelihood for that edge of belonging to a specific edge class is maximum. Other processing stages have to cope with the problem of ambiguous edges. Each of the two previously stated choices simply carries different assumptions on the subsequent stages of the edge recognition process. However, the main goal we wanted to achieve with our study, was to test the performance of the Gabor EF's based model for visual detection at the task of discriminating between different edge types, that is, if, and to which extent, a relation between image edges and physical edges can be made explicit on the basis of the combined spatial and spectral signatures alone. Therefore, in spite of the mentioned findings, we decided to include representatives from all three semantically defined edge classes in the training set. Accordingly, a number (that is, fifteen) of *edge* points for each class were manually collected and their (x, y) coordinates were given as training set to the classifier.

All points in the image were classified using a varying number of Gabor filters. To this end, another class, *background*, was added to the previous three. Therefore, each point in the classified output was either labeled as belonging to one of the edge classes, or was labeled as no-edge.

Results of classification are shown in fig.2b when eight Gabor filters are used[3]. In this images, convex and occluding edges (steps), shadow edges (ramps) and concave edges (roofs) are shown in decreasing tonalities of gray. The no-edge points (background) are shown in white. The large band corresponding to the blurred part of the shadow edge, which represents the *ramp* aspect of this edge, is surrounded by two thinner bands: a *step* band on its

---

[3] Specifically, σ_x was set to the values $\sigma_x = 4$ and $\sigma_x = 8$ and the ratio r was set to the values $r = 2$ and $r = 4$. As usual, both the sine and the cosine part of each Gabor EF were considered.

external side, and a *step* or a *roof* band on its internal side. What the Gabor filters have captured in this case corresponds to the Mach bands illusion.

This is not surprising. An explanation of the Mach band perceptual phenomenon was accounted for by Marr's multiresolution model of early vision. Basically, the perception corresponds to a composition of the responses by larger scale filters, which signal the linear change of luminance in the middle of the ramp, and the smaller scale filters, which respond to the discontinuities of the luminance gradient at the two extremes of the ramp but are insensitive to the linear change which characterizes the ramp. This is graphically illustrated in fig.5, which shows the response profiles to an ideal ramp edge (fig.5a), when Gabor EF's with increasing values for σ are used (figs.5b to 5d)[4].

Mach band effect is strong along the shadow border on the opposite side of the wedge, but becomes less noticeable as the shadow border approaches the illuminated surface of the wedge. As previously noted, this corresponds to the fact that the image edge becomes steeper in this region until its shape becomes rather similar to the shape of a step edge. As for the concave edge, it gets consistently classified as roof along the darker side of the image, although its step aspect becomes increasingly detected by the classifier as one moves toward the illumination source. Again, as one get close to the source of illumination the rate of luminance change becomes very high on one side of the roof (fig.3d). The Gabor EF centred on that edge therefore, classifies it as a step edge.

The results obtained by the application of our algorithm on two more test images are displayed in fig.7 and fig.8. The first image (fig.7a) is the digitized version of a black and white photograph representing a scene from the word of polyhedra. The *ramps* and the *roofs* captured by our technique are shown in fig.7a and fig.7b, respectively. A white coffe cup on a uniform background is pictured in fig.8a. This time our algorithm was requested to identify the shadow edges and the convex edges (that is, the borders of the cup). Results are displayed in fig.8b and fig.8c.

4 Conclusions

In this paper we have presented an edge classification scheme aimed at uniquely assigning, to each pixel in an image, a label identifying the specific edge type category to which the pixel belongs.

A main goal of this work was to evaluate a widely accepted hypothesis in the image analysis literature that image edge classification is possible by means of a local, context-free analysis of image edges properties only. Our intention was to build a computational tool for looking into the relationship between image edges and their 3D counterparts. To this end, the most successful local image descriptors and the most powerful context-free classification paradigm were employed. Specifically, a point by point maximum likelihood classification algorithm was used to assign each candidate edge point to a specific edge class

---

[4] The fact that a roof pattern appears mixed to a step pattern on the inner (low luminance) side of the shadow profile finds its explanation on the specific shape of test data at hands. This is best shown graphically in fig.6 where a one dimensional cross-section of the image is displayed. Instances of the three types of edge profiles appear in this image, namely, step, roof and ramp from left to right. As this figure reveals, the left side of the ramp is shaped much like a roof edge.

on the basis of the spatial frequency characteristics of that point detected at different spatial scales by a set of Gabor EF's.

The classifier was trained to recognize image edges belonging to three distinct semantics classes, namely, convex edges (steps), concave edges (roofs) and shadow edges (ramps), in real polyhedral images.

Results show that the classifier classifies successfully (that is, in agreement with the assigned semantics) a great deal of image edges. In fact, most pixels on shadow, concave and convex edges are labelled as ramps, roofs and steps, respectively. Many image points, however, which also clearly belong to the shadow, concave and convex boundaries, are given labels which deviate from the previous assignment. A most interesting point here is that a visual screening of these *misclassified* edges shows that in many cases their 1D geometrical shape matches the shape of the edge class to which they have been assigned. In other words, it seems as if the grouping of these edges in feature space is still correctly performed based on their shape while the *a-priori* interpretation superimposed on them does not univocally matches the edge classes thus found on the image. For example, shadow edges and obscuring edges tend to be step shaped when close to the source of illumination (or to the strongly illuminated surface). Consequently, they tend to be labelled as *steps*. The classifier was trained to recognize image edges on the basis of their physical interpretation. However, edge labelling is performed on the basis of edge *syntactic* characteristics. These results suggest that a complete recovery of the intrinsic nature of image edges is not possible based only on local cues. In fact, as we have tried to show, even in the simple image under study, the given edge semantics appear to have an inherent ambiguity which cannot be solved by a context-free approach only.

Acknowledgements

This work was partially supported by CNR Progetto Finalizzato Robotica No:92.01113.PF77.

References

Bonifacio, A., Catanzariti, E. & Di Martino, B. (1992). Unsupervised segmentation of textured images using Gabor functions. To appear in *Proceedings of the 2nd International Conference Dedicated to Image Communication*. Bordeaux, France.

Bovik, A.C., Clark,M. & Geisler, W.S. (1990). Multichannel texture analysis using localized spatial filters. *IEEE Transactions on PAMI*, **12**(1), 55-73.

Canny, J. (1986). A Computational approach to edge-detection. *IEEE Transactions on Pattern Analysis and Machine Intelligence*, **8**(6), 679-698.

Catanzariti, E. (1992). Maximum-likelihood Classification of Image Edges Using Spatial and Spatial-frequency Features. *Proceedings of the 11th International Conference on Pattern Recognition* (Vol. 1, pp. 725-729). The Hague, The Netherlands: IEEE Computer Society Press.

Cheng, J-C., & Don, H-S. (1991). Roof edge detection by mathematical morphology. *Proceedings of Vision Interface'91* (pp. 89-96). Calgary, Canada.

Daugman,J.G. (1985). Uncertainty relation for resolution in space, spatial frequency, and orientation optimized by two dimensional visual cortical filters. *Journal of Optical Society of America*, **2**(7), 1160-1169.

Daugman, J.G. (1988). Complete discrete 2-d Gabor Transforms by Neural Networks for image analysis and compression. *IEEE Transactions on Acoustic, Speech and Signal Processing*, **36**(7), 1169-1179.

Fleet,D.J. & Jepson,A.D. (1989). Hierarchical construction of orientation and velocity selective filters. *IEEE Transactions on Pattern Analysis and Machine Intelligence*, **2**(3).

Herskovitz, A., & Binford, T. (1970). *On Boundary Detection* (Memo 183). Cambridge, MA: M.I.T. Artificial Intelligence Laboratory.

Horn B. (1975). Obtaining shape from shading information. In P. H. Winston (Ed.) *The Psychology of Computer Vision* (pp. 115-155). McGraw-Hill.

Lee, D. (1989). Edge detection, classification and measurement. *Proceedings of the IEEE Conference on Computer Vision and Pattern Recognition* (pp.2-10). San Diego, California: IEEE Computer Society Press.

Marcelja, S. (1980). Mathematical description of the responses of simple cortical cells. *Journal of the Optical Society of America*, **70**(1), 1297-1300.

Marr, D.C. (1976) *Early processing of visual information*, Phil. Trans. R. Coc. London, **B-275**, 483-524

Marr,D. & Hildreth,E. (1980) *Theory of edge detection*, Proc. R. Soc. Lon, **B-207**, 187-217

Perona, P., & Malik, J. (1990). Detecting and localizing edges composed of steps, peaks and roofs. *Proceedings of the 3rd International Conference on Computer Vision* (pp. 52-57). Osaka, Japan: IEEE Computer Society Press

Petrou,M. & Kittler,J. (1991) Optimal edge detectors for ramp edges. *IEEE Transactions on Pattern Analysis and Machine Intelligence*, **13**(5), 483-491

Porat,M. & Zeevi,Y.Y. (1989). Localized texture processing in vision: Analysis and synthesis in the Gaborian Space. *IEEE Transactions on Biomedical Engineering*, **36**(1), 115-129

Shirai Y. (1975). Analyzing intensity arrays using knowledge about scenes. In P. H. Winston (Ed.) *The Psychology of Computer Vison* (pp. 93-113). McGraw-Hill.

Watson.A.B. (1987). Detection and recognition of simple spatial forms. In A. Braddik and A.C. Sleigh (Eds.) *Physical and Biological Processing of Images* (pp. 100-114). Springer-Verlag.

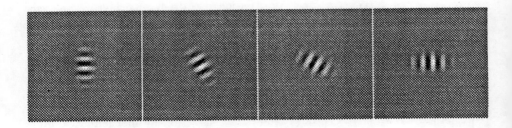

Figure 1: Gabor EF's (sine) when r=1: a) $\theta = 0°$ b) $\theta = 30°$ c) $\theta = 60°$ d) $\theta = 90°$.

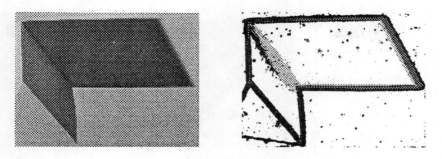

Figure 2: a) *Wedge* b) Classified edges in *wedge*

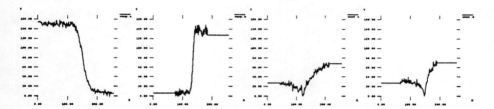

Figure 3: a) and b) Shadow edge intensity profiles, c) and d) Concave edge intensity profiles

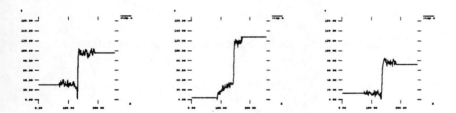

Figure 4: Convex edge intensity profiles

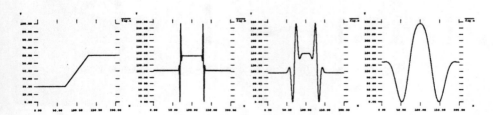

Figure 5: Mach band effect: The ideal ramp in a) is convolved with Gabor filters of increasing spatial extent: b) $\sigma = 2$ c) $\sigma = 8$ d) $\sigma = 32$

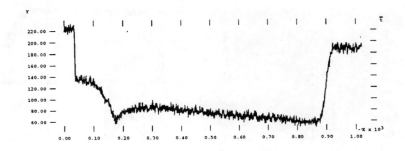

Figure 6: Horizontal cross-section intensity profile of test image

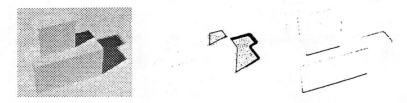

Figure 7: a) *Blocks* b) shadow edges c) concave edges

Figure 8: a) *Coffee cup* b) convex edges c) shadow edges

Computer-based iconic communication

Colin Beardon
Faculty of Art, Design & Humanities, University of Brighton
Brighton, U.K.

Abstract

How can two people who have no common language communicate with each other? Invented languages have not been popular and machine translation is unlikely to be practical. Perhaps the use of drawings and animated gestures is more promising. Computer-based icons, with an operational as well as a representational definition, provide a more powerful medium than do printed icons. Insights from artificial intelligence combined with good graphic design could enhance the ability of people to communicate without a common language. Some initial research in this direction has led to a general purpose communication system based almost entirely on the use of icons, using natural language text only for proper names.

The CD-Icon system is based around a message structure derived from Schank's Conceptual Dependency formalism, with the ultimate referential objects being self-explaining icons. The recipient receives a set of interconnected and animated screens representing the message. Messages are composed through the use of four specially designed screens. As these screens can be accessed recursively, the resulting medium is very powerful. The system is theoretically significant because it provides independent testing of semantic formalisms (such as Conceptual Dependency) that claim to represent the "real" meaning of natural language sentences.

1 Introduction

In *The Hitch-Hiker's Guide to the Galaxy* (Adams, 1979) we are introduced to the Babel fish. This is a remarkable creature that you place in your ear and, no matter what language a person uses when speaking to you, the biological properties of the fish convert that speech into your own native tongue. Alas, the Babel fish is purely fictional, and while the technology of communication has become more widespread and people need to communicate across language barriers, we still lack the device to make this possible.

Machine translation might be one approach but, whatever claims there have been concerning automatic translation of formal texts, the possibility of real-time translation of informal conversation is very remote. Neither have attempts to create international languages, such as Esperanto, been very successful. At the macro-level Esperanto is a clever solution but there is something not quite right about its micro-economics. A lot of time and effort has to be invested in learning the language but it is unwise to do so unless it is known that the person to be communicated with has already made a similar investment. Being unsure of this, Esperanto is not learned. The lesson from this is that practical communication systems must be extremely simple to learn and use, appearing almost intuitive, or else they are doomed.

It is, however, possible to understand other people across language barriers in different ways. There are internationally recognised symbols and icons which one can find in public places. They are typically used to directly denote the place where one might find trains or change money, but they may also contain a number of "meta-icons" such as the red diagonal and the arrow. Much work has been done on iconic languages (Bliss, 1965; Neurath, 1978) but their transformation to the computer has yet to be fully explored. Also it is possible to make oneself understood, at least at some basic level, by means of gestures and mime. People who find themselves in a situation where they know little or nothing of the local language can often communicate simple ideas by pointing or indicating their intentions through actions, which may even border on acting.

The objective of the iconic communication project (Beardon, Dormann, Mealing & Yazdani, 1992; Mealing 1992; Yazdani & Mealing, 1991) is to explore the potential of building computer-based systems that enable cross-language[1] communication by relying upon graphical symbols, avoiding where possible the use of words. This paper discusses the principles behind such systems and describes one particular project, called CD-Icon.

2 Icons

According to the picture theory of meaning (Wittgenstein, 1961), a sentence is ultimately composed of atoms that refer directly to objects. Figure 1 contains various atoms which refer to the object "men" and the way they do this varies from the arbitrary (Figure 1a), through the iconic (Figures 1b - 1d), to the pictorial (Figure 1e).

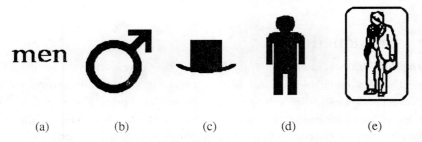

Figure 1 Five different ways of referring to men

Words are essentially arbitrary in the way they refer and there is no alternative but to learn their meaning. With pictures there is little to learn because the relationship to the referent is direct, but it would be quite wrong to see pictures as an ideal method of referring. There are a number of things pictures are not good at expressing, for example "love" or "the class of all mammals". For the purpose of pictorial communication, pictures can also reveal too much information. Figure 1e might stand for "business men" or have a number of other potential interpretations.

Whilst an icon describes its referent, it does not describe it precisely. We can

---

[1] "Cross-language communication" means communication irrespective of the languages spoken by the participants. It is distinguished from "cross-cultural communication" which raises a number of new problems. Whilst we can speculate how cross-cultural issues might be addressed, there has to date been little research undertaken in this area.

summarise the mapping between an icon and its referent by identifying three cases.

1. Image-based, or based on a direct physical correspondence between the image and the object (e.g. Figure 1d). Bliss calls this a *pictogram* which represents directly by means of an outline of the object (Bliss, 1965).

2. Aspect-based, or based on the representation of a particular example, part or property of the object (e.g. Figure 1c). Bliss calls this an *ideogram*, and adds that this may also exploit a metaphor (Bliss, 1965).

3. Convention-based, or based on an arbitrary relationship that is created by a particular or systematic convention (Figure 1b). Bliss further distinguishes between an *abstraction* that is essentially arbitrary, and a *symbol* which implies an accepted cultural meaning (Bliss, 1965).

What advantages do icons give us over words? They can be easier to learn because their visual form suggests a set of possible referents. They can also be easier to remember because, if we cannot simply recall what they stand for, we may recall the process whereby the meaning was constructed from the symbol. By the introduction of conventions (e.g. Figure 1b) both abstract and material objects can be represented by icons, though the price we pay is the need for some prior learning.

Compound icons can be constructed according to the principle of compositionality (Charniak, 1981). That is to say, an icon can be composed of a number of more elementary icons in a configuration, and the meaning of the compound is a function of the meaning of the elements and the configuration. There are examples of this in Chinese characters, for example the use of the Chinese character for "tree" to represent "a single tree" (once), "a group of trees" (twice) and "thick forestation" (three times). Other examples are the use of the red diagonal to represent negation or prohibition when used in conjunction with another icon, or the superimposition of a national flag onto an icon for a person to represent a citizen of that country.

3 Computer-based iconic communication

In previous iconic languages the meaning of messages depended upon our learning the conventions and interpreting correctly the pictorial images. In the language Bliss, messages are read as a sequence of icons with sub- and super-scripts (Bliss, 1965), while in Isotype messages are two-dimensional "texts" in which a whole page may be interpreted simultaneously (Neurath, 1978). Such two dimensionality can be exploited to express certain elements of a message by the relative position of icons rather than by the use of a separate icon.

A computer with a video monitor provides a significant new environment for the use of icons. The computer image can vary over time, meaning that the temporal dimension can also be exploited to convey meaning, through animation for example. Interactivity can also be exploited, requiring an icon in a computer environment to be defined both representationally and operationally. In addition to asking what an iconic picture stands for, one can also ask what happens when the user performs various operations on the icon. For example, one operation may be the command, "Explain yourself!", which could give rise to any multimedia representation of what the icon stands for.

From a philosophical standpoint, much work on iconic languages has been from a logical positivist perspective. Otto Neurath saw Isotype as a practical manifestation of logical positivism and was associated with the Vienna Circle which was strongly influenced by Wittgenstein's *Tractatus* (Wittgenstein, 1961). By adding new dimensions to meaning, through extending the concept of use, we are providing a new orientation for iconic languages much closer to the later period of Wittgenstein's thought (Wittgenstein, 1953).

There are many directions in which research into iconic communication could take place. It would be possible, for example, to develop purely iconic interfaces to existing software. The task of such an interface would be to translate iconic messages into statements in a formal language that the existing software understands. It would also be possible to attempt to develop systems that translate between iconic messages and corresponding sentences in natural language or vice versa. Whilst not being against such developments, this research follows neither of these directions. It is concerned with direct person-to-person communication through an iconic medium with no linguistic reference point. We do not see success as being the ability to communicate what is already expressible in natural language, but rather wish to explore a new medium to see what it tells us about the possibility of non-linguistic communication. Users may therefore have to learn how to think and express themselves in terms of the iconic system, rather than in terms of language.

The particular project that is the subject of the remainder of the paper is concerned with a far more flexible view of meaning than that adopted by the logical positivists and this is reflected in the major principles that guide it. These are,

1. The objective of the research is to see to what extent people can communicate (i.e. make themselves understood to others) using iconic systems.

2. The system should enable users to understand messages created by other users irrespective of the languages they know.

3. Natural language text is to be used only for proper names.

4. New users will require some guidance on how to use the system, but this should be brief (no more than 10 minutes) and the details should be self-taught.

4 CD-Icon: an experimental system

CD-Icon is an iconic communication system that has been developed at the University of Brighton. Messages composed within CD-Icon have a structure that is based upon

Schank's Conceptual Dependency (CD) representation. Demonstration models have been developed using HyperCard II on Apple Macintosh equipment.

There is a sense in which CD-Icon has turned natural language processing on its head. Computational linguists have sought to devise canonical representations that have the expressive power of natural language, but avoid its ambiguity or diversity of forms. In Chomsky's case, the canonical representation was syntactic ("kernel sentences" or "deep structure"), for Winograd it was procedures in Planner, for Schank it was conceptual dependency diagrams, for Montague it was IL, while others have used first-order logic, frame notations, messages, and many more besides. Within such approaches, the meaning of a natural language sentence is defined in terms of a canonical system, which Woods called a Meaning Representation Language (MRL) (Woods, 1978). It is this canonical representation that, linguists claim, clearly reflects the true meaning of a sentence and it is by means of this representation that we can extract some benefit from the processing of natural language. An MRL, for example, may provide direct access to a database or similar system.

The intention of the work reported here is quite different. It uses the MRL as the surface structure of a new language in which the atoms are not words but icons. Furthermore, we aim to exploit the full power of multimedia computing and knowledge of graphic design to make the system as easy to use and as self-explanatory as possible. From this perspective we can provide new insights into particular MRLs. If we can develop an interface for these formal notations that is graphical and self-explanatory then their direct use may become attractive. We choose to use an underlying canonical representation that is very close to Schank's conceptual dependency (Schank, 1973). This was chosen because it is inherently diagrammatic and it proved to be of some practical value in a number of projects (e.g. Schank, 1975; Wilensky, 1981). This does not mean that we are not critical of this notation, but prefer any weaknesses to show themselves in the endeavour of getting people to use it directly for communication.

4.1 Schank's conceptual dependency

Schank's conceptual dependency claims to be a well-defined notation for representing the meaning of sentences in "an unambiguous language-free manner" (Schank, 1973). The description of CD diagrams involves four basic types of construct:

1. a message - a simple conceptualisation, or a compound conceptualisation containing a conceptual relation (e.g. time, place, causality, etc.);

2. a conceptualisation - typically based around one of fourteen primitive ACTs and their related cases, but may also describe a state or change of state;

3. a picture - based around a Picture Producer and various modifiers which are Picture Aiders or other Picture Producers with specified relations (e.g. location or poss-by);

4. a lexical entry - an atomic symbol for a Picture Producer or Picture Aider.

Lexical entries are, in CD-diagrams, simply words in natural language. The lexicon is a list of such words with sufficient semantic information necessary to implement various semantic constraints (e.g. MOVE requires an animate agent). In addition, there a number of "conceptual tenses" that can be applied to messages and conceptualisations to indicate tense, negation, etc.

4.2 Understanding messages in CD-Icon

CD-Icon has been constructed as four modules, each processing one type of construct, with clearly defined interactions. We will demonstrate the system by first showing what the recipient of a message will see. The nearest English equivalent of the message we will describe is,

"*A girl called Jane will go to the Devon countryside by train.*"

The CD form of the message is shown in Figure 2.

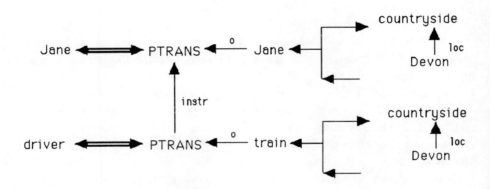

Figure 2   CD representation of the sample message

This will be presented as a message screen indicating the main ACT is a PTRANS, two conceptualisation screens representing the two simple conceptualisations, and three picture screens representing Jane, the train and the Devon countryside.

The first thing that the recipient sees is the message screen shown in Figure 3 which indicates that the message contains a single conceptualisation. The icon suggests that this is a PTRANS but identification is not crucial at this stage.

Clicking on the PTRANS icon in Figure 3 will reveal the animated event shown in Figure 4. The icon representing a small girl will move slowly across the screen from left to right.

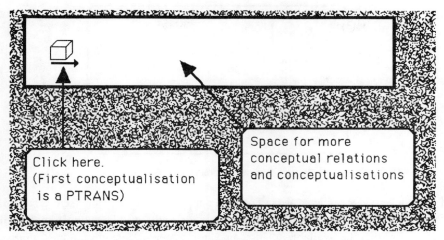

Figure 3 Message screen for recipient (with annotations)

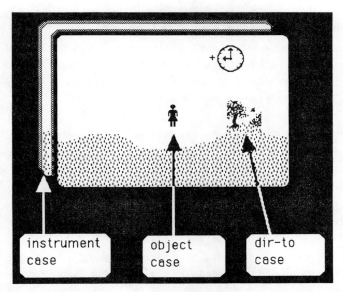

Figure 4 Conceptualisation screen (PTRANS) for recipient (with annotations)

When the animation stops the icons can be accessed. Clicking on most of the cases will result in a picture screen (as in Figure 5), but the instrument case is represented in CD not by a picture but by another conceptualisation. It is shown as another event stacked behind the current one and clicking on it will reveal another animated event screen with its own case slots.

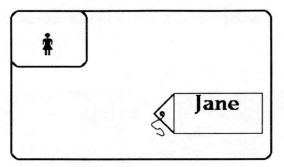

Figure 5 Picture screen for recipient

If one clicked on the object case in Figure 4, one would see the picture screen shown in Figure 5. Clicking on the icon for a small girl will reveal whatever multimedia explanation may be available. This could, for example, contain pictures, animated diagrams, sounds, or even video. The same would apply if one clicked on the clock-based icon in Figure 4 (which represents the future tense).

4.3 Generating messages in CD-Icon

The creator of the message is confronted with a slightly different set of screens, for these are presented as a structured set of the options available, in a manner similar to systemic grammar. Because of this need to represent options, as well as instances, this part of CD-Icon has three conventions which have to be learned.

1. There are two types of icon. Grey icons can be thought of as representing types or classes of objects, such as *physical objects, places* or *sizes*. Wherever a grey icon is seen, the act of clicking on it will produce a set of options from which a selection can be made. Black and white icons represent specific objects, such as *dog* or the size *large*. When a black and white icon is selected then it will replace the grey icon which led to it being offered.

2. When a picture or a conceptualisation has been completed it can be referred to from elsewhere by means of a single icon. All black and white icons effectively have more information associated with them which can be obtained by clicking on the icon.

3. Clicking on any icon with the Option key held down is the system convention for "Help". If it is a black and white icon then it will result in revealing its full description. Otherwise it will result in a short animated explanation in the form of "self-explaining icons" (Mealing, 1992).

With these three conventions in mind we can see how CD-Icon allows the composition of a message in four stages: the message, the conceptualisation, the picture and the lexicon.

The system is entered at the message level, with each level calling the level below until the lexicon is reached at which point icons begin to be passed back up the various levels.

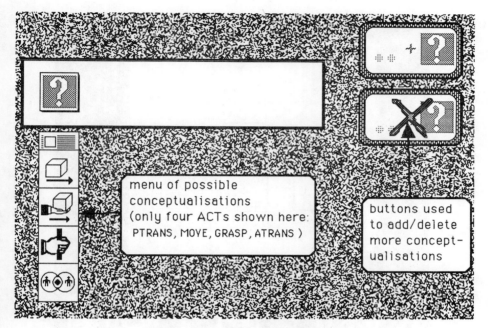

Figure 6 Message screen for sender (with annotations)

At the topmost level the user must decide whether the message consists of a single conceptualisation or is compound. Clicking on a conceptualisation icon will result in a menu of possible ACTs being offered, as in Figure 6. If the message is compound, the relationship between the two component conceptualisations is indicated by a grey plus-sign representing the class of conceptual relations (which can be logical, temporal or spatial). These options will appear as a menu if the grey plus-sign is clicked.

Conceptualisations typically consist of a primitive action (an ACT) and a number of defined case slots. (This is not always true as conceptualisations can be purely descriptive and consist of a picture and a Picture Aider or two pictures.) The fourteen ACTs listed by Schank will be presented as a menu and the user is invited to make a selection. When the user selects one of these control will pass to the appropriate screen; in our example we require the PTRANS option which is shown in Figure 7.

The PTRANS screen contains a basic background with grey icons representing the object, origin and destination cases. The stacked screen indicates the instrument case and we offer the conceptual tenses of time, negation and assertion, interrogation or imperative. (We are experimenting by not using the agent case on the grounds that PTRANS can be seen as purely indicating a change of location.) The grey house represents the class of places, the grey question mark represents the class of objects and the grey clock represents the class of times. By clicking on the icons for the origin, object or destination slots control is passed to the picture screen and a single icon will be returned.

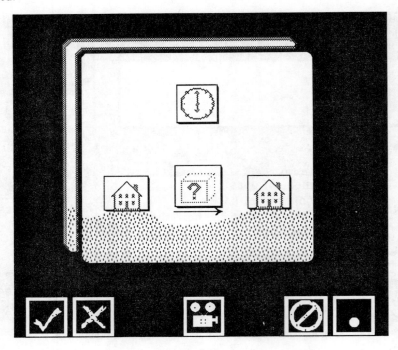

Figure 7 Conceptualisation (PTRANS) screen for sender

The instrument case is handled differently because it always points to a conceptualisation. This means that control must pass recursively to the conceptualisation screen, and not to the picture screen. As there is never more than one instrument case for any conceptualisation, the stacked screen form of representation is possible.

Pictures are made up of a Picture Producer, which we call the 'head', and various modifiers (Picture Aiders and Picture Producers) (see Figure 8). Pictures originate from conceptualisations or other pictures and the head must be selected from the lexicon before the picture screen is entered. When the head icon is chosen any preferred modifier types are added (as they are types they will be in grey). The user can click on any of these modifier icons to be taken to the appropriate page of the lexicon to select a specific modifier (colour, size, location, etc.) These are not strictly semantic constraints but merely recommendations, as users are permitted to browse through the lexicon and select any icon they wish to use as a modifier.

The lexicon contains icons that denote both "Picture Producers" and "Picture Aiders". It appears as a series of pages, with each page representing a type (Figure 9). The icon for the type appears in the top right corner of the page, with the centre of the page containing both type and specific icons. At the bottom of the page there are arrows to

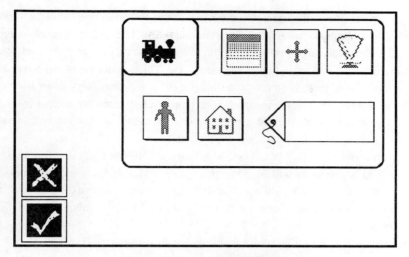

Figure 8 Picture screen for sender

allow browsing through the lexicon, a "Picture" icon to allow reference to an existing picture and the query icon for the formation of questions.

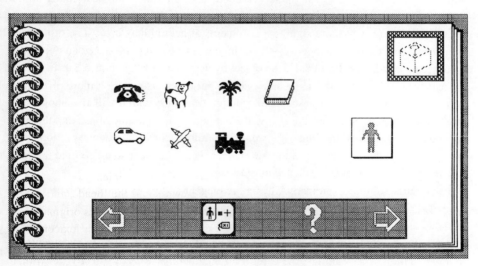

Figure 9 Lexicon (available only to sender)

5 Testing

Testing is taking place in three stages. The first stage aims to see whether iconic messages can be understood by users after a minimal introduction to the system. A simple instruction sheet is issued and participants are given a disk which contains eight iconic messages, varying from quite simple to complex and involving questions, instrument cases, different tenses, and compound messages. For each message the participant writes down the nearest English equivalent and an indication of how easy or difficult they found it to understand. A preliminary round of testing is nearly complete and all users to date have been able to operate the system satisfactorily and reach a reasonable level of understanding of the messages. Three major problems have shown up. The handling of tense is far too crude and not as "self-explanatory" as we had hoped; some users tend to guess at the meaning of an icon and will often not look at its picture; and users are unable to distinguish between references to the same object and references to similar objects.

Stage two of testing will be concerned with the abilities of users to create their own messages and participants will be given the system and asked to create six messages, writing down the nearest English equivalent of their message. Two experienced users of the system will read the messages and record their own English equivalents and the two will be compared.

Stage three of testing will involve iconic dialogues between two users.

6 Future plans

The first two stages of testing are concerned with iconic messages which are the equivalent of the natural language sentence, and they will suggest many minor amendments that will improve the usability of the system. The third stage will require us to address a number of more significant problems of natural language. The problem of anaphora is one example (Singer, 1990). Using text and having introduced an object, say a house, we can later refer to the same object by using "it", or "the house", or in several other ways. This makes language more varied, but it can also make it more difficult to comprehend. In a computer-based iconic communication system it will also happen that on several different occasions the user will want to refer to the same object and there are two possible solutions: the linguistic solution to allow icons for pronouns (adopted by Bliss), and the iconic solution to allow several instances of an icon to all point to the same full description of the object (adopted in CD-Icon).

The latter solution is in many ways more elegant and gives computer-based iconic systems a distinct advantage over natural language, but there is a price to pay. Firstly, as we have seen, users need the means to determine whether the current object is the same as one they have seen before. Secondly, linguistic text is not just a series of statements but involves, among many things, the careful arrangement of what is already known and what is new. There is a potential problem in these systems that it will be impossible to distinguish between "the white house" and "the house is white". This problem can be avoided in iconic systems, but only by the explicit introduction of focus, perhaps through the temporal sequencing of information with the initial screen showing what is given and the later screen adding what is new.

Two pronouns that no iconic system could avoid are those for "me" and "you". These can be defined within the context of a message by means of an iconic envelope saying who is sending the message to whom. The sender and recipient will be associated with icons which can then be used throughout the message. Furthermore, we are exploring the extent to which we can graphically indicate the type of speech act that is being offered. Hence the envelope might show that, say, Mary is sending a story to Brian, or Jill is asking a question of Jane.

A final requirement of a dialogue system will be the ability to refer to the text itself. This is particularly necessary if a message is not understood, where there needs to be some way to request clarification or recomposition of the message.

7 Conclusion

A system has been described which makes it technically possible for people to communicate quite complicated messages without the use of words. Initial testing suggests that single-sentence messages can be understood by recipients and we are confident that a workable system for the transmission of such messages could be made to work. Though CD-Icon has been based upon the CD formalism, the experience of users questions the necessity of the Agent case in PTRANS, and the handling of tense in this notation.

The current stage of development suggests that it is technically feasible to have a non-textual communication medium with surprisingly few conventions. The system is expressively very powerful, due its recursive facilities. The next major development will be to embed the system within a system for handling dialogue. If such a system can be developed then the feasibility of using it as an interpersonal communication system can be assessed.

In the long term, a range of practical applications could follow from a successful implementation and the possibilities of language independent instruction manuals, aids to language learning and translation, and writing devices for the severely disabled have been proposed. Our present concern, however, is to develop a sound theoretical approach and we find inspiration for this in the active combination of graphic design and artificial intelligence.

References

Adams, D. (1979). *The Hitch-Hiker's Guide to the Galaxy*. London: Pan Books.

Beardon, C., Dormann, C., Mealing, S., & Yazdani, M. (1992). Talking with pictures: exploring the possibilities of iconic communication. *AISB Quarterly*, forthcoming.

Bliss, C.K. (1965). *Semantography*. Australia: Semantography publications.

Charniak, E. (1981). The case-slot identity theory. *Cognitive Science*, 5, 285-292.

Mealing, S. (1992). Talking pictures. *Intelligent Tutoring Media*, 2 (2), 63-69.

Neurath, O. (1978). *ISOTYPE: International picture language*. Reading, UK: University of Reading, Department of Typography and Graphic Communication.

Schank, R.C. (1973). Identification of conceptualisations underlying natural language. In R.C.Schank & M.C.Colby (Eds.) *Computer models of thought and language* (pp.187-247). San Francisco, CA: W H Freeman.

Schank, R.C. (1975). *Conceptual information processing*. New York, NY: North-Holland.

Singer, R.A. (1990). *Human-computer graphical dialogue* (CITE Report No 104). Milton Keynes, UK: Open University, Institute of Educational Technology.

Wilensky, R. (1981). PAM. In: R.Schank & C.Reisbeck (Eds.) *Inside computer understanding* (pp.136-179). Hillsdale, NJ: Lawrence Erlbaum.

Wittgenstein, L. (1961). *Tractatus logico-philosophicus*. Trans. D.F.Pears & B.F.McGuiness. London, UK: Routledge & Kegan Paul.

Wittgenstein, L. (1953). *Philosophical Investigations*. Oxford, UK: Blackwell.

Woods, W. (1978). Semantics and Quantification in Natural Language Question Answering. In M.Yovits (Ed.) *Advances in Computers, Vol 17* (pp.2-64). New York, NY: Academic Press.

Yazdani, M. & Mealing, S. (1991) A computer-based iconic language. *Intelligent Tutoring Media*, **1** (3), 133-36.

Toward Non-Algorithmic AI

Selmer Bringsjord
Rensselaer Polytechnic Institute
Troy, USA

Abstract

In his *The Emperor's New Mind* Roger Penrose propounds and defends the thesis that human cognition is beyond the algorithmic. I tackle herein this question concerning the Penrosian position:

QUEST Since "quantum computing," when rigorously unpacked, doesn't provide a formal foundation for the Penrosian position, and since the same is true of the connectionist's analog "neural computing," is there perhaps reason to think the Penrosian position simply cannot be founded upon formal results (in which case many AIniks would dismiss the position out of hand), or is there something *else* which might provide a mathematical underpinning for the claim that human thought is uncomputable?

1 Introduction

In his *The Emperor's New Mind* Roger Penrose (1991) propounds and defends the thesis, put roughly for now, that human cognition is beyond the algorithmic.[1] Others, e.g., Hofstadter (1979), Kugel (1986), and Smolensky (1988) — the last an orthodox, arch-connectionist —, have advanced and affirmed essentially the same thesis,[2] and there's little doubt many *outside* the Artificial Intelligentsia find the thesis tempting.[3] What rationale does Penrose offer in support of this view, which certainly at least *seems* inimical to AI and Cognitive Science? Penrose's main argument for the thesis in question is an enthymematic resurrection of the old Lucas-made (1964) claim that Gödel's incompleteness results imply, in conjunction with "facts" about the human assimilation of these results, that humans can do things machines in principle can't. I'm not interested herein in evaluating such arguments.[4] Nor am I interested in evaluating "mysterian"[5] claims that the brain's "causal powers" produce non-algorithmic cognition of a sort forever beyond the reach of "fleshless"

---

[1] See esp. pages 416-422 of (Penrose, 1991).
[2] Though sometimes for non-Penrosian reasons.
[3] Many creative writers, for example, firmly believe that what they do is *in principle* beyond the capacity of a computer. This is an attitude I've faced as co-director of *Autopoeisis*, a research project in story generation which is discussed below.
[4] For such a detailed evaluation, see "Chapter VII: Gödel" of (Bringsjord, 1992).
[5] The term "mysterian" may be Daniel Dennett's. See "The Brain," in *Newsweek*, April 20, 1992.

computers (Searle, 1980-a, 1980-b, 1982, 1983, 1984, 1990).[6] Rather, I'm interested in tackling this question concerning the Penrosian position:

> QUEST Since "quantum computing," when rigorously unpacked, doesn't provide a formal foundation for the Penrosian position, and since the same is true of the connectionist's analog "neural computing," is there perhaps reason to think the Penrosian position simply cannot be founded upon formal results (in which case many AIniks would dismiss the position out of hand), or is there something *else* which might provide a mathematical underpinning for the claim that human thought is uncomputable?

My plan is as follows. Section 2 is a review of the rudiments of recursion theory needed to state the Penrosian position and to mathematize both standard and "extravagant" computing machines. In section 3, after briefly explaining why neither quantum nor neural computing gives Penrose & Co. what they need, I proceed to tackle QUEST. To tip my hand a bit, my response to QUEST, in a nutshell, will be: "Yes, there are indeed machines out there which, unlike quantum and neural ones, could formally underpin the Penrosian thesis — but since (the simplest of) these machines ('Zeus' machines) apparently cannot be sensibly identified with human brains (or parts thereof, or even with anything physical), they don't do the job, unless Penrose is willing to wed some form of dualism to his claim that human thought is at least in part non-algorithmic. Besides, ZMs are not particularly fertile, mathematically speaking. On the other hand, other machines more powerful than TMs, corresponding to Π_1-and-beyond points in the Arithmetic Hierarchy of recursion theory (if the Arithmetic Hierarchy is new to you, relax and bear with me: it's explained below), provide not only a glimmer of hope for rendering the Penrosian thesis formally respectable, but perhaps reveal, *via* plausible supporting deductive arguments that center around story generation and recognition, the nature of human cognition, and hence the heart of (non-algorithmic) AI."

2 Background

I assume, on the part of the reader, familiarity with standard elementary recursion theory — naïve set theory, string theory, the standard automata hierarchy [from finite automata to Turing machines (TMs)], simulation proofs [which, e.g., equate k-tape TMs with cellular automata (CAs)], and logical systems primarily of the first-order, extensional type.[7] The architecture of ZMs and machines corresponding to Π_1-and-beyond points in the Arithmetic Hierarchy will for us be the architecture of a standard TM. I'll assume that we also have in hand some standard formalization of the

---

[6]For such a detailed evaluation, see "Chapter V: Searle" of (Bringsjord, 1992).

[7] Readers wanting a detailed look at elementary recursion theory can turn to a quartet of books I have used in teaching logic and computability: (Lewis & Papadimitriou, 1981), (Ebbinghaus, Flum et al., 1984), (Boolos & Jeffrey, 1980), and (Hopcroft & Ullman, 1979). For a comprehensive , mature discussion of such matters that includes succinct coverage of *un*computability, including the Arithmetic Hierarchy, which figures centrally in my answer to QUEST, see (Davis & Weyuker, 1983), (Soare, 1980).

notion of a **program** for a TM — but I call upon a generalized version of the Church-Turing Thesis[8] to sanction the description of these programs in simple, austere English. We need the generic concept of a set of instructions which is broader than that of a program, since we need to give instructions to machines beyond TMs: let's call a set of instructions for a machine of *any* type a **procedure**. (For cognoscenti: procedures are programs augmented by way of things like oracles.) We then reserve 'program' for procedures which drive TMs in the normal (Σ_0 and Σ_1) sense. [Again, if the Arithmetic Hierarchy is new for you, simply ignore, for now, references (such as Σ_0) to it.]

Though what TMs can *do* defines the class of computable functions (the μ-recursive functions), for this paper it's more important to know what they *can't* do. Two functions TMs can't compute are the **busy beaver function** (f_{bb}) and the **full halting problem**. The first function is $f_{bb} : \mathbb{N} \to \mathbb{N}$ (i.e., a function from the natural numbers to the natural numbers) defined as follows. For every $n \in \mathbb{N}$, $f_{bb}(n)$ is the largest number p such that, for some Turing machine $\mathcal{M}$ with alphabet $\{1, \#\}$ (or $\{0, 1\}$, etc.) and with exactly n states, $\mathcal{M}$ goes from a starting state s in which $\mathcal{M}$'s tape is blank, to a halting state h in which there are p 1s on $\mathcal{M}$'s tape. Informally put, the busy beaver function takes in a natural number n, and outputs the number of 1s which is the maximum an n-state TM can write on its tape. Now, what about the full halting problem? We write $\mathcal{M}_P : u \to \infty$ to indicate that machine, or automaton, $\mathcal{M}$ goes from input u through a computation, directed by procedure P, that never halts. We write $\mathcal{M}_P : u \to$ halt when machine $\mathcal{M}$, directed by P, goes from input u to a computation that does halt. We can also harmlessly suppress mention of inputs and talk only of machines halting or not halting *simpliciter*. Furthermore, let the traditional property of decidability be handled by way of the symbols **Y** ("yes") and **N** ("no"). Finally, let $n^{\mathcal{M},P}$ be the gödel number of pair machine $\mathcal{M}$ and program P. The full halting problem is classically unsolvable, i.e., when, again, the machines we're talking about are Turing machines, and P and P^* are programs, there is no machine-program pair $\mathcal{M}, P$ such that for every pair $\mathcal{M}^*, P^*$:

$$\mathcal{M}_P : n^{\mathcal{M}^*,P^*} \to \mathbf{Y} \text{ iff } \mathcal{M}^*_{P^*} \text{ halts}$$
$$\mathcal{M}_P : n^{\mathcal{M}^*,P^*} \to \mathbf{N} \text{ iff } \mathcal{M}^*_{P^*} \text{ doesn't halt}$$

---

[8]Two recent, interesting AI-and-Cog-Sci-relevant papers on CTT are (Nelson, 1987) and (Mendelson, 1990). CTT, for our purposes, says "a function f is effectively computable if and only if f is Turing-computable." To see the indispensability of CTT for sanctioning the informal and enormously time-saving technique of expressing programs in simple English, see either (Davis & Weyuker, 1983) or (Ebbinghaus, Flum et al., 1984). Those skeptical that CTT plays a central role in AI and Cognitive Science need to keep in mind, among other things, that the thesis is presupposed whenever a formal result about what (say) Turing machines can't do (such as solve the full halting problem) is taken to imply that there is a corresponding something which no humanly created computer system (whether AIish or not) can do.

3 The Quest

Let's spend a moment disposing of processing — of the quantum and "neural" sort — which *can't* undergird the Penrosian position. [We'll truly spend but a *moment* on this processing, since processing that is at least a contender for formally grounding the Penrosian thesis is our focus in this paper. For details on quantum computing in the context of foundations of AI and Cognitive Science, see (Bringsjord & Zenzen, 1992). For details on neural computing, connectionism, and AI, see (Bringsjord 1991). In what immediately follows, I draw upon these two papers.]

Pagels (1989), pp. 306-7, claims to have effortlessly defined 'quantum computer,' and at first glance it appears that he succeeds.[9] Unfortunately, upon closer inspection, his account proves to be maddeningly inchoate: it's ambiguous between at least three different construals, each of which is itself reducible to an ordinary TM. Each construal corresponds to a different choice about how to delineate, in Pagels' prose account (see footnote 9), the boundaries of that entity which is to qualify as a quantum computer. And no matter how the boundaries are marked off, the result is just an ordinary TM. At most, Pagels specifies how TMs could play a trivial role in the sort of *calculation* that is at the heart of quantum mechanics.

Deutsch (1985), on the other hand, does rigorously describe a quantum computer, which we'll call 'Q^*.' [Specification and illustration of Q^*, due to space constraints, cannot be included here. Again, for a detailed discussion, see (Bringsjord & Zenzen, 1992).] Deutsch's description requires that one affirm the Everett ontology and interpretation of quantum mechanics, which is somewhat controversial, but the real problem is that since the class of functions computable by Q^* is precisely equivalent to the μ-recursive functions, those computable by TMs, it's hard to see how Q^* can provide any "mathematical meat" for the Penrosian position. Perhaps the rebuttal will come back that there are nonetheless things Q^* can do which Turing Machines can't, and that it's by virtue of *these* things that Q^* undergirds the Penrosian thesis. As Deutsch (1985) shows, there *are* things Q^* can do which TMs can't. For example, Q^* — to put it roughly here — can generate "true random numbers," while TMs can only "fake it." The problem is that it's exceedingly hard to see how one could believe, with any confidence, that

---

[9] He says: "[Creating a quantum computer] is easily done by using the long-range correlations among quantum states. One imagines a source of photons (light quanta) that emits pairs of spin-correlated photons, each one of the pair going in opposite directions. These pairs of photons are then detected at two distinct stations after going through some polarizers. The pattern of detected photons — whether they get through a polarizer or not — can be represented as a random sequence of 0's and 1's (0 = no detection, 1 = detection). The two sequences of 0's and 1's at each station, where each is random, are correlated in a nonrandom way. Furthermore, that correlation cannot be accounted for by classical laws of physics; it is intrinsically quantum mechanical. These random sequences can be used as partial input for two ordinary Turing machines at each station, and these machines use the inputs for some computation. Now the whole apparatus consisting of both Turing machines can be viewed as a single computer. This computer is not reducible to a universal Turing machine for the simple reason that the correlations of quantum mechanics cannot be accounted for by classical mechanics." (Pagels, 1989, pp. 306-307)

true random number generation offers a model of that cognition alleged to be non-algorithmic. In fact, it's hard to find *any* intimate, principled connection between true (as opposed to pseudo) random number generation and human cognition — so Penrose and those of like mind are hardly likely to be heartened by Q^*.

If Q^* and its relatives don't provide a formal foundation for the Penrosian position, what does? It might be said that connectionist systems promise to be, or perhaps point the way toward, such a foundation. Smolensky (1988), a prominent connectionist who has spent much time trying to distinguish connectionist systems from traditional, symbolicist systems (which, at bottom, are Turing machines in action), tells us

> I believe that ... there is a reasonable chance that connectionist models will lead to the development of new somewhat-general-purpose self-programming, massively parallel analog computers, and a new theory of analog parallel computation: They may possibly even challenge the strong construal of Church's Thesis [= our CTT] as the claim that the class of well-defined computations is exhausted by those of Turing machines. (Smolensky 1988, p.3)

Unfortunately, Smolensky's views constitute a leap of faith: we have yet to see a shred of evidence against CTT, and we continue to build up overwhelming evidence *for* the thesis. Moreover, there is every formal and philosophical reason to think that connectionist and symbolicist systems are mathematically equivalent (Bringsjord 1991). A large part of the support for this equivalence is this instance of hypothetical syllogism: neural networks are equivalent to CAs, which are equivalent to k-tape TMs, which are equivalent to ordinary TMs, which are equivalent to logics (an alphabet, a grammar for generating well-formed formulas, and a proof theory); hence neural networks are, in a real sense, nothing more than what symbolicist AIniks (like John McCarthy) have been developing and refining since the late 50's. It's worth noting, in this regard, that CAs are widely used by the Artificial *Life* community — a community which, like connectionist AI, insists on fidelity between computational structures and actual *physical* organisms and organism-parts (Langton, 1989, 1992; Bringsjord, 1993).

Where does this leave QUEST? Should it be answered in the negative? I believe such an answer would be premature. There is, after all, an entire, mature (but difficult, even intimidating) branch of theoretical computer science devoted to uncomputability (= unsolvability = non-algorithmicity), and it's known that that there *are* machines which "compute" functions Turing machines can't. The simplest machines having this property, as far I know, are "Zeus" machines (ZMs). ZMs have the architecture of TMs, but (in rather naïve and free-wheeling fashion) are allowed to work ever faster and faster during some "computation." Consider the sets $\mathbb{N}$ and $\mathbb{R}$ (reals). Suppose that $Ex\Phi$ iff x enumerates set Φ over a finite interval $[t_i, t_k]$; and that Zx iff x is a ZM. Then we know (Boolos & Jeffrey, 1980) that (where '$\diamond$' is 'logically possibly') $\diamond \exists x [Zx \wedge Ex\mathbb{N}]$. How can this be? A ZM $\mathscr{Z}$ could print out 0 in .5 seconds, 1 in .25 seconds, 2 in .125 seconds, ..., until, after one second, all of $\mathbb{N}$ has been printed out! Lest is be thought that a ZM can do anything, it should be noted that (Boolos & Jeffrey, 1980, pp. 14, 15, 16, 40, 122) $\neg \diamond \exists x [Zx$

∧ ɛxR]. What do these results imply, exactly? Apparently that there is a *well-defined* grade of power automata can have, having nothing at all to do with quantum mechanics (but, it must be confessed, perhaps having to do with something even more bizarre than the "microworld" of quantum mechanics!), which puts them beyond TMs. ZMs can, for example, "compute" the busy beaver function (Bringsjord 1992, chap. VIII).

Unfortunately, ZMs don't provide much of a formal foundation for the Penrosian position — unless one is prepared to say either that (i) human thought takes place in non-physical entity(ies) (since *physical* ZMs are, we can all agree, rather hard to identify with brains, which, with respect to space and time, are assumed to be "normal" physical objects), or (ii) the human brain capitalizes on some mysterious quirks which it enable it to do an infinite amount of work in a finite amount of time. I think that taking the Penrosian position seriously will inevitably involve clarification and serious consideration of (i) and (ii) — but that this work ought to be postponed for the moment, and in fact for the duration of this paper. Why? Because, in a nutshell, ZMs aren't worth it: they aren't all that fertile, mathematically speaking: whatever else their virtues, they don't lead to the fine-grained analysis that gives rise to the Arithmetic Hierarchy, and, more generally, degree theory. (One quick example: ZMs blur the distinction between the *full* halting problem and the version of this problem coinciding with Σ_1, in which a solution consists solely in an answer of **Y** if and only if the machine-program pair under scrutiny halts.)

4   The Right Stuff?

It's time to get serious about making formal sense of the Penrosian position, by appealing to that part of recursion theory which, I daresay, most in AI and Cognitive Science, let alone most outside-the-community detractors (like Penrose), don't know about, or don't understand. In order to make the following discussion tractable, I'm going to focus on just one speck in the Arithmetic Hierarchy (AH), the first uncomputable point in it (Π_1); and with respect to that one speck I will build up to and end with what I think is a formidable *deductive* argument for the proposition that human story recognition is "characteristic" of Π_1. This deductive argument is related to a compressed, off-the-cuff *abductive* argument given by Kugel (1986), as we shall see.

I begin with a *very* brief introduction to the AH in a purely "syntactic" manner. (This introduction is in no way a substitute for any of the accomplished introductions to AH; see note 7.) Suppose we have some **totally computable** predicate $S(P, u, n)$ iff $\mathcal{M}_P : u \to$ halt in exactly n steps. (Remember that our machines, architecturally speaking, are always simply TMs.) Predicate S is totally computable in the sense that, given some triple (P, u, n), there is some program P^* which, running on some TM $\mathcal{M}^*$, can give us a verdict, **Y** or **N**, for whether or not S is true of this triple. (P^* could simply instruct $\mathcal{M}^*$ to simulate $\mathcal{M}$ for n steps and see what happens.) This implies that $S \in \Sigma_0$, the starting point in AH, and composed of **totally computable predicates**. But now consider the predicate H, defined by

$$H(P, i) \text{ iff } \exists n S(P, i, n).$$

Since the ability to determine, for a pair (P, i), whether or not H is true of it, is equivalent to solving the full halting problem, we know that H is not totally computable. Hence $H \notin \Sigma_0$. However, there is a program which, when asked whether or not some TM $\mathcal{M}$ run by P on u halts, will produce **Y** iff $\mathcal{M}_P : u \to$ halt. For this reason H is **partially computable**, and in Σ_1. To generalize, informally, the syntactic representation of the landscape here is:

Σ_n set of all predicates definable in terms of totally computable predicates using at most n quantifiers, the first of which is *existential*

Π_n set of all predicates definable in terms of totally computable predicates using at most n quantifiers, the first of which is *universal*

Δ_n $\Sigma_n \cap \Pi_n$

We have, based on this scheme, the Arithmetic *Hierarchy* because, where $\subset$ is proper subset,

$$\Sigma_0 \subset \Sigma_1 \subset \Sigma_2 \dots$$
$$\Pi_0 \subset \Pi_1 \subset \Pi_2 \dots$$
$$\text{for every } m > 0, \Sigma_m \neq \Pi_m$$
$$\Pi_m \subset \Sigma_{m+1}$$
$$\Sigma_m \subset \Pi_{m+1}$$

It's possible to devise a more procedural view of (at the least the lower end of) AH. Σ_0 and Σ_1 have already been viewed procedurally. How, then, could Π_1, the first genuinely uncomputable stop in AH, be viewed procedurally? Kugel (1986) has aptly called Π_1 procedures **non-halting procedures**; here's how they essentially work. Let R be a totally computable predicate (a crucial assumption for the following); then there is some program P which decides R. Now consider a corresponding predicate $G \in \Pi_1$, viz.,

$$G(x) \text{ iff } \forall y R(x, y)$$

Here's a non-halting procedure P^+ [*not* a program; and note, also, that we count P^+'s last output (if there is one) as its result] for solving G, in the sense that a **Y** is the result iff Gx:[10]

- Receive x as input
- Immediately print **Y**
- Compute, by repeatedly calling P, $R(x, 1)$, $R(x, 2)$, ..., looking for a **N**
- If **N** is found, erase **Y** and leave result undefined

Okay, okay, you say, but what the heck does all this have to do with the Penrosian position? Why have we gone through all this trouble? What about AH, and specifically Π_1, provides what quantum computing couldn't?

---

[10] The reader should satisfy herself that the following procedure *does* decide G.

These questions can be answered *via* my deductive argument that human story recognition necessarily corresponds to at least a non-halting procedure. Unfortunately, before articulating the argument, I need to introduce another formal notion, that of a **productive set**:

A set A is **productive** iff
  (i) A is classically undecidable (= no program can decide A);
  (ii) there is a computable function f from the set of all programs to A which, when given a candidate program P, yields an element of A for which P will fail.

Put informally, a set A is productive iff it's not only classically undecidable, but also if any program proposed to decide A can be counter-exampled with some element of A.

Now, Kugel (1986) has suggested that perhaps the human ability to recognize certain inscriptions as the letter 'A' (or certain objects as 'beautiful') makes the set of all letter 'As' (or the set of all beautiful objects) a productive set. (Predicates can be identified with corresponding sets, *viz.*, the set of objects of which the predicate is true.) I say that Kugel has *suggested* this position because he really doesn't give an argument for it. And if his suggestion is read as an enthymeme, then the argument in question is **abductive** in form. More concretely, if we charitably unpack what he's up to, then Kugel is seemingly attracted to arguments like

If cognitive process C involves a productive set, then ϕ.
ϕ.
∴ Cognitive process C involves a productive set.

Such arguments have a rather unimpressive pedigree, at least when stacked against deductive arguments. This is because abductive arguments actually embody a well-known deductive fallacy ("affirming the consequent"), and as such they lose what makes sound deductive arguments compelling, *viz.*, that if such an argument's premises are true, its conclusion must be also, so if a rational agent accepts the premises, that agent must also accept the conclusion. Here, then, is a question: *Are there deductive arguments for the view that non-halting procedures are at work in human cognition?* I think there are; and in fact the least complicated one I know of can be derived from some of Kugel's (1986) vague but suggestive reasoning:

We seem to be able to recognize, as beautiful, pieces of music that we almost certainly could not have composed. There is a theorem about the partially computable sets that says that there is a uniform procedure for turning a procedure for *recognizing* members of such sets into a procedure for *generating* them. Since this procedure is uniform — you can use the same one for all computable sets — it does not depend on any specific information about the set in question. So, if the set of all beautiful things were in Σ_1, we should be able to turn our ability to recognize beautiful things into one for generating them.... This suggests that a person who recognizes the Sistine Chapel Ceiling as beautiful knows enough to paint it. Even though this claim applies to the knowledge involved, and not the painting skills, it still strikes me as somewhat implausible. (Kugel 1986, pp. 147-148)

I used to think Kugel's remarks here were hopelessly vague and hasty, that AH was only mathematically, not "cognitively," significant; but after reflecting on these remarks, and after toiling for a year as a researcher in *Autopoeisis*, an AI project in story generation, I now think there's something to Kugel's views — or at least that what he says can be reconstructed. In order to carry out this reconstruction, I suggest a focus on the ability of humans to recognize those stories which are interesting. The property of "interestingness" is one which all stories generated by machines have hitherto lacked. [E.g., this complaint has been justifiably leveled against Meehan's (1981) story generator, TALE-SPIN.] I have become increasing tempted to regard the set of interesting stories — call it $\mathscr{S}$ — to be a productive set. Why? Recall that productive sets must have two properties; let's take these properties in turn, in the context of $\mathscr{S}$. First, $\mathscr{S}$ must be classically undecidable; i.e., there is no program which answers the question, for an arbitrary story in $\mathscr{S}$, whether or not it's interesting. Second, there must be some computable function f from the set of all programs to $\mathscr{S}$ which, when given as input a program P which purportedly decides $\mathscr{S}$, yields an element of $\mathscr{S}$ for which P fails. It seems to me that $\mathscr{S}$ does have both of these properties — because, in a nutshell, the *Autopoeisis* research group seems to invariably and continuously turn up these two properties "in action." Every time someone suggests an algorithm-sketch P for deciding $\mathscr{S}$, it's easily shot down by a counter-example consisting of a certain story which is clearly interesting despite the absence in it of those conditions P regards to be necessary for interestingness. (It has been suggested that interesting stories must have inter-character conflict, but monodramas can involve only one character. It has been suggested that interesting stories must embody age-old plot structures, but some interesting stories are interesting precisely because they violate such structures ....) But my aim is not merely to share intuitions which have sprouted in the process of team-building a story generator. My aim is to move to a deductive argument, reminiscent of the Kugel quote above, for the proposition that $\mathscr{S} \in \Pi_1$ (or higher). And here is this argument:

(1) If $\mathscr{S} \in \Sigma_1$ (or $\mathscr{S} \in \Sigma_0$), then there exists a procedure P which reverses programs for *deciding* members of $\mathscr{S}$ so as to yield programs for *enumerating* members of $\mathscr{S}$.

(2) It's false that there exists a procedure P which reverses programs for *deciding* members of $\mathscr{S}$ so as to yield programs for *enumerating* members of $\mathscr{S}$.

∴ (3) $\mathscr{S} \notin \Sigma_1$ (or $\mathscr{S} \notin \Sigma_0$). (by *modus tollens*)

∴ (4) $\mathscr{S} \in \Pi_1$ (or above in the AH). (disj. syll. on the rest of AH)

This is actually a powerful little argument, for the following reasons. Clearly, it's formally valid, i.e., the inferences go through (the natural deduction rules are standard). Next, premise (1) is not only true, but necessarily true, since it's part of the canon of unsolvability theory. This leaves just premise (2). If (2) is true, then our QUEST is ended: the Penrosian position turns out to have a formal foundation, and is furthermore supported by some fairly careful reasoning (of a sort that Penrose, with all due respect, doesn't himself envisage). *Is* (2) true? I'm inclined to think it is. What sort of evidence can be offered in support of (2)? In this regard, I think the ability of students with little knowledge of, and skill for, creating interesting stories to recognize interesting stories is significant. It does appear that while students in the *Autopoeisis* project can recognize elements of $\mathscr{S}$, they can't enumerate them. That is, students who are, by their own admission, egregious creative writers, are nonetheless discriminating critics. They can single out stories which are interesting, but *producing* such stories is another matter. These would be, necessarily, the same matter if the set of all interesting stories, $\mathscr{S}$, was in either Σ_0 or Σ_1, the algorithmic portion of AH.

Now perhaps this is all an illusion. Perhaps the students involved have a hidden ability to enumerate $\mathscr{S}$. But it certainly *seems* that they don't; and thus it would be safe to say, I think, that (2) is, though not self-evident, quite plausible. And saying this results in something that may give pause to those who habitually and unreflectively reject Penrose-like views: a clear, formidable, deductive argument for such views.

A number of objections, of course, need to be rebutted (e.g., is 'interesting story' a well-defined class?). Unfortunately, I haven't the space here to detail the ensuing dialectic. I'll pause for just a moment to consider *one* objection that will inevitably arise: "But look, stories are just strings over some finite alphabet A (which alphabet will depend on which natural language we're talking about). We know that A*, the set of all strings over A, is enumerable; A* can, for example, be lexicographically ordered. This implies that $\mathscr{S}$ can be enumerated, since one could simply enumerate A*, and at each step along the way one could decide whether or not the most recent output is interesting." The problem with this objection is that it begs the question at issue, since it identifies stories with finite strings, and such an identification does indeed, by the reasoning just indicated, imply that $\mathscr{S}$ is enumerable. It's not at all obvious that stories, in the real world, *are* strings. Authors often think about, expand, refine, ... stories without considering any strings whatever. They may "watch" stories play out before their mind's eye, for example. It's not at all obvious that there aren't an uncountable number of stories in this extended, real world sense. Hence the objection here is far from fatal.

Finally, however, it must be conceded that the issue, postponed above, of how a Π_1-or-beyond procedure runs on the finite human brain (which presumably doesn't have an infinite "tape," something such procedures capitalize on) must be confronted. That's another paper, though.

References

Boolos, G. S., & Jeffrey R. C. (1980). *Computability and Logic.* Cambridge, UK: Cambridge University Press.

Bringsjord, S. (1991-a). Is the Connectionist-Logicist Clash One of AI's Wonderful Red Herrings? *Journal of Experimental and Theoretical Artificial Intelligence,* 3 (4), 319-349.

Bringsjord, S. (1991-b). *Soft Wars.* New York, NY: New American Library, an imprint of Penguin USA (ISBN 0-451-17041-5).

Bringsjord, S. (1992-b). *What Robots Can & Can't Be.* Dordrecht, The Netherlands: Kluwer (ISBN 0-7923-1662-2).

Bringsjord, S. (1993-in press). Artificial Life: AI's Immature Brother? *Computing & Philosophy.*

Bringsjord, S. & Zenzen M. (1991). In Defense of Hyper-Logicist AI. *IJCAI 1991*, 1066-1072.

Bringsjord, S. & Zenzen M. (1992). Quantum Computing. *Sixth International Conference on Philosophy & Computing.* University of West Florida, Orlando, FL, August 1992.

Davis, M. & Weyuker E. (1983). *Computability, Complexity, and Languages: Fundamentals of Theoretical Computer Science.* New York, NY: Academic Press.

Deutsch, D. (1985). Quantum Theory, the Church-Turing Principle and the Universal Quantum Computer. *Proceedings of the Royal Society of London,* **400**, 97-117.

Ebbinghaus, H. D., Flum J. & Thomas W. (1984). *Mathematical Logic.* New York, NY: Springer Verlag.

Hofstadter, D. R. (1979). *Gödel, Escher, Bach: An Eternal Golden Braid.* Hassocks, UK: Harvester.

Hopcroft, J. E. & Ullman J. D. (1979). *Introduction to Automata Theory, Languages and Computation.* Reading, MA: Addison-Wesley.

Kugel, P. (1986). Thinking May Be More Than Computing. *Cognition,* **22**, 137-198.

Kugel, P. (1977). Induction Pure and Simple. *Information and Control,* **35**, 276-336.

Langton, C. G. (Ed.) (1989). *Artificial Life.* Reading, MA: Addison-Wesley.

Langton, C. G. (Ed.) (1992). *Artificial Life II,* Reading, MA: Addison-Wesley.

Lewis, H. & Papadimitriou C. (1981). *Elements of the Theory of Computation,* Englewood Cliffs, NJ: Prentice-Hall.

Lucas, J. R. (1964). Minds, Machines, and Gödel. In A. R. Andersen (Ed.) *Minds and Machines,* (43-89). Englewood Cliffs, NJ: Prentice Hall.

Meehan, J. (1981). TALE-SPIN. In R. C. Schank R. & C. Reisbeck (Eds.), *Inside Computer Understanding: Five Programs Plus Miniatures* (pp. 197-226). Hillsdale, NJ: Lawrence Erlbaum Associates.

Mendelson, E. (1990). Second Thoughts about Church's Thesis and Mathematical Proofs. *Journal of Philosophy,* **87** (5), 225-233.

Messiah, Albert (1966). *Quantum Mechanics.* New York, NY: John Wiley & Sons.

Nelson, R. J. (1987). Church's Thesis and Cognitive Science. *Notre Dame Journal of Formal Logic*, **28**, 581-614.
Pagels, H. R. (1989).*The Dreams of Reason*. New York, NY: Simon & Shuster.
Penrose, R. (1991). *The Emperor's New Mind*. New York, NY: Penguin.
Searle, J. (1980-a). Author's Response. *Behavioral & Brain Sciences*, **3**, 450-457.
Searle, J. (1980-b). Minds, Brains, and Programs. *Behavioral & Brain Sciences*, **3**, 417-424.
Searle, J. (1982). Review of Hofstadter and Dennett. *New York Review of Books*, pp. 3-10.
Searle, J. (1983). *Intentionality*. Cambridge: Cambridge University Press.
Searle, J. (1984). *Minds, Brains and Science*. Cambridge, MA: Harvard University Press.
Searle, J. (1990) Is the Brain's Mind a Computer Program? *Scientific American*, **262** (1), 25-31.
Smolensky, P. (1988). On the Proper Treatment of Connectionism. *Behavioral & Brain Sciences*, **11**, 1-22.
Soare, R. (1980). *Recursively Enumerable Sets and Degrees*. New York, NY: Springer-Verlag.

Poster Session

A Unified Approach To Inheritance With Conflicts

R. Al-Asady
Computer Science Department, Exeter University,
Exeter, U.K

Abstract

Nonmonotonic reasoning is reasoning which allows both a statement and its negation simultaneously to attain some degree of reliability. Such situations are called *ambiguous*. In this paper we investigate and describe a new approach for multiple inheritance with conflict, which concentrates on finding the degree of plausibility of the statement and of its negation. We identify a new implicit default measure simultaneously between the conflict classes and their shared property which we call *default correlation* (DC). To measure the strength of one class with respect to the other we compare their DC levels. The class with higher DC is the extension that the network will follow. This approach produces unambiguous results when applied to any inheritance hierarchy with exceptions.

1 Introduction

In order to represent the monotonic aspect of inheritance in AI, and to infer any explicitly represented knowledge, most of the mechanisms use the usual *isa* and *not-isa* relations directly. They deal with single inheritance hierarchies with or without exception successfully. However these systems are not capable of handling multiple inheritance with exception [Touretzky, 1984]. They lack the ability to represent the nonmonotonicity aspect of inheritance. Multiple inheritance with exception addresses two problems; the redundant problem and the conflict classes (the interested readers can refer to [Touretzky, 1986] for details). In this paper we describe Default Correlation (DC), a new approach to knowledge representation systems for multiple inheritance with conflicts.

Within inheritance hierarchies, our approach is capable of solving the problems of ambiguity and redundancy as well as that of masking [Al-Asady, 1992].

2 Default Correlation

2.1 Mechanics of default correlation

We treat any inheritance hierarchy as a network containing two parts:

1. Inheritance part.

2. Instantiation part.

The intuition underlying our approach is that people as intelligent entities deal with relations between properties by employing a kind of default reasoning. We call this Default Correlation (DC).

Definition 1 *Default Correlation (DC) is the method of analyzing conflicting classes (properties) in terms of properties or classes further down the hierarchy (factors). Each factor is assigned a correlation level linked it to the superclass directly above it.*

Definition 2 *Default Correlation Rule:Higher correlation levels override lower correlation levels when choosing a path through a network.*
$DC[weight_i \rightarrow property_0] > DC[weight_j \rightarrow property_0]$.
If an instance (a) has two inconsistency relations, which related to each other in a way such as the relation above, then (a) will prefer the extension to the property $property_0$ via the weight $weight_i$.

We recognise that it is not always possible to know exact correlation values. With this in mind we have developed a qualitative approach to reasoning in such circumstances based on four different kinds of network link which we call *qualitative defaults*. We now give an informal description of each of these.

1. $\odot$ links represent positive partially linked correlation between a class and its superclass.
2. $\neg \odot$ links represent negative partial exception correlation between a class and its superclass.
3. $\otimes$ links represent positive fully linked correlation between a class and its superclass.
4. $\neg \otimes$ links represent negative full exception correlation between a class and its superclass.

There is an ordered set of weights (W_i levels) attached to these links eg. $\otimes W_1$ or $\odot W_n$. Links are compared with each other according to the following procedure:
$W_1 > W_2 > ... > W_n$ Similarly $\neg W_1 > \neg W_2 > ... > \neg W_n$, where W_i and $\neg W_i$ are incomparable for a very specific level.

$\otimes$ or $\neg \otimes$ override both $\odot$ and $\neg \odot$ relations. For specific uses of DC the reader is referred to [Al-Asady, 1992].

2.2 DC framework

In this approach, DC allows us to predict the performance of a hierarchy by analyzing one ambiguous property regarding its conflict classes into its conflict factors.
I.e., $property_i$ may constructed from $\sum_{i=1}^{n} W_i$ subtheories (classes) where i can be either smaller than $n/2$ or greater than that.
Wi=$\otimes$ if $i > n/2$ otherwise Wi = $\odot$.
In other words we use different levels of a theory to represent different degrees of reliability.

A DC theory T is a set, $(T_1...T_n)$, where each T_i is a set of weights, $(W_1...W_n)$, inside a property , i.e $W_i > W_j$. The information in $\otimes W_1 T_i$ is more reliable than $\odot W_n T_j$.

We now show how Default Correlation deals with the ambiguity problem. The essence of the problem is that in multiple inheritance, when a class is a specialization of other classes, conflicts may occur. In Default Correlation the Correlation levels acts rather like weights in a balance allowing us to select the correct extension. The difference will determine if there is a (stronger) inheritance relation between one of the conflict classes and the superclass.

2.3 Formal relationship between the link types and DC

According to Definition 2 of DC and the informal description of the link types, consider an instance (x) related to two conflict classes R and Q, where the first class has a negative relation with a property P (i.e., $R \not\to P$) and the second has a positive relation with P (i.e., $Q \to P$). We can formalise the relation between these links and their DC independently using Reiter's [Reiter, 1980] default reasoning as follows:-

$$\frac{R(x) : \neg P(x)}{\neg P(x)} \quad and \quad \frac{Q(x) : P(x)}{P(x)}.$$

Analyzing the property P to it's class factors [Al-Asady, 1992], assuming that $\otimes[R_x W_1] > \odot[Q_x W_n]$ regarding the property P, (i.e the weight for W_1 regarding the class R is greater than that of W_n regarding the class Q), we can substitute the preferred weight factors instead of the original classes as follows:-

$$\frac{\otimes[R_x W_1] : \neg P(x)}{\neg P(x)} \quad and \quad \frac{\odot[Q_x W_n] : P(x)}{P(x)}$$

in which

$$R_{x \neg P} \text{ will override } Q_{xP} .$$

3 Ambiguity revisited

Reiter and Criscuolo [Reiter, 1981] described a variety of settings in which they discussed interacting normal defaults (conflicting default assumption). They claimed (page 97) "if a pair of defaults have contradictory consequents C but whose prerequisite (A and B) may share common instances (a), then the typical A which is also B leads to no conclusion".

Brewka [Brewka, (1989, 1991)] introduced a modification of DL called prioritized default logic PDL, in which he introduced the notion of preferred maximal consistent subsets of the premises. His approach subsumes Poole's [Poole, 1988] approach based on hypothetical reasoning.

Touretzky identified two approaches to the ambiguity problem : the skeptical reasoning approach (as used in TMOIS) and the credulous reasoning approach [Touretzky and Horty, 1987]. Both of these approaches didn't identify a unique theory to follow.

All these approaches which deal with default reasoning in an explicit way neither allow a system to represent priorities between classes in an ambiguous situation nor have the ability to formalise any relation between two pairs of defaults that have contradictory consequents.

4 Conclusion

In this paper we presented a new approach to nonmonotonic inheritance with conflicts. We dealt with default correlation levels, both qualitative and quantitative, in inheritance relations without recourse to the Instantiation relations. We used this approach to solve the ambiguity problem in particular and the masking problem in general.

References

Al-Asady, R. (1992). *A unified approach to inheritance with conflicts. Technical Report no CS-222,* Exeter, UK: University of Exeter, Department of Computer Science.

Brewka, G. (1989). Prefered subtheories: An extended logical framework for default reasoning. *IJCAI, 1043-1048.*

Brewka, G. (1991). *Nonmonotonic reasoning: Logical foundations of commonsence.* Cambridge, UK: Cambridge University Press.

Poole, D. (1988) A Logical framework for default reasoning. *AI,* **36**.

Reiter, R. & Criscuolo, G. (1981). On interacting defaults. *IJCAI-81, 94-100.*

Reiter, R. (1980). A logic for default reasoning. *AI,* **13**(1,2), 81-132.

Touretzky, D. (1984). Implicit ordering of defaults in inheritance systems. *Proceedings of AAAI-84, 332-325.*

Touretzky, D. (1986). *The mathematics of inheritance.* UK: Morgan Kaufmann.

Touretzky, D., Horty, J. & Thomason, R. (1987). A clash of intuitions: The current state of nonmonotonic inheritance systems. *IJCAI-87, 476-482.*

From Propositional Deciders to First Order Deciders: a Structured Approach to the Decision Problem

Alessandro Armando[*] Enrico Giunchiglia[*] Paolo Traverso[†]

Abstract

In this paper a hierarchical structure of highly specialized decision procedures for subclasses of FOL of increasing complexity is presented. Each decision procedure is specialized on a recursive class of formulae and it can be exploited in an effective way by higher level procedures whose design and implementation is thus greatly simplified. The overall structure turns out to be highly modular, extendible and allows to face the decision problem at the "right level" of detail. This results in an architecture of well integrated and very efficient decision procedures.

1 Introduction

In this paper a hierarchical architecture of many special-purpose deciders for (some) known decidable fragments of FOL is presented. We present (in order of increasing complexity) complete deciders for the propositional calculus (PTAUT), the quantifier-free theory of equality without function symbols (PTAUTEQ), the class of first order formulae provable using only the propositional inference machinery and the properties of equality (TAUT and TAUTEQ) and the class of prenex universal-existential formulae not containing function symbols (UE-decider).

The whole structure results high modular, extensible and easily maintainable. In fact, it can be easily extended adding more powerful decision procedures on the top of the hierarchy. The effort in building such procedures is restricted to the implementation of a procedure mapping a formula of the new class into a formula belonging to a class for which a decider is already available. The procedures described in this paper have been implemented and are currently available inside the GETFOL system (Giunchiglia 1992). GETFOL provides the user with a set of inference rules close to Prawitz' (1965) Natural Deduction rules. Natural deduction rules and decision procedures can be both applied to prove theorems.

Due to the lack of space the paper is not self-contained: the proposed architecture is sketched and the internal behavior of the deciders is only summarized. A detailed description of the deciders and a formal presentation of the technical results can be found in Armando and Giunchiglia (1992).

2 The Tautology Deciders

At the bottom of the hierarchy there is a decider for the propositional calculus (PTAUT). Actually it is able to deal with quantifier-free first order formulae but it exploits only their propositional structure and therefore it make sense to regard it as a "propositional" decider.

---
[*]DIST, University of Genoa, Genova - Italy
[†]Istituto per la Ricerca Scientifica e Tecnologica, Trento - Italy

PTAUT implements the technique of the *Truth Value Analysis* devised by Quine (1959). Such method (i) does not require the formula to be in any normal form, (ii) suitably optimized it is very efficient in the average case, having a linear cost on some recursive classes of formulae and (most important) (iii) its algorithmical structure is such that it can be easily specialized to deal with decidable quantifier-free theories.

As already pointed out in Murray (1982), we want to emphasize that translating a formula into disjunctive (conjunctive) normal form often obscures the structure of the formula and may yield formulae of intractable length. The improved translations suggested in Plaisted and Greenbaum (1986) succeed in yielding normal formulae of tractable length but at the cost of the extra-complexity due to the introduction of new propositional letters.

We thus suggest a decision procedure generalizing Davis-Putnam-Loveland's since it is not restricted to clausal formulae. Let α, β and w be propositional formulae. We take for granted the notion of *positive (negative) polarity* (Murray, 1982) and define the notion of *Top Level Disjunctive Occurrence* (TLDO) in the following way:

1) if $w = \alpha$ then α is a TLDO of w;

2) if $\neg \alpha$ is a TLDO of w then also α is a TLDO of w;

3) if $\alpha \vee \beta$ ($\alpha \wedge \beta$) is a TLDO of w and occurs with positive (negative) polarity in w then both α and β are TLDO of w;

5) if $(\alpha \rightarrow \beta)$ is a TLDO of w and occurs with positive polarity in w, then both α and β are TLDO of w;

It is easy to check the following facts:

Theorem 1 *If ϕ^* is an assignment assigning F (T) to every propositional constituent of w occurring only with positive (negative) polarity then w is a tautology if and only if the formula w^*, obtained by partially evaluating w w.r.t. ϕ^*, is a tautology.*

Theorem 2 *If P is a propositional constituent of w with a positive (negative) TLDO in w then w is a tautology if and only if the formula w^*, obtained by partially evaluating w w.r.t. the partial assignment assigning F (T) to P, is a tautology. If P occurs twice in w as TLDO but with opposite polarity, then w is a tautology.*

and verify that, in a clausal language, they correspond to the *Affirmative-Negative Rule* and the *One-Literal Clause Rule* respectively of the Davis-Putnam-Loveland procedure, respectively.

If no propositional letter occurs either as a TLDO or only positively or only negatively, as already noticed in Quine (1959) the algorithm can be made faster in the average case by choosing the propositional letter occurring the greatest number of times. Then two assignments (one assigning T the other assigning F to the chosen propositional letter) are generated and finally the formula is partially evaluated with respect to such assignments. Such steps, are recursively applied to the simplified formulae.

Experimental results confirm the effectiveness the optimizations. We have implemented an algorithm randomly generating formulae on the base of connectives $\{\neg, \wedge, \vee, \rightarrow\}$ given the number of propositional constituents and the total number of occurrences. Submitting 5000 randomly generated formulae with 20 propositional constituents and 1000 occurrences, PTAUT takes 24 recursive calls to solve the hardest formula, against the 3195 recursive calls needed by the non optimized version. Repeating the test with 5000 randomly generated formulae with 20 propositional constituents but with 3000 occurrences, PTAUT takes 208 recursive calls in the worst case against the 16221 needed by the algorithm if the optimizations are not applied.

Besides complexity arguments, PTAUT can be extended into a decision procedure for any decidable quantifier-free theory. In fact, given a quantifier-free theory Γ, it is possible to build a decision procedure for the quantifier-free formulae deducible from Γ by providing an algorithm to test whether any given assignment of truth values to the atomic subformulae satisfies the axioms in Γ (obviously the existence of such an algorithm guarantees the decidability of the class).

We have built a decider for the *quantifier-free theory of equality without function symbols*. The algorithm for testing the admissibility of an assignment we have implemented is a simplification of that described in Shostak (1977) for testing the satisfiability of a set of equalities. The main difference is the ability to explicitly deal with predicates, which in Shostak (1977) are substituted with equalities containing new function symbols, so obscuring the form of the original formula.

The following fact provides a criterion for determining whether a given assignment satisfies the properties of equality. Let $\mathcal{C}$ be the set of individual constants occurring in the input formula and the binary relation "$\simeq$" be defined as the smallest equivalence relation over $\mathcal{C} \times \mathcal{C}$ such that if $\phi(\text{``}c_1 = c_2\text{''}) = T$ then $c_1 \simeq c_2$. In a similar way, if $\mathcal{A}$ is the set of atomic subformulae occurring in the input formula, let us define "$\cong$" as the smallest equivalence relation over $\mathcal{A} \times \mathcal{A}$ such that if $P(r_1, \ldots, r_n) \in \mathcal{A}$, $P(s_1, \ldots, s_n) \in \mathcal{A}$ and for $i = 1, \ldots, n$ $r_i \simeq s_i$ then $P(r_1, \ldots, r_n) \cong P(s_1, \ldots, s_n)$.

Theorem 3 *An assignment ϕ does not satisfy the equality properties if and only if there exist individual constants $c_1, c_2 \in \mathcal{C}$ such that $\phi(\text{``}c_1 = c_2\text{''}) = F$ and $c_1 \simeq c_2$ or there exists atomic formulae $A_1, A_2 \in \mathcal{A}$ such that $\phi(A_1) \neq \phi(A_2)$ and $A_1 \cong A_2$.*

3 The UE-decider

The developed decision procedure efficiently expands the formula over the Herbrand universe. In order to simplify the description, we make the further assumption that the propositional connectives occurring in the formula are only "$\wedge$" and "$\vee$" and that the negation symbols "$\neg$" (if any) are applied only to atomic formulae.

The finiteness of the Herbrand universe allows to compute the expansion of the formulae before applying a propositional decider. The tricky point is to make in the most efficient way the substitution of the members of the domain for the existential variables in the formula. In particular while substituting an existential variable not occurring in one of the two sides of the formula, it is not necessary to copy the other side of the formula. More formally, for any formula α, define:

- D the set of constants in α

- $E(\alpha)$ the set of existential variables in α;

- $H_D^S(\alpha)$ to be the disjunction of all the formulae obtained from α substituting in all possible ways the elements in D for the members in S.

Then the following fact holds:

Theorem 4 *Let* $\circ \in \{\wedge, \vee\}$, $\alpha = (\beta \circ \gamma)$, D *a set of constants.*

$$\vdash H_D^{E(\alpha)}(\alpha) \iff \vdash H_D^{E(\alpha)}(H_D^{E(\beta)\backslash E(\gamma)}(\beta) \circ H_D^{E(\gamma)\backslash E(\beta)}(\gamma))$$

This fact is a trivial consequence of the distributivity of "∨" with respect to "∧". If you get D to be the set of constants and free variables in α, then both $H_D^{E(\alpha)}(\alpha)$ (*standard Herbrand expansion*) and $H_D^{E(\alpha)}(H_D^{E(\beta)\backslash E(\gamma)}(\beta) \circ H_D^{E(\gamma)\backslash E(\beta)}(\gamma))$ (*partitioned Herbrand expansion*) are logically equivalent to α by Herbrand's theorem. Calculating the size of the partitioned expansion, it turns out that they are quite relevant (Armando and Giunchiglia, 1992). In the special case of some *monadic* formula, it turns out that the size of the partitioned Herbrand expansion is linear with the cardinality of the Herbrand domain.

4 Conclusions and acknowledgements

We have described a hierarchical architecture of decision procedures each of them can be used individually or jointly with the others to solve a set of decision problems of increasing complexity. Each procedure is specialized and optimized with respect to a particular decision problem and the whole structure results highly extendible and efficient.

This work has been done with the support of the Italian National Research Council (CNR), Progetto Finalizzato Sistemi Informatici e Calcolo Parallelo (Special Project on Information Systems and Parallel Computing).

References

Armando A. and Giunchiglia E. (1992). Embedding complex decision procedures inside an interactive theorem prover. To appear in the *Annals of Artificial Intelligence and Mathematics*.

Giunchiglia F. (1992) *The GETFOL Manual - GETFOL version 1*. Technical Report 9204-01, DIST - University of Genova, Genoa, Italy, IRST-Manual, forthcoming 1992, IRST, Trento, Italy.

Murray N.V. (1982). Completely non-clausal theorem proving. *Artificial Intelligence*, **18**:67-85.

Plaisted D.A. and Greenbaum S. (1986) A structure-preserving clause form translation. *Journal of Symbolic Computation*, 2:293-304.

Prawitz D. (1965). *Natural Deduction - A proof theoretical study*. Almquist and Wiksell, Stockholm.

Quine W.V.O. (1959). *Methods of Logic*. Henry Holt and Co., New York.

Shostak R. (1977). An algorithm for reasoning about equality. *Proc. of the 7th International Joint Conference on Artificial Intelligence*, 526-527.

Modelling Auditory Streaming: Pitch Proximity

Kevin L. Baker
Departments of Psychology and Computer Science, University of Sheffield
Sheffield, UK

Sheila M. Williams
Departments of Psychology and Computer Science, University of Sheffield
Sheffield, UK

Phil Green
Department of Computer Science, University of Sheffield
Sheffield, UK

Roderick I. Nicolson
Departments of Psychology, University of Sheffield
Sheffield, UK

The auditory system has evolved so that the acoustic signals which hit the ear drum are normally analysed with great efficiency. Auditory scene analysis (Bregman, 1990) attempts to explain the processes by which the listener decodes the single sound pressure wave that arrives at the ear into meaningful descriptions of the events which may have produced it.

Auditory stream research seeks to discover the principles behind the grouping and segregation of sound, which are frequently investigated through the use of repetitive loops of stimuli. One of the principles by which the auditory system separates sounds, identified in this way, is frequency (or pitch) proximity. The Baroque composers, such as Bach, exploited the grouping effect of pitch proximity in the use of counterpoint to imply a polyphony from a single instrument. However, the metric by which such proximity is determined by the auditory system has not yet been fully described.

The work presented here forms part of the 'Mapping the Auditory Scene' project (Williams, Nicolson & Green, 1990). The project has two aims: (1) to further investigate the processes which take part in grouping complex sound patterns into auditory streams; and (2) to further develop computational models of auditory grouping, in order to demonstrate and test our theories concerning these processes. The objective of the computational modelling approach is to explore different possible methods and algorithms for creating a predictive model of auditory scene analysis and in so doing, to identify

---

Requests for reprints should be addressed to Sheila M. Williams, Department of Computer Science, University of Sheffield, PO Box 600, Sheffield, UK S1 4DU. email: s.williams@dcs.sheffield.ac.uk
This research was supproted by the SERC/ESRC/MRC joint Council Initiative into Cognitive Science/HCI. We would like to thank Al Bregman and Pierre Ahad for their continued support and Proffessor David Huron for his suggestions, also Rob Waltham for his help with the music theory literature.

further areas of fundamental psychoacoustic research necessary to the solution of the auditory scene analysis problem. Using a variety of computational approaches including object oriented programming, it is hoped that once a suitable level of representation for primitive auditory objects has been established, the heuristics used in partitioning the auditory scene can be applied to predict the percepts which are most likely to arise from any given scene. The psychoacoustic programme is directed towards filling the gaps in the (sometimes sparse) descriptions of the many grouping processes which have been identified as participating in auditory scene analysis.

Within this project the frequency proximity metric is being explored. In this paper we consider data from a series of experiments designed to explore stream organization due to frequency proximity and relate this to other work in the area, suggesting that any model of auditory streaming needs to address the definition and relevance of pitch and pitch proximity in a new way.

Some of the earliest experiments in auditory streaming stemmed from the observation by Miller and Heise (1950) that when two tones of a certain frequency separation were played in a repeated loop, they were heard as either a trill or as a distinct 'jumping' pattern depending on the *speed of repetition* and *frequency separation* of the tones. From this beginning, the effect of frequency proximity on stream organisation has been widely explored but as yet there have been few attempts to describe the possible metric by which such proximity can be measured and applied in a simple computer model. One attempt, by Beauvois and Meddis (1991), demonstrates that an enhanced physiology-based auditory model may provide a sufficient transformation of the sound signal for streaming of simple two-tone stimulus patterns such as those employed by Miller and Heise. In contrast, the project reported here (Williams et al., 1990) takes a perceptual approach. Williams builds on Bregman's (1984, 1990) conviction that auditory scene analysis is based on Gestalt-like perceptual principles such as good continuation, similarity, and proximity, etc. as fundamental but higher order grouping mechanisms. Such principles have been represented computationally through an object-oriented approach (Williams, 1989). These two approaches may not be incompatible since each approaches stream segregation at a different level of explanation.

In the physiological account, the peripheral auditory system is often modelled as a bank of overlapping band-pass filters giving a description of the sound spectrum at each moment in time, which implies the need for a higher order frequency tracking mechanism to account for the effects of sequential grouping. This type of explanation, which often refers to psychophysiological studies of critical bandwidth, explains streaming in terms of a *breakdown*, or failure of the frequency tracking mechanism (eg van Noorden, 1975).

Bregman's thesis (1984, 1990), however, is based on a contention that the splitting of tones into streams, or perceptual groups, cannot be due to a frequency-tracking mechanism alone, for the context in which a tone is presented significantly affects the ability to discriminate aspects of the stream (Bregman & Pinker, 1978). Thus, Bregman explains the auditory system as being designed not "to integrate successive sounds willy-nilly, but to integrate those that probably arose from the same source" (1978, p.392). This approach treats stream formation as an accomplishment, not a breakdown.

Another factor, widely identified in music theory (Terhardt, Stoll & Seewann, 1982; Parncutt, 1988; Huron, personal communication, 1991) but which appears to have been overlooked in psychoacoustic studies, is the problem that when the frequency of a pure tone is manipulated in a measured way, the perceptual change of the tone varies along more than one dimension. Terhardt, Stoll and Seewann (1982) have explained the timbral change of pure tones (i.e. those whose acoustic form is a perfect sine-wave) by positing the existence of a spectral dominance region either side of which such tones lose their quality of perceptual purity and begin to sound like sizzles (very high frequencies) and rumbles (very low frequencies). For pure tone partials of complex tones, he has identified the region around 700 Hz as having the 'clearest' pitch perception, and has devised a model of pitch weighting for tones either side of the dominance region to account for their relative *salience*.

The empirical investigation at Sheffield began by applying the streaming paradigm to two different stimulus patterns (see Fig. 1) in order to determine the streaming mid-point between two target tones (Baker, Williams & Nicolson, 1992). This was intended to confirm that the frequency proximity metric in auditory streaming was indeed the same as that calculated by other methods. Thus the target tones, and a captor selected from a range of values varying between the target tone values, were presented in a repeating cycle at rates sufficiently rapid to induce streaming such that two of the tones formed a rapidly beating sequence and the remaining tone formed a slower beating and independent stream. The midpoint was defined as the point at which streaming with the upper or the lower target was equally likely. It was expected that this would manifest itself by either a breakdown in the streaming process, leading to greater uncertainty in the subjects responses, or by a reduction in predictability, leading to different perceptual partitions of the same stimulus set at different presentations.

Fig. 1.   (a) the sequential pattern   (b) the synchrony pattern

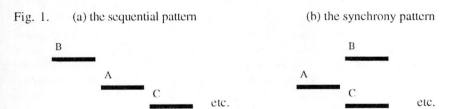

For both patterns, the capturing tone A is placed between the two target tones B&C. At the rate of repetition used in the experiments, the middle tone *always* tends to group with either the top or bottom tone, producing two perceptual streams. One stream is fast in rhythm, and the other is slow in comparison. By varying the frequency position of the middle tone, and asking for judgments of the streaming organizations, the midpoint of the frequency proximity metric may be ascertained.

In the sequential pattern all three tones occupy a separate time-space. This stimulus pattern represents a simple auditory scene in that frequency proximity is the only dimension which may determine the grouping characteristics, due to all other factors being held constant. In the synchrony pattern, however, the capturing middle tone is 'competing' against the synchronous grouping of the two tones as a tonal dyad.

For both tone patterns, tones B and C were always played at the same frequencies marking the limits of the range between which A was varied. All frequencies were chosen to be prime numbers and not within 5% of any whole-number multiple of any other tone present. These values were chosen to avoid the possibility of simple harmonic relationships between captor and targets from influencing the stream organization.

Our first group of experiments, covering the frequency range from 929-1972Hz, did not confirm our hypothesis that existing scales would be appropriate to model frequency proximity in streaming. The midpoints arising from both stimulus patterns were substantially lower than the midpoints predicted by either physiology-based scales or simple perceptual scales or by a simple log relationship and, furthermore, the midpoints derived from the experimental data for each of the two stimulus patterns differed significantly from each other. Further experimentation has shown similar effects with stimulus patterns in different frequency ranges.

Applying the empirical results to modelling such behaviour, it seems difficult to ascertain whether the variation in the pitch metric is due to pitch dissonance and should be accounted for in the initial representations derived from the acoustic signal or belongs to the interaction of complex grouping processes and as such should be represented in the weighting factors required at the level of evidence combination.

Our results demonstrate that stream organization is due to an interaction of the time and frequency information present in the acoustic signal, although this is not a simple relationship being dependent upon whether elements of the pattern co-occur and the particular frequency range under consideration, among other factors.

References:

Baker, K. L., Williams, S. M., & Nicolson, R. I. (1992). *Evaluating frequency proximity in stream segregation.* Manuscript submitted for publication.

Beauvois, M. W., & Meddis, R. (1991). A computer model of auditory stream segregation. *The Quarterly Journal of Experimental Psychology*, **43**, 517-541.

Bregman, A. S. (1984). Auditory scene analysis. *Proceedings of the Seventh International Conference on Pattern Recognition.* Montreal, 168-175.

Bregman, A. S. (1990). *Auditory scene analysis.* Cambridge, MA: MIT Press.

Bregman, A. S., & Pinker, S. (1978). Auditory Streaming and the Building of Timbre. *Canadian Journal of Psychology*, **32**, 19-31.

Miller, G. A., & Heise, G. A. (1950). The trill threshold. *Journal of the Acoustical Society of America*, **22**, 637-638.

Parncutt, R. (1989). *Harmony: a psychoacoustical approach.* Berlin: Springer-Verlag.

Terhardt, E., Stoll, G., & Seewann, M. (1982). Algorithm for extraction of pitch and pitch salience from complex tonal signals. *Journal of the Acoustical Society of America*, **71**, 679-688.

van Noorden, L. P. A. S. (1975). *Temporal coherence in the perception of tone sequences.* Unpublished doctoral dissertation. Eindhoven University of Technology.

Williams, S. M. (1989). *STREAMER: A prototype tool for computational modelling of auditory grouping effects* (Research Report No: CS-89-31). University of Sheffield, Department of Computer Science.

Williams, S. M., Nicolson, R. I., & Green, P. D. (1990). STREAMER: Mapping the auditory scene. *Proceedings of the Institute of Acoustics Autumn Conference on Speech and Hearing*, **12**(10), 567-575.

A Simulated Neural Network for Object Recognition in Real-time

Bernhard Blöchl and Lampros Tsinas

Institut für Meßtechnik, Universität der Bundeswehr München
Neubiberg, Germany

Abstract

To build a hybrid system which unites the approach of symbol manipulation and connectionism the existing system is enhanced by recognition modules on the basis of neural networks. A first network which has the task to detect and classify vehicles is introduced here. The network itself as well as training strategies and first results of real-time experiments are described. In general it can be said that, second only to the type of network used, the method of learning is pivotal for the successful performance.

1 Introduction

Two classes of AI research are commonly distinguished today. The traditional variety, also called good old fashioned AI (or GOFAI as Haugeland (1985) coined) and connectionism. Both have the same root, the paper "A logical calculus of the ideas immanent in nervous activity" from the psychologist/psychiatrist Warren McCulloch and the mathematician Walter Pitts (McCulloch & Pitts, 1943).

Usually it is assumed, that since Turing machines are symbol manipulators and Turing had proved that they could compute anything, he had proved that all intelligence could be captured by logic. Therefore, a symbolic approach unlike an holistic (connectionist) approach does not need justification. The uninterpreted symbols of a Turing machine (zeros and ones) were confused with the semantically interpreted symbols of the kind of symbols of a vision system.

The dominant role of symbol manipulation obscured the fact that the symbols (which represent physical objects) receive their semantics from the programmer, rather than from cognitive processes (which had been the original intention). In vision systems the symbol extraction from the relatively unstructured image data stream will be performed by connectionist systems with more power and less effort. Although computer speed and capacity have increased steadily in the last years, the levels necessary for a complete, monolithic, purely connectionist implementation of a real-time system have still not yet been reached.

2 Towards a hybrid system

For the interpretation of traffic scenes high system performance is achieved by a system architecture optimally designed for the task of machine vision. The concept of the layered model is described in Graefe (1989) and shown in Figure 1. The latest state of the hardware is described in (Blöchl 92). The system is based on loosely coupled parallel object processors. A system bus is used for interprocessor communication. A wide band video bus distributes video data from four cameras.

For each object or class of object one recognition module exists (Blöchl et al., 1992). A cognitive process extracts a set of significant object features from the video data. The task of the situation processor is context evaluation through symbol manipulation.

The modular structure of the existing multiprocessor system not only makes the system easier to develop and test. It also allows a productive mix between traditional methods and an implementation of neural-network techniques on the object level without detrimental impact on the proven system design. Besides the "classical" methods of object recognition in image processing, easily simulated artificial neural networks (SANN) may be introduced on the recognition module (object processor) level. The neural network is simulated, that is to say, is developed in a simulated environment on the PC and runs in real-time. The implementation and integration into the system is on the way.

In no way is this concept or its proposed goals to be compared to those of the ALVINN project of the CMU Navlab, where the goal is an exclusive adaptation to a monolithic SANN system (Pommerleau, 1991).

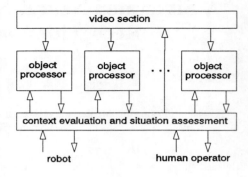

Figure 1. Conceptual structure of object oriented vision systems (Graefe 89).

3 Vehicle SANN

For the vehicle SANN a back propagation network led to the best results. The network has 16x16 input elements, 98 hidden elements as a vector in a single layer, and one output element. The number of hidden elements used here has been proved in experiments (trial-and-error) to be appropriate for optimal efficiency.

As road traffic and on-site transport always take place in the plane, there is no need to take rotational invariance into account. This allows correspondingly smaller neural networks to be employed, and reduces both development efforts and computer processing time. The vehicle module's task is not to determine the position of a vehicle within a frame, but to simply be able to recognize whether or not a vehicle lies within the frame (the image excerpt being surveyed). The single output element delivers a binary statement of the vehicle's whereabouts.

Figure 2. Detail of a traffic scene; typical shape of a vehicle.

Vehicles are more or less flat shapes in the image (Figure 2). As long as the resolution is sufficient, objects must be recognizable, regardless of their distance from the vehicle. The goal of this approach is a recognition process independent of the size of the object in the frame. This would also achieve independence from focal length.

3.1 Training

For the training an input matrix (or training vector) must be prepared. A total picture is read into the PC by means of a frame-grapper. To produce static training files a detail is manually selected to ensure that the entire vehicle is visible.

In the starting trial phase, tests with grey values were performed. This first tests produced no positive results at all. Even by varying the number of training patterns, no usable results were achieved.

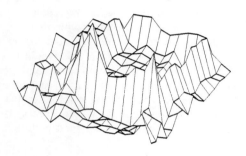

The process of interpreting images is more difficult because vehicles can appear not only in positive, but also in negative contrast. To alleviate this problem, the absolute value of the mean of the gradient in horizontal and vertical direction is produced. In a gradient image, all the vertical and horizontal edges, especially the corners, are prominently highlighted, as Figure 3 illustrates. By this method, pictures with positive and negative contrast produce identical patterns.

Figure 3. 3-D mesh surface of the absolute value of the gradients of the vehicle shown in Figure 2.

Only through cyclical training (the simultaneous re-learning from previous input matrixes and increasing the number of training patterns) was it possible to bring the network to a measurable status and increase its recognition rate.

4 Experiments

In one attempt in which 50 input vectors were presented as input pattern, 50% from scenes with vehicles and 50% from scenes without vehicles, the network achieved 100% accuracy through recognition of the trained pattern. The network also was able to recognize patterns that were similar or close to the learned pattern.

In a different experiment, the same type of network was trained by presenting it with only one out of the following three, distinctly different patterns, namely an image with vehicle with strong contrast, an image with vehicle with weak contrast, and an image with weak contrast without vehicle. After training, the network could classify unknown images with up to 78% accuracy. This shows that the higher the variety of scenes with which the network has been trained, the higher its tolerance is. Consequently, the network could easily surpass the first trial run results by introducing broader standards for the proposed expectations. A further improvement of accuracy would be achieved by increasing the number of training patterns.

5 Real-time results

For real-time experiments a PC (80846, 50 MHz) was used together with a frame-grapper card. Traffic scenes were recorded on video tape and read into the PC in real time via the frame-grapper. The reading-in of the recorded scenes, and thus the experiments, can be repeated. The advantage of this procedure is obvious.

For real-time video scenes the above described trained networks were used. In each step a predefined search area of 192 x 192 pixels was evaluated. The input data were normalized and gradients were calculated. The processing of one area needs less than 500 ms including normalization and calculation of the gradients. Most of the time is used to transfer the pixel data via the PC bus. This "minimal" system is not optimal for real-time applications but ideal for program testing and debugging.

The trained network was able to recognize most of the trained objects (vehicles) when confronted with them in dynamical traffic scenes. An increased number of neurons would lead to a higher resolution and thus increase the recognition rate. On the other hand this would lead to a higher cycle time if the computing power remains unchanged. The present cycle time of less than 500 ms, however, is fully acceptable considering the moderate velocity of autonomous vehicles. The cycle time is considerably lower after the implementation of the network into the real-time vision system, whose access to image data is much more faster. The implementation is on the way and will be described in one of the next papers.

Acknowledgement

Part of the work reported was supported by the Ministry of Research and Technology (BMFT) and by the German automobile industry within the EUREKA project PROMETHEUS.

References

Blöchl, B. (1992). Coarsely Grained Multiprocessor Image Processing System Coupled with a Transputer Network. In R. Trappel (Ed.), *Cybernetics and Systems Research '92-Vol.2* (pp. 1423-1429). London: World Scientific.

Blöchl, B., Berends, J.-U., Tsinas, L. (1992). Objekterkennung in Verkehrsszenen auf Transputern [Object Recognition in Traffic Scenes on Transputers]. *4. Transputer-Anwender-Treffen TAT'92.* Aachen, 128-131. In print by Springer.

Graefe, V. (1989). Dynamic Vision System for Autonomous Mobile Robots. *Proceedings of IEEE/RSJ International Workshop on Intelligent Robots and Systems. IROS'89.* Tsukuba, 12-23.

Haugeland, J. (1985). *Artificial Intelligence; The Very Idea.* Cambridge, MA: MIT Press/Bradford Books.

McCulloch, W. S., and Pitts, W. H. (1943). A logical calculus of the ideas immanent in nervous activity. *Bulletin of Mathematical Biophysics,* **5**, 115-133.

Pommerleau, D. A. (1991). Neural network-based vision processing for autonomous robot guidance. *Proceedings of SPIE, Applications of Artificial Neural Networks II,* **1469**, 121-128.

Constraints and First-Order Logic

James Bowen

North Carolina State University

Raleigh, USA

Dennis Bahler

North Carolina State University

Raleigh, USA

Abstract

We present a constraint-based approach to making the <u>full</u> FOPC available for knowledge representation. The resulting inference technique is sound and terminating. Although the technique is necessarily incomplete in general, it is refutation-complete for many practical applications.

1 Introduction

We agree with Nilsson (1991) that "For the most versatile machines, the language in which declarative knowledge is represented must be at least as expressive as first-order predicate calculus" (FOPC). Since the full FOPC is undecidable, the inference engines for such languages cannot be sound, complete and terminating. Although termination is a *sine qua non*, neither of the other two attributes is indispensable. Unsound inference has long been accepted; it has been codified as various forms of non-monotonic logic. Acceptance of incomplete inference is less widespread, but we argue that there are many situations in which full FOPC expressiveness is more important than completeness. In this paper, we present a constraint-based approach to making the full FOPC available for knowledge representation purposes. Although this approach cannot guarantee completeness, inference is refutation-complete for many practical applications.

Formally, a constraint network can be defined as follows:

<u>Constraint Network:</u>
A constraint network is a triple $\langle \mathcal{U}, \mathbf{X}, \mathbf{C} \rangle$ where $\mathcal{U}$ is a universe of discourse, $\mathbf{X}$ is a finite tuple of q non-recurring parameters, and $\mathbf{C}$ is a finite set of of r constraints. Each constraint $C_k(T_k) \in \mathbf{C}$ imposes a restriction on the allowable values for the a_k parameters in T_k, a sub-tuple of $\mathbf{X}$, by specifying that some subset of the a_k-ary Cartesian product $\mathcal{U}^{a_k}$ contains all acceptable combinations of values for these parameters. □

The overall network constitutes an intensional specification of a joint possibility distribution for the values of the parameters in the network. This joint possibility distribution, called the network *intent* (Bowen and Bahler, 1991a), is a q-ary relation on $\mathcal{U}^q$.

The Intent of a Constraint Network:
The intent of a constraint network $\langle \mathcal{U}, \mathbf{X}, \mathbf{C} \rangle$ is $\Pi_{\mathcal{U},\mathbf{X},\mathbf{C}} = \overline{C}_1(\mathbf{X}) \cap ... \cap \overline{C}_r(\mathbf{X})$, where, for each constraint $C_k(T_k) \in \mathbf{C}$, $\overline{C}_k(\mathbf{X})$ is its cylindrical extension into the Cartesian space defined by $\mathbf{X}$. □

In a first-order language $\mathcal{L} = \langle \mathcal{P}, \mathcal{F}, \mathcal{K} \rangle$, $\mathcal{P}$ is the vocabulary of predicate symbols, $\mathcal{F}$ is the vocabulary of function symbols, and $\mathcal{K}$ is the vocabulary of constant symbols. A total interpretation for a language $\mathcal{L}$ is a set of mappings, such that each symbol in $\mathcal{P} \cup \mathcal{F} \cup \mathcal{K}$ is mapped. A partial interpretation for $\mathcal{L}$ is a set of mappings in which at least one symbol is not mapped. Consider the class of problem in which, given a theory Γ written in $\mathcal{L}$ and a partial interpretation function $\mathcal{I}_p$ for $\mathcal{L}$ in terms of a universe $\mathcal{U}$, one has to find a total interpretation $\mathcal{I}$ for $\mathcal{L}$ such that $\mathcal{I}_p \subset \mathcal{I}$ and $\langle \mathcal{U}, \mathcal{I} \rangle \models \Gamma$. It can be shown (Bowen and Bahler, 1991b) that, for instances in which only constant symbols of $\mathcal{L}$ are uninterpreted by $\mathcal{I}_p$, this class of problem is equivalent to finding one tuple in the intent of a constraint network $\langle \mathcal{U}, \mathbf{X}, \mathbf{C} \rangle$ where the components of $\mathbf{X}$ are the uninterpreted constants of $\mathcal{L}$ and the constraints in $\mathbf{C}$ are derived from the sentences in Γ.

2 Example 1: A Practical Application

To see the practical utility of this constraint-based approach to logic, consider the following theory, which describes the relationships between the specifications for a computer and a model chosen to meet the specifications.

$\Gamma = \{cost = price(chosen) * 1.06, \quad slots(chosen) \geq reqd\_slots,$
$\quad speed(chosen) \geq reqd\_speed, \quad capacity(chosen) \geq reqd\_disk,$
$\quad \neg(\exists X)(slots(X) \geq reqd\_slots \wedge speed(X) \geq reqd\_speed \wedge$
$\quad\quad capacity(X) \geq reqd\_disk \wedge price(X) < price(chosen))\}.$

This theory is written in a first-order language
$\mathcal{L} = \langle \{=, \geq, <\}, \{*, speed, capacity, slots, price\},$
$\quad \mathcal{R} \cup \{m1, m2, m3, reqd\_speed, cost, reqd\_disk, reqd\_slots, chosen\} \rangle$

where $\mathcal{R}$ is the set of strings formed from the characters +, -, . and 0..9 according to a grammar for well-formed numeric strings. In this paper, to distinguish between symbolic constants and entities in the universe of discourse, we use typewriter font for the latter. Thus, 1.06 is a constant symbol in $\mathcal{R}$, while 1.06 is a real number in $\Re$. The universe of discourse for $\mathcal{L}$ is $\mathcal{U} = \Re \cup \{\texttt{m1}, \texttt{m2}, \texttt{m3}\}$. Using $\mathcal{I}_\mathcal{R}$ to denote the standard interpretation (with mappings such as $1.06 \mapsto \texttt{1.06}$) from $\mathcal{R}$ to the finitely expressible rationals Q_f, we have the following partial interpretation $\mathcal{I}_p$ for $\mathcal{L}$:
$\mathcal{I}_p = \mathcal{I}_\mathcal{R} \cup \{= \mapsto \{\langle X, Y \rangle | \text{EQUALS}(X, Y)\}, \geq \mapsto \{\langle X, Y \rangle | \text{GEQ}(X, Y)\},$
$\quad < \mapsto \{\langle X, Y \rangle | \text{LESS}(X, Y)\}, * \mapsto \{\langle X, Y, Z \rangle | \text{EQUALS}(Z, \text{TIMES}(X, Y))\},$
$\quad m1 \mapsto \texttt{m1}, m2 \mapsto \texttt{m2}, m3 \mapsto \texttt{m3}, price \mapsto \{\langle \texttt{m1}, \texttt{4500} \rangle, \langle \texttt{m2}, \texttt{3000} \rangle, \langle \texttt{m3}, \texttt{3500} \rangle\},$
$\quad capacity \mapsto \{\langle \texttt{m1}, \texttt{628} \rangle, \langle \texttt{m2}, \texttt{128} \rangle, \langle \texttt{m3}, \texttt{140} \rangle\}, slots \mapsto \{\langle \texttt{m1}, \texttt{3} \rangle, \langle \texttt{m2}, \texttt{6} \rangle, \langle \texttt{m3}, \texttt{5} \rangle\},$
$\quad speed \mapsto \{\langle \texttt{m1}, \texttt{20} \rangle, \langle \texttt{m2}, \texttt{25} \rangle, \langle \texttt{m3}, \texttt{25} \rangle\}\}.$

In Γ, there are five uninterpreted constant symbols: *reqd_speed*, *reqd_disk*, *reqd_slots*, *chosen*, and *cost*. The first three of these symbols represent the required functionality, the fourth represents the appropriate machine, and the fifth represents the cost of this machine. The first sentence in Γ specifies that *cost* is the price of *chosen* plus six percent tax; the second, third and fourth fourth sentences specify that *chosen* must provide the required functionality, and the fifth sentence specifies that there should not be anything cheaper than *chosen* which offers this functionality. In the network $\langle \mathcal{U}, \mathbf{X}, \mathbf{C} \rangle$ corresponding to this situation,

the joint possibility distribution for $\langle chosen, cost, reqd\_disk, reqd\_slots, reqd\_speed \rangle$ is the network intent
$\Pi_{\mathcal{U},\mathbf{X},\mathbf{C}} =$
$(\{\texttt{m1}\} \times \{4770\} \times \{X|\textsc{gt}(X,140) \wedge \textsc{leq}(X,628)\} \times \{X|\textsc{leq}(X,3)\} \times \{X|\textsc{leq}(X,20)\}) \cup$
$(\{\texttt{m2}\} \times \{3180\} \times \{X|\textsc{leq}(X,128)\} \times \{X|\textsc{leq}(X,6)\} \times \{X|\textsc{leq}(X,25)\}) \cup$
$(\{\texttt{m3}\} \times \{3710\} \times \{X|\textsc{gt}(X,128) \wedge \textsc{leq}(X,140)\} \times \{X|\textsc{leq}(X,5)\} \times \{X|\textsc{leq}(X,25)\}).$

By interacting with this network, it is possible to solve a wide variety of problems. To see some of this versatility, we will consider various "what if" scenarios.

Scenario 1 Suppose that, with a limited budget, we want to buy a computer which will provide a lot of disk space and sufficient slots to connect several peripheral devices. Suppose we estimate the disk space needed to be 200MB, thereby producing $\Gamma' = \Gamma \cup \{reqd\_disk = 200\}$. A unique machine and cost is determined, since
$$\Pi_{\mathcal{U},\mathbf{X},\mathbf{C}'}^{chosen} = \{\texttt{m1}\} \text{ and}$$
$$\Pi_{\mathcal{U},\mathbf{X},\mathbf{C}'}^{cost} = \{4770\}.$$

That is, according to the intent of the new network $\langle \mathcal{U}, \mathbf{X}, \mathbf{C}' \rangle$ that corresponds to the augmented theory Γ', the set of possible values for *chosen* is $\{\texttt{m1}\}$ and the set of possible values for *cost* is $\{4770\}$. In other words, the only possible machine is the m1 and the cost is $4770.

Scenario 2 Suppose that $4770 is more than we want to spend. We retract $reqd\_disk = 200$ and assert $cost < 4000$, to produce $\Gamma'' = \Gamma \cup \{cost < 4000\}$. We now find that
$$\Pi_{\mathcal{U},\mathbf{X},\mathbf{C}''}^{chosen} = \{\texttt{m2},\texttt{m3}\},$$
$$\Pi_{\mathcal{U},\mathbf{X},\mathbf{C}''}^{reqd\_disk} = \{X|\textsc{leq}(X,140)\},$$
$$\Pi_{\mathcal{U},\mathbf{X},\mathbf{C}''}^{reqd\_slots} = \{X|\textsc{leq}(X,6)\} \text{ and}$$
$$\Pi_{\mathcal{U},\mathbf{X},\mathbf{C}''}^{reqd\_speed} = \{X|\textsc{leq}(X,25)\}.$$

Suppose we assert $reqd\_slots = 6$, to produce $\Gamma''' = \Gamma'' \cup \{reqd\_slots = 6\}$. We now find that
$$\Pi_{\mathcal{U},\mathbf{X},\mathbf{C}'''}^{chosen} = \{\texttt{m2}\},$$
$$\Pi_{\mathcal{U},\mathbf{X},\mathbf{C}'''}^{reqd\_disk} = \{X|\textsc{leq}(X,128)\},$$
$$\Pi_{\mathcal{U},\mathbf{X},\mathbf{C}'''}^{reqd\_slots} = \{6\} \text{ and}$$
$$\Pi_{\mathcal{U},\mathbf{X},\mathbf{C}'''}^{reqd\_speed} = \{X|\textsc{leq}(X,25)\}.$$

That is, the only machine which is within our budget cap of $4000 and which provides enough peripheral slots is the m2. We may consider buying this even though, at 128 MB, it does not provide as much disk space as we would have liked.

Scenario 3 Suppose we are interested in checking out the m3. We can check its functionality by returning to the original network and asserting $chosen = m3$, to produce $\Gamma'''' = \Gamma \cup \{chosen = m3\}$. We now find that
$$\Pi_{\mathcal{U},\mathbf{X},\mathbf{C}''''}^{chosen} = \{\texttt{m3}\},$$
$$\Pi_{\mathcal{U},\mathbf{X},\mathbf{C}''''}^{reqd\_disk} = \{X|\textsc{leq}(X,140)\},$$
$$\Pi_{\mathcal{U},\mathbf{X},\mathbf{C}''''}^{reqd\_slots} = \{X|\textsc{leq}(X,5)\} \text{ and}$$

$$\Pi_{\mathcal{U},X,C''''}^{reqd\_speed} = \{X|\text{LEQ}(X, 25)\}.$$

If we check on the cost of getting this machine, we find that, although it offers only 5 slots, it is within our budget, since

$$\Pi_{\mathcal{U},X,C''''}^{cost} = \{3710\}.$$

We can investigate other "what if" situations by asserting and retracting appropriate sentences and observing the resultant possibility distributions.

3   Galileo3: A Constraint-based language

Galileo3 is a programming language based on the above ideas. A program in Galileo3 provides a declarative specification of a constraint network, analogous to the problem specification in Example 1. The Galileo3 program for Example 1 is:

```
function price =::= datafile(qdb,$PRICES.DBE$).
function capacity =::= datafile(qdb,$DISKCPTY.DBE$).
function slots =::= datafile(qdb,$NUMSLOTS.DBE$).
function speed =::= datafile(qdb,$SPEED.DBE$).
cost = price(chosen) * 1.06.
slots(chosen) >= reqd_slots.
speed(chosen) >= reqd_speed.
capacity(chosen) >= reqd_disk.
not(exists X : slots(X) >= reqd_slots and speed(X) >= reqd_speed and
    capacity(X) >= reqd_disk and price(X) < price(chosen)).
```

The program reads external files, that is PRICES.DBE, DISKCPTY.DBE, NUMSLOTS.DBE and SPEED.DBE, to get the extensions for the function symbols *price*, *capacity*, *slots* and *speed*. Thus, this program is more general than Example 1, in that it can handle whatever types of computer are described in the external files, which can be changed at any time.

Galileo3 provides richer expressive power than any other constraint language, because it allows the theory used to specify a network to contain arbitrary FOPC sentences. Essentially, the CLP languages, for example CLP($\Re$) (Jaffar and Michaylov, 1986) and CHIP (Dincbas et al., 1988), restrict the theory to ground sentences since they treat free variables as uninterpreted constant symbols. A full discussion of Galileo3 and its implementation is beyond the scope of this paper.

References

Bowen, J. and Bahler, D. (1991a). Conditional Existence of Variables in Generalized Constraint Networks *Proc. 9th Conf. of the Amer. Assoc. for Artif. Intel.*, 215-220.

Bowen, J. and Bahler, D. (1991b). *Full First-Order Logic in Knowledge Representation - A Constraint-Based Approach* (Research Report Number TR-91-29). Raleigh, NC: North Carolina State University, Department of Computer Science.

Dincbas, M., van Hentenryck, P., Simonis, H., Aggoun, A., Graf, T. and Berthier, F. (1988). The Constraint Logic Programming Language CHIP *Proc. Fifth Gen. Comp. Sys. '88*.

Jaffar, J., and Michaylov, S., (1986). Methodology and Implementation of a CLP System *Proc. of the 4th Int. Conf. on Logic Programming.*

Nilsson, N., (1991). Logic and artificial intelligence *Artificial Intelligence*, **47**, 31-56.

Programming Planners With Flexible Architectures *

Alessandro Cimatti, Paolo Traverso
IRST
Trento, Italy

Luca Spalazzi
University of Ancona
Ancona, Italy

Abstract

In this paper we discuss some of the features of a domain independent development environment for planning systems, called MRG. Within MRG, the user can program the control mechanism of a planning system depending on the application requirements: detection and reaction to failure, interleaving plan formation and plan execution, ad hoc strategies execution and so on. MRG can thus be used as a flexible and powerful tool to build real systems that work in complex, real world domains.

1 Introduction

In this paper we discuss some of the main features of a domain independent development environment (called MRG) for planning systems. MRG provides a programming language for the development of planners which must deal with the complexity of the real world. Besides the ability to represent goals, actions and plans of actions, MRG has the following two fundamental features:

- in the language, failure is explicitly represented and constructs are provided to build plans that detect and react to failure. The control of the planner can thus be programmed to react to failure depending on the application domain;
- MRG provides a set of libraries containing various basic planning steps, like, for instance, different plan formation, execution and replanning mechanisms. These planning steps are explicitly represented in the programming language

---

*This work has been conducted as part of MAIA, the integrated AI project under development at IRST and has been done with the partial support of the Italian National Research Council (CNR), Progetto Finalizzato Robotica (Special Project on Robotics). Many of the underlying intuitions have been provided by Fausto Giunchiglia. He has also provided useful help and encouragement during the development of the whole project. The Mechanized Reasoning Group at IRST is also thanked.

itself (Traverso, Cimatti & Spalazzi, 1992); they can be defined, reused, composed, modified and activated in the most proper way according to the particular application domain.

Plans and planners are uniformly represented in MRG as **tactics**[1], explicit data structures that can involve a variety of control structures (*e.g.* conditionals and recursion). A tactic can represent a real world action and a complex plan, as well as a planning strategy that controls the activation of different activities of the planning system. Unlike in most planning systems, in MRG, the user can not only define goals and operators, but can use tactics to define the whole planner.

Tactics are briefly explained in section 2. In section 3 we show how failure handling is performed within MRG. Section 4 gives an outline of some of the MRG tactic libraries. Section 5 gives some conclusions.

2 Planners as programs: tactics

Tactics (see Traverso *et al.* (1992) for a formal definition of tactics) are programs in the MRG language. They are inductively defined on the base of *primitive tactics*. Primitive tactics are defined on the basis of *tactic identifiers* and *constants*: A primitive tactic is an expression of the form $(t_i \; c_1...c_n)$, where $c_1, ..., c_n$ are constants and t_i an n-ary tactic identifier. Tactic identifiers have associated an action type. Constants denote entities over which actions are executed. Technically, MRG keeps track of the link between the tactic identifier and the action in a *tactic-action pair*, a data structure $\langle t_i, a \rangle$ where t_i is the tactic identifier of a primitive tactic and a a piece of code that can be executed. MRG keeps track of the link between the constants of the language and the elements of the domain in a *c-d pair*, a data structure $\langle c_i, d_i \rangle$ where c_i is a constant of the MRG language and d_i a data structure.

MRG has a tactic interpreter, whose task is to activate and control the execution of the actions represented by tactics. When primitive tactics are interpreted, the corresponding actions are executed: according to their success or failure, the tactic is said to succeed or fail. When a tactic fails, the interpretation of a tactic is said to be `failure`. Given the set of primitive tactics, the user can define *defined tactics* as expressions of the language of MRG. Defined tactics allow function composition and lambda abstraction. Furthermore, tactics can be composed to build more and more complex ones by a set of general combination operations over tactics, called *tacticals* (see Traverso *et al.* (1992) for a formal and complete definition of defined tactics and tacticals). Tacticals are functions mapping tactics onto tactics. Some of the tacticals implemented in MRG are `then`, `if`, `orelse`. Intuitively, `then` constructs sequential tactics, `if` builds conditional tactics, `orelse` detects failure. For instance, when the tactic (`then t1 (if cond-tac then-tac (orelse t2 t3))`) is executed, `t1` is executed first. If it succeeds, then the conditional tactic is executed. `cond-tac` is a tactic whose evaluation returns, when it succeeds, a truth value. If it succeeds, `then-tac` or (`orelse t2 t3`) are executed according to the result. The tactical `orelse` is explained in section 3.

---

[1] The terms *tactic* and *tactical* are borrowed from ML (Gordon, Milner & Wadsworth, 1979), a procedural metalanguage for the specification of theorem proving strategies. There actually exist some similarities between the two languages, but there are rather strong conceptual and technical differences, mainly due to the different domain of application.

3 Failure handling in MRG

orelse is the tactical for failure handling. Tactics built without orelse fail any time the execution of a primitive tactic fails. This is completely in disagreement with the intuition that after the failure of an action, some alternatives should be tried, and the overall process might succeed anyway. Accordingly, (orelse t_1 t_2) may be thought of as a retrial point: if t_1 fails, then t_2 is tried.

Definition 1 (Orelse interpretation) *Let t_1 and t_2 be tactics. Let t be a tactic defined as* (orelse t_1 t_2). *If the interpretation of t_1 succeeds (with value v_1), then the interpretation of t succeeds (with value v_1). If the interpretation of t_1 is* failure, *then the interpretation of t is the interpretation of t_2.*

Information can be acquired by controlling failure. Failure, in complex and unpredictable application domains, is frequent and difficult to prevent. Furthermore, in some applications, failures are not critical nor unrecoverable: on the contrary, failure can be thought of as a particular way to get information. Success and failure correspond to the result of an "attempt" to solve the problem in a certain way. Failure while executing a particular action tells us to try a different approach. Furthermore, depending on the application domain, some strategies and actions are more likely to fail than others. Then, a good strategy must be able to specify which actions might probably fail and how failure recovery must be undertaken. orelse provides the proper mechanisms to control failure: that is, it provides the means to "capture" failure and to specify alternative actions (or strategies) to be executed in the case of failure.

4 Planning libraries

MRG is provided with a set of libraries containing tactics that perform different planning activities. For instance, the plan generation library contains different tactics to generate plans that can be used according to the application requirements. The libraries include but are not limited to plan generation: tactics to execute plans, to monitor execution, to handle failure and to replan in different ways are available. In this way, various planning steps can be programmed in MRG to represent various planning phenomena. As an example of this capability, we describe briefly how plan formation, execution and information acquisition are represented in the MRG libraries and can be combined and interleaved.

Plan formation library: in MRG, plan formation activities are represented by tactics whose argument is a goal and returning tactics. We call this kind of tactics "plan-for tactics". We have several plan-for tactics in the MRG library. They can be either primitive or defined tactics. A *primitive* plan-for tactic represents a *fixed* process searching over tactics to find a solution to the given goal and is not modifiable at the moment of the installation. A *defined* plan-for tactic is built from primitive ones. Its definition is accessible by the user and is modifiable according to the application domain at the moment of installation. The simplest plan-for tactic that we have in the library is a function that, given a goal, returns the corresponding tactic in a goal-tactic table. On the other hand, the library provides a set of complex plan-for tactics, each of them implementing a different general purpose planning search. Other `plan-for` tactics represent various special purpose plan generation tasks.

Plan execution library: in `MRG`, plan execution activities are represented by "exec tactics" whose argument is a *tactic*. An `exec` tactic, when interpreted by the `MRG` interpreter, executes the argument tactic. In case of success it returns the result of the execution, otherwise it fails. The execution of a tactic representing a plan can be performed in different ways by various `exec` tactics. The simplest execution modality is performed by a tactic that calls the `MRG` interpreter. In this way the whole plan is executed till the end or till a failure occurs that is not captured by `orelse`. Other execution tactics that we have implemented allow us to monitor execution.

Data acquisition library: these tactics acquire information from the real world. Examples of acquisition tactics are tactics returning the information acquired from robot sensors and through a user interface module.

Data-driven library: reactive systems are systems able to react to events and changes in the external environment. In `MRG`, this task is performed by "data-driven" tactics. These tactics acquire information from the external world by means of data acquisition tactics.

Solve library: the global task of a planner is to carry out actions in the real world to achieve some defined purpose. This functionality is represented in `MRG` by "solve tactics", whose argument is a goal. Simple `solve` tactics are compositions of `plan-for` and `exec` tactics, where plan formation is followed by plan execution; this is what usually happens in classical planners. Note that the two fundamental planning steps, plan formation and plan execution, are represented explicitly and kept separated by `plan-for` and `exec` tactics (Traverso *et al.*, 1992). We can program the planner to undertake any action between plan formation and plan execution. All of this depending on the application requirement.

5 Conclusions

`MRG` is a domain independent system for the development of planners. `MRG` allows the user to program the basic control mechanisms of a planning system in a flexible way. `MRG` provides a language with the basic constructs (i) to define basic planning actions and compose them into flexible architectures; (ii) to activate plan formation and execution and to interleave them when necessary; (iii) to handle and to control failure in a flexible way.

References

Gordon M.J., Milner A.J. & Wadsworth C.P. (1979). *Lecture Notes in Computer Science: Vol. 78. Edinburgh LCF - A mechanised logic of computation.* New York, NY: Springer Verlag.

Traverso P., Cimatti A. & Spalazzi L. (1992). Beyond the single planning paradigm: introspective planning. In *Proceedings ECAI-92* 643–647. IRST-Technical Report 9204-05, IRST, Trento, Italy.

On Seeing that Particular Lines could Depict a Vertex Containing Particular Angles.

G. Probert
Department of Pure Mathematics, Queen's University
Belfast, Northern Ireland

R. Cowie
School of Psychology, Queen's University
Belfast, Northern Ireland

1. Introduction

We consider the geometry of deciding what kinds of corner three given lines in a picture could represent. We are specifically concerned with the angles that may occur between the edges which form the corner: this reflects evidence that human vision favours particular configurations of angles (Probert & Cowie 1990, Cowie & Walsh this volume).

Previous work includes well known results for special cases involving right angles (Perkins, 1968; Attneave & Frost, 1969), and work which shows that in the general case fitting a given vertex to particular lines involves solving a quartic with complicated coefficients (Dhome et al, 1989; Probert & Cowie, 1990). One of the issues which emerges we call the limits problem: that is, distinguishing between a range of trihedral corners that particular lines could represent and a range that they could not. Perkins described rules for checking whether lines could represent a cubic corner (i.e. one where all three angles were right) without undertaking full 3-D reconstruction, but it remains unknown whether comparable tests exist for less specialised configurations.

Our approach to the limits problem uses spherical geometry to identify the ranges of vertices which could generate particular projections. Mapping the range exactly is computationally expensive, but we identify approximations which are easy to compute. For machine vision, these results suggest ways of pre-filtering before matching vertices in a model to junctions in an image. For psychology, they raise a range of empirical questions about humans' ability to judge what simple line configurations could represent.

2 Spherical geometry : the basis of the analysis

Figure 1 represents a trihedral vertex, i.e. three edges meeting at a point O, and its projection onto a plane. What are the possible combinations of the angles **U**, **V**, and **W** between the edges that produce projection angles **A** and **B**?

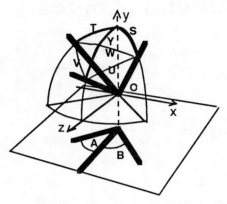

Figure 1

A fundamental relationship in spherical trigonometry leads directly to the basic equations

Cos **U**= Cos **S** Cos **Y** + Sin **S** Sin **Y** Cos **A**
Cos **V**= Cos **T** Cos **Y** + Sin **T** Sin **Y** Cos **B**
Cos **W**= Cos **S** Cos **T** + Sin **S** Sin **T** Cos (**A** + **B**)

Our approach is to consider specific values of **A**, **B**, **U**, **V** and **Y**. The value of **S** is defined as follows:

$$\text{Sin } S_1 = \frac{\text{Cos } U \cdot \text{Sin } Y \cdot \text{Cos } A - \text{Cos } Y \sqrt{\text{Sin}^2 U - \text{Sin}^2 Y \cdot \text{Sin}^2 A}}{1 - \text{Sin}^2 Y \cdot \text{Sin}^2 A}$$

$$\text{Sin } S_2 = \frac{\text{Cos } U \cdot \text{Sin } Y \cdot \text{Cos } A + \text{Cos } Y \sqrt{\text{Sin}^2 U - \text{Sin}^2 Y \cdot \text{Sin}^2 A}}{1 - \text{Sin}^2 Y \cdot \text{Sin}^2 A}$$

The equations for **T** give parallel alternatives. Given **S** and **T** it is straightforward to recover **W**: clearly it has up to 4 values for each combination of **A**, **B**, **U**, **V** and **Y**.

3 YW curves, bar codes, and limit objects

The spherical equations mean that for a given **A**, **B**, **U** and **V**, up to 4 graphs can be plotted for **W** against **Y**. We will call these **YW** curves. Figure 2 shows the **YW** curves associated with one image. These **YW** curves lead directly to conclusions about the range of values **W** can take. Take the highest point on each **YW** curve, and the lowest. The range between them defines a range of values which **W** can have. It is shown by a dark bar on the right of the graph. Doing that for all four **YW** curves specifies the range of **W** for these particular values **A, B, U** and **V**. We will call this description **W**'s bar code.

A single bar code is just part of a much bigger object. Suppose we calculate a bar code for every possible combination of **U** and **V**. Together they form a three-dimensional object. We call this a limit object. Figure 3 shows a slice through one. It is formed by taking particular values of **A** and **B**, then holding **U** constant and varying **V**. The complexity of the limit objects we find argues strongly that neat analytic solutions to the limits problem are most unlikely to exist.

Computing all the **YW** curves behind Figure 3 is very time-consuming. However most the points on its boundary can be easily computed by evaluating **W** at the end points of **YW** curves and introducing a straight line (upper left in figure 3) which represents the situation where all three edges of the vertex lie in a single plane (so **U+V=W**).

Figure 4 illustrates our work on approximation more fully. It considers the symmetric case (**U=V**), where the whole limit object can be shown because it is two-dimensional. Psychologically, equality of angles seems to be an important kind of regularity (Clements & Cowie 1991). Panel (a) shows the object computed exhaustively. Panel (b) shows key curves which lie behind it. The solid curves have simple formulae, and the upper left straight line represents plane vertices. The problem is to understand the other points on the periphery. They are related to the other dotted curves.

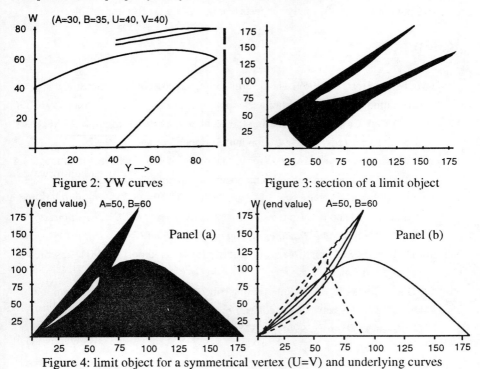

Figure 2: YW curves   Figure 3: section of a limit object

Figure 4: limit object for a symmetrical vertex (U=V) and underlying curves

A critical issue is whether the angles **A** and **B** are acute or obtuse. When both are acute or both obtuse there is a point we call spaghetti junction where the dotted lines meet, and it is about this point that the relevant extrema of the **YW** curves occur. We are exploring ways to compute **W** at these extrema or make a good approximation. The double acute case is an exact reflection of the double obtuse about the vertical **U=V**=90°.

When **A** and **B** are mixed (one acute and one obtuse) options tend to be very limited. This is because of the two curves which form a "smile" on the left hand side of 4(b). Each is a simple function of one of the picture angles, and they create the "hole" that is visible in 4(a). When **A** and **B** are mixed these curves lie on opposite sides of the vertical **U=V**=90°, giving a large "exclusion zone". This points to a significant general strategy. It is possible to identify junctions which severely restrict the range of vertices that could be fitted, and capitalising on these may be a powerful way of filtering options.

4 Conclusion

It seems clear that there is no short and perfect way to answer the question "could *this* projection represent a vertex containing *these* angles?". On the other hand there are short ways which involve some level of approximation. Exactly parallel conclusions apply to the related questions "if *this* is the projection of a vertex V, what range of angles could V contain?". This illuminates options which research on either machine or biological vision can sensibly pursue - and strongly suggests that others are (sadly) impossible.

References

Attneave, F. and Frost, R. (1969) The determination of perceived tridimensional orientation by minimum criteria. *Perception and Psychophysics* **6**, pp. 391 -396.

Clements, D and Cowie, R. Measuring the 'Rubber Rhomboid' effect. In M.McTear & N. Creaney (eds) *AICS 90* London: Springer Verlag. pp 193-208.

Cowie & Walsh (this volume) Why go for virtual reality if you can have virtual magic? A study of different approaches to manoeuvring an object on screen.

Dhome, M., Richetin, M., Lapreste, J-T., and Rives, G. (1989) Determination of the Attitude of 3-D objects from a single perspective view. *IEEE transactions of Pattern Analysis and Machine Intelligence* **11**, 1265-1278.

Perkins, D. N. (1968) *Cubic Corners*. Quarterly Progress Report 89, MIT Research Lab in Electronics.

Probert, G and Cowie, R. (1990) Interpreting trihedral vertices by using assumptions about the angles between the edges. *Proceedings of the British Machine Vision Conference: Oxford, 1990.* pp. 373-378.

Terminological Representation and Reasoning in TERM-KRAM[*]

Daniela D'Aloisi
Fondazione Ugo Bordoni
Rome, Italy

Giuseppe Maga
IP-CNR, National Research Council
Rome, Italy

Abstract

TERM-KRAM is a fully implemented terminological language, endowed with the direct representation of n-ary predicates, the consequent elimination of the role as an autonomous entity, the use of an INCORPORATES link, and the introduction of a STRUCTURAL DESCRIPTION for each defined concept to describe the internal structure of terms. TERM-KRAM emphasizes the intensional aspects of concepts. Furthermore, it provides their meaning on the ground of previously defined terms by assigning them a CONCEPTUAL SCHEME that reflects their lexical decomposition, i.e., their meaning with respect to natural language. These features make TERM-KRAM particularly suitable for describing verbs and complex terms.

1. Introduction

TERM-KRAM is the terminological component (Tbox) of a knowledge representation system, called KRAM (Knowledge Representation for Agency Modeling) (D'Aloisi & Castelfranchi, 1992), that follows a hybrid approach (Nebel, 1990): KRAM also includes an assertional component (Abox), PROP-KRAM (D'Aloisi et al., 1989), connected with a Truth Maintenance System. KRAM is part of a project whose task is to build a realistic model of how agents interact with each other and with the external world. Even if the system does not support the use of natural language, it allows a user to have at his/her disposal all the expressive richness that natural interaction permits, i.e., the system should offer a formalized language that is as communicative as natural language can be. As a consequence, TERM-KRAM's task is not limited to the representation of a taxonomy of objects connected by means of relations, but includes the provision of a language able to describe the internal structures of terms usually used by agents in a dialogue, the image of a term as it exists in the mind of cognitive agents. As a matter of fact, in a natural interaction attributing a unique meaning to the words is important since a communicative act implies more knowledge than explicitly expressed. Moreover, it is necessary to represent not only objects, but also other types of terms as verbs, actions, and complex entities. TERM-KRAM provides a language capable to represent the meaning of all the entities that are necessary to model interactions between agents. It uses n-ary predicates instead of only unary and binary terms (respectively called concepts and roles in most hybrid systems) which allow for a more direct and efficient representation that results in an increase in the expressive power of the language. The use of the n-ary predicates has caused the suppression of the role as an autonomous entity. Each term is associated to a Conceptual Scheme (CS) which is the representation of the term's internal structure based on its lexical decomposition. As regards the inferential capacity, TERM-KRAM attempts to mostly exploit the knowledge structures by defining query functions capable to retrieve information from the semantic connections among different concepts. The knowledge

---

[*] Correspondence should be sent to the first author at Fondazione Ugo Bordoni, Via B.Castiglione 59, I-00142 Rome, Italy (e-mail: fubdpt5@itcaspur.bitnet).

structure connected to a description can improve the inferential capabilities of the system: as a matter of fact, when the descriptive knowledge is poor and/or little structured, the inferred knowledge may end up being too limited and not very useful; if the definition of a term is expressively rich, the results of the reasoning process can augment the applicability of the system.

2. The Language

TERM-KRAM defines a language that allows a terminological network to be built, in which the nodes are concepts and the edges state relationships between concepts. It also offers tools to set up the network and to reason on it. The concepts are represented by n-ary terms: *man(x)* and *tell(x y z)* are examples of concepts. The edges represent SUPERC and INCORPORATES relationships: the SUPERC link implements a hierarchical nexus between concepts, and the INCORPORATES link allows concepts to be embodied in the description of other concepts. The well-formed terms of TERM-KRAM have to obey the following syntax (for its semantics see (D'Aloisi & Maga, 1991)):

<term-entry>	::=	***THING*** /
		Top^n / with $n \geq 1$
		<concept> /
		<disjoint-concept>
<concept>	::=	<primitive-concept> / <complex-concept>
<primitive-concept>	::=	*(<k-predicate-symbol> ((<var>$_1$ <type-restriction>$_1$)*
		... (<var>$_k$ <type-restriction>$_k$))
		***NIL*)**
<complex-concept>	::=	*(<k-predicate-symbol> ((<var>$_1$ <type-restriction>$_1$)*
		... (<var>$_k$ <type-restriction>$_k$))
		<Conceptual Scheme>)
<type-restriction>	::=	*<1-predicate-symbol> / **Top^1***
<Conceptual Scheme>	::=	*([<ConceptConjunction>] [<StructualDescription>])*
<ConceptConjunction>	::=	***(AND** <k-predicate-symbol>*)*
<StructualDescription>	::=	***(INCORPORATES** <terminological-statement>*)*
<terminological-statement>	::=	*(<h-predicate-symbol> <arg>$_1$...<arg>$_h$) /*
		*(**NOT** <h-predicate-symbol> <var>$_1$...<var>$_h$))*
<k-predicate-symbol>	::=	**a name of predicate**
<arg>	::=	*<var> / <terminological-statement>*
<var>	::=	*x / y / z /....*
<disjoint-concept>	::=	***(DISJOINT** <k-predicate-symbol>$_1$*
		<k-predicate-symbol>$_2$)

The language can represent primitive and complex concepts, the disjointness between concepts, and the special concepts *Top^n* and THING. THING represents the most general term, i.e., the *Universal Concept*, and *Top^n* represents the most general concept of arity *n*. A concept -both primitive and complex- is denoted with an n-ary predicate, where *n* is the number of its arguments. Concept arguments are denoted with variables, which are just internal symbols, and are characterized by a conceptual role by means of a 1-place predicate, which limits the domain of the possible values that the variable can take on, i.e., it defines a *sort* for the variable.

Primitive concepts describe entities lacking definition; complex concepts refer to partially or completely analyzed entities. A complex concept *C* is characterized by its definition, which represents its meaning, designated in the syntax with the non-terminal symbol <*ConceptualScheme*>. It consists of two parts: the list of the concepts whose conjunction forms *C*, introduced by the AND operator, and the Structural Description (SD), introduced by the INCORPORATES operator. The first list allows the user to explicitly state the set of concepts the term is made up from. The Structural Description consists of a network of structured *terminological statements* that relate a concept to other terms on the ground of semantic relations --i.e., cause-effect or conditional links-- in order to reflect the fact that a term t_1 can concur in the formation of the meaning of a term t_2 without transmitting its properties to it. The INCORPORATES link relates two concepts so that the meaning of the described concept is specified by means of the second one, and can be thought of as a sort of implication between two terminological statements. From the inheritance's point of view, the incorporation is a poorer relation than it is generalization: in fact, it relates two concepts simply by connecting two *terminological propositions*, while a hierarchical link, i.e., C_1 subsumes C_2, claims that C_2 inherits all the knowledge structure of C_1. Each terminological statement consists of a previously defined predicate applied to arguments that can also be propositions in their turn.

The language also offers the possibility of declaring a disjointness between concepts. A concept inherits the propriety of disjointness from its predecessors. As a matter of fact, if $Subc(C_a)$ and $Subc(C_b)$ are respectively the set of the concepts subsumed by C_a and the set of the concepts subsumed by C_b, then if C_a and C_b are disjoint then

$$\forall C_i \in Subc(C_a), \forall C_j \in Subc(C_b) \ (\varepsilon[C_i] \cap \varepsilon[C_j] = \varnothing).$$

Let us show a brief example of a description in TERM-KRAM (Fig.1). The definition of the verb *to tell* connects an actor, who is speaking, a recipient, who is listening, and an argument, what the actor is telling. The definition makes the system aware of the underlying facts that the act of telling semantically implies. The structured list of incorporated propositions allows the system to share the mental schemes of an agent, and also to deduce what the execution of an action causes: to a certain degree, the Structural Description might be thought of as the *add list* of a planning operator even if it can comprise different kinds of information. TERM-KRAM offers deduction functionalities, that are strongly related to the representational structures of terms, that are briefly discussed in the next Section. For a more detailed discussion about the properties of TERM-KRAM see (D'Aloisi & Maga, 1991).

```
(Tell
    ((Agent x) (Patient y) (Object z))
    ((AND
        Communicate)
    (INCORPORATES
        (Person x)
        (Person y)
        (Fact z)
        (Know x z)
        (Speak x y)
        (Cause (Speak x y) (Know y z))
        (Listen y x))))
```

Fig.1

3. Functionalities of TERM-KRAM

TERM-KRAM offers a collection of functions to build the knowledge base and to perform queries on it: it embodies a graphical interface, which displays the terminological network and permits a straightforward interaction with the system. The primitive functions, that the system puts at the users' disposal, can be divided into two groups, the TELL functions and the ASK functions. The TELL functions are used for updating the knowledge base, i.e., for introducing new concepts, for incrementally defining concepts, and for deleting concepts. In addition, they check the entries so that the information stored in the knowledge base is always consistent. The ASK functions retrieve information or deduce implicitly stored information from the knowledge base.

Another important functionality of TERM-KRAM is the subsumption calculus, i.e., finding out if concept C_1 is more general than concept C_2. As a consequence, also the classification, i.e., automatically inserting a concept in the terminological network, retains an important place. Most of the last works in this field focused on finding a language that is both expressive and efficient for the performance of the subsumption calculus: in our opinion, it is also important to stress the expressive power of the representation: sometimes it is better to have a general and expressive -possibly incomplete- language suitable to a large class of different applications, than a complete language impoverished with respect to its representational competences. In TERM-KRAM, two different subsumption algorithms are defined, a *subsumption algorithm* (SUBSUME?) and a *strong-subsumption algorithm* (STRONG-SUBSUME?) in order to consider the particularity of the SD. The function SUBSUME? uses weaker conditions in order to make the algorithm computable. As matter of fact, the subsumption procedure is sound even if incomplete in some cases, but the strong-subsumption procedure is sound and complete. The complexity of the subsumption algorithm is $O(n^4)$, where n is the number of the concepts in the terminological network; currently, we are verifying the complexity of the strong-subsumption algorithm. TERM-KRAM also defines a classification algorithm, whose task is to automatically insert new concepts in the terminological network. The features and functionalities of TERM-KRAM are more largely described in (D'Aloisi & Maga, 1991).

TERM-KRAM is implemented in Common Lisp on a Symbolics Lisp Machine. A graphic interface facilitates the interaction by showing the terminological network and by providing menus that implement its basic functionalities. TERM-KRAM has been applied with good results in practical applications, for instance as the representational component of an intelligent help system (D'Aloisi, 1992).

Acknowledgement
We thank Cristiano Castelfranchi who contributed to the progress of the KRAM project.
This work has been partially supported by "Progetto Finalizzato Sistemi Informatici e Calcolo Parallelo" of CNR under grant n.104385/69/9107197; Daniela D'Aloisi carried out her work in the framework of the agreement between the Italian PT Administration and the Fondazione Ugo Bordoni. Giuseppe Maga's present address is: Tecsiel, Via Ponte di Piscina Cupa 43, I-00128 Rome, Castel Romano, Italy.

References
D'Aloisi, D., Castelfranchi, C., & Tuozzi, A. (1989). Structures in an Assertional Box. In H. Jaakkola & S. Linnainmaa (Eds.), *Scandinavian Conference on Artificial Intelligence - 89* (pp.131-142). Amsterdam: IOS Press.
D'Aloisi, D., & Maga, G. (1991, November). *Terminological Representation and Reasoning in TERM-KRAM (Extended Version)*. (FUB Tech.Rep. 5T05091) Rome: Fondazione Ugo Bordoni.
D'Aloisi, D. (1992). A Terminological Language for Representing Complex Knowledge. In F. Belli & F.J. Radermacher (Eds.), *Industrial and Engineering Applications of AI & ESs* (pp.515-524). Berlin: Springer-Verlag.
D'Aloisi, D., & Castelfranchi, C. (1992). Propositional and Terminological Knowledge Representations. *Proceedings of the AAAI Spring Symposium on "Propositional Knowledge Representation"* (pp.57-67). Stanford, CA: AAAI (An extended version will appear in *The Journal of Experimental & Theoretical AI*, **5** (3), 1993).
Nebel, B. (1990). *Reasoning and Revision in Hybrid Representation Systems*. Berlin: Springer-Verlag.

Integrating Planning and Knowledge Revision: preliminary report

Aldo Franco Dragoni

Istituto di Informatica, Università di Ancona
Ancona, Italy

Abstract

We've examined various thinkable ways in which plan formation, execution, monitoring and replanning could take advantage from Truth Maintenance techniques. The creation of a plan is based on a set of assumptions about the external environment and task goals. The inconsistency of these sets or the falsehood of some among their elements can cause failures both in plan generation or execution. On failure it becomes important to understand which assumptions were responsible for the flaw in order to get a more updated knowledge base or a consistent goal. Consistency restoration, multisensor information integration and updating knowledge bases, all need knowledge revision techniques. In this paper we report some possible schemes for the interleaving between the activities of an Assumption Based Truth Maintenance System (ATMS) and one or more (STRIPS-like) Planners. We distinguish four cases of useful adoption of an ATMS:
1. to check and eventually resolve inconsistencies during a single planner planning,
2. to update the knowledge base of a single planner from multisensor information integration or on failure during the execution,
3. case 1 but in a multi-planner architecture,
4. case 2 but in a multi-planner architecture.

1 Preliminaries

Given a first order language L, we describe the state of the world by means of a set S of atomic ground sentences of L. Let R be a definite set of sentences of L representing the causal theory of the world. We define $D = S \cup R$ a *description of the world*.

The *Reviser* is a couple ATMS/Chooser, that is a typical knowledge revision system (Martins, 1988). It is an automaton that accepts as input a description of the world, eventually inconsistent,[1] and gives as output a consistent and anyhow preferable subset of that description. A *context* is a subset of D that is *consistent* (i.e. it has no inconsistent subsets) and *maximal* (it becomes inconsistent if augmented with whatsoever else assumption in D). The ATMS takes as input D and gives as output the set of all the contexts of D. The task of the Chooser is that of selecting one of the contexts supplied by the ATMS as the preferred context. The Chooser is a plausibility/preferability meta-function; it is an opportune domain dependent algorithm that adopts some arbitrary domain dependent selecting criteria. We adopt the simple STRIPS-like definition for actions (Nilsson, 1982). They are atomic means to modify the state of the world. An action A changes the state of the world from S to A(S). A goal statement G is a set of sentences of L. A planner is an automaton that takes as input a description D of the world and a goal statement G, and gives as output a plan, that is an acyclic labelled graph of states. The arcs are labelled with action instances. Each leaves of the graph is a final description SF_i of the world in which the goal statement is satisfied. That is, for each complete ordered action sequence in the plan $A_1..A_n$ it holds

$$A_n(A_{n-1}(..(A_1(D))..)) \models G$$

---

[1] We give to the term "Inconsistency" a broad and indefinite meaning. It could mean strictly "logical inconsistency" (i.e. the description has no models), but in this case we would have troubles to check for it in a full first order description. It could mean "strong-inconsistency", that is the contradiction already belongs to the set of sentences, not only it is derivable from that set. It could also mean simply that, for whatever arbitrary reasons, the sentences in the set - "nogood sets" in (de Kleer, 1986) - can't stand together. What is important to use an ATMS is that if a set of sentences is inconsistent, then every superset of it is inconsistent too.

If the actions are represented by sound operators and a situation S is consistent, then A(S) should be consistent too. So, if the initial description of the world D is consistent and all the operators available to the planner are sound, then each SF_i shall be consistent too and needs not to be checked for consistency. The output of a planner is the overall graph. Hence, the output of the planner and the input of the ATMS are quite incongruent while the output of the Reviser and both the two inputs of the planner are congruent.

2 Integrating a Reviser and a single Planner

If the initial state description is inconsistent, then the final state descriptions may be inconsistent too. If the situation is *logically* inconsistent the problems are more serious; in this case every sentence of L is a logical consequence of the initial situation. A classical planner, before planning an action, checks for the action's preconditions to be verified in the current situation. Suppose that the method used by the planner to verify a sentence is complete, that is, if a sentence is a logical consequence of a situation D then the planner is able to prove it. If the situation is logically inconsistent then such a planner will verify every precondition of every operator. It will be so possible the planning of every action and the whole planning process will be vanished. In such a scenario it is justified the use of a Reviser to check and eventually resolve in input to the planner the inconsistencies of the initial situation. The resulting initial situation will be a consistent maximal subset of the previous one. In practice, the occurrence of an inconsistent initial situation is not unlikely, especially in robot planning domain, where the information about the environment come from a multisensor apparatus. In a multi-agent domain, the initial situation is also built upon information coming from the various other agents in the world. This information sources' multiplicity is being considered as the main cause of inconsistency. Another cause for the inconsistency of the initial situation is the dinamicity of the world. If the world changes during the time needed to the system to sinthesize the plan, then the new information coming from the sensors may be inconsistent with the previous ones. Before starting the execution of the plan it will be necessary to recheck for and eventually to restore the consistency. It will mean to retain as much of the previously held knowledge is consistent with the new information about the world. Once detected the contradictions and calculated the set of contexts, we need a good Chooser to select the preferred context. In (Dragoni, 1991) we've examined a Chooser for Knowledge Revision in a multi-agent environment, presenting some general algorithms and criteria that can be adapted to the specific case under consideration. If the goal statement G is inconsistent then every final situation verifying it shall be inconsistent too. If the initial situation is consistent and all the operators available to the planner are sound then we can't obtain an inconsistent final situation, therefore there not exists a plan to reach that goal statement. However, this seems to be an acceptable behaviour because it would be strange the planner's being able to sinthesize plans for goals that do not have a model. Goal inconsistency can appear, for instance, in multi-agent cooperative domain, when an agent adopts the goals of more then one other agent at the same time.

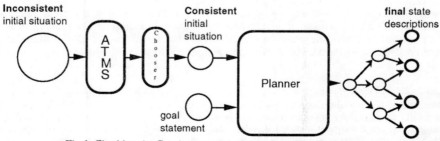

Fig 1. Checking the Consistency of the Initial State Description.

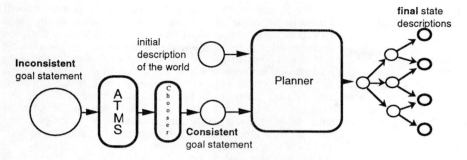

Fig 2. Checking the Consistency of the Goal.

During the execution of a plan, a robot could find the state of world different from the one expected. May be the expected initial situation was different from the real world's one. May be the state of the world changed because of casual or unforeseen events. Most of the differences and of the changes cause inconsistencies with the previous expected or unchanged situation. The general frame problem is that of specifying what doesn't change when an event occurs (McCarthy, 1969). In our implementations we have addressed this problem simply retaining from the previous situation as much knowledge as possible that is consistent with the new information. This task has been accomplished by a Reviser. In this case, we've added a special important criterion to those presented in the previous section, that is simply:

1. new coming information from an agent/sensor are always the strongest assumptions

In this way the preferred context will always contain the latest information about the world and as much knowledge, from the previously held one, is consistent with them and preferable according to the other criteria presented. May be the changes don't affect the plan execution. The "interesting" case is that in which the removing of inconsistencies causes the invalidity of the preconditions of an action in the plan. There are various possible strategies. At least two extremely positions are worth notice:

1. the planner could stop the execution and replan from the current situation (the situation in which the robot has perceived the change), or

2. the planner could continue the execution until the unperformable action is encountered and then replan from that new situation (the situation it has reached).

Both strategies have their rationales and pitfalls.

3 Integrating a Reviser in a Multi-Planner domain

In distributed planning (Bond, 1988) a single plan is produced and executed by the cooperation of several planners. Each planner produces a subplan, but there may be conflicts among subplans that need to be reconciled. Given a goals conjunction, each planner develops a plan for a subgoal. May be each planner is specialized on a class of problems, may be they work at different levels of abstraction. As usual, we must check for and, possibly, resolve, negative interactions between the various subplans (a subplan is incompatible with another one). We are going to study the use of a Reviser to check and eventually resolve these incompatibilities. Each planner produces its plan and then gives the final situation in which the world would be from its point of view (i.e. as if its plan were the only to be executed). We define *global final situation* the union of each final situation. This final situation may be inconsistent. It can happen even if the goals conjunction is consistent. In fact, if MG is the set of the models of a goal and MFin is the set of the models of the final situation produced by a plan achieving G, then MG⊇MFin. In such a scenario, the Reviser could select a preferable context CFin of the global final situation. At this point, the idea is that of accepting only the plans whose final situation belongs to CFin. The planners whose plan is not accepted must replan taking as initial situation the union of the final situations of the accepted plans, i.e. CFin minus the set of the sentences introduced only by the rejected plans. Obviously, the plans that will substitute those rejected must be executed after the end of the execution of the plans accepted initially. This procedure is iterated until all the planners produce an accepted plan. Alternatively, a planner could replan using a dependency directed backtracking strategy (may be using a TMS-like system). If we record the dependencies between actions and effects, the planner whose plan is rejected may be notified with the earliest action in its plan that introduced the problematic effects. Then it could replan from that action forward. These strategies are not necessarily alternative.If there is at least one planner that in unable to sinthesize a plan to reach its goal, the only solution is that of backtracking over the choice of the rejected plans. It means that we now accept a plan previously rejected (may be the plan whose planner has been unable to sinthesize another one) and reject some plans (in a minimal number) among them previously accepted.

Suppose that all the planners were able to produce their plan and all the final incompatibilities were solved. Suppose that one (or more) of the planners fails during execution because of the fact the world is different from what expected. May be the global initial situation was not correspondent to the real world situation, may be something is changed due to casual or unforeseen events, may be the plans were "final compatibles" (compatibles from the final situation point of view) but not compatibles in their intermediate steps. Because of the fact that we can't know a priori the actions' duration we can't resolve the intermediate step's incompatibility simply by synchronising the actions. The drastic solution is that of making a preliminary cross compatibility test among every action in every plan and all the other actions of all the other plans. On detecting a cross incompatibility one of the two planners will have to replan (may be only the subsequence of the plan starting from the incriminate action). When the cross compatibility test is successful the execution starts. At the moment we can't say much on this argument because we are still discussing the subject. However, it seems that the use of an ATMS during the execution in a distributed architecture is not much useful!

4 Conclusions

We've studied four cases of integration between the classical planner's and the ATMS's paradigms.

1. During a single planner planning stage we've justified the employment of a Reviser to preserve the consistency of the initial description of the world and of the task goal.

2. During a single planner execution stage we've found useful a Reviser to restore the consistency of the description of the world, lost because of unexpected changes of the world.

3. During a distributed planning stage we've justified the employment of a Reviser to preserve the final consistency of the distributed plan.

4. During a distributed execution stage, at the moment, we've found the employment of a Reviser not more useful than in the second case.

5 References

Bond, A. H., & Gasser, L. (Eds.) (1988). *Readings in Distributed Artificial Intelligence*. San Mateo, CA: Morgan Kaufmann Publishers.

De Kleer, J. (1986). An Assumption-based TMS. *Artificial Intelligence*, **28**, 127-162.

Dragoni, A. F. (1991). A Model for Belief Revision in a Multi-Agent Environment, In E. Werner & Y. Demazeau (Eds.), *Decentralized A. I. 3*. Amsterdam, The Netherlands: North Holland Elsevier Science Publisher.

Martins, J. P., & Shapiro, S. C. (1988). A Model for Belief Revision. *Artificial Intelligence*, **35**, 25-79.

McCarthy, J., & Hayes, P. (1969). Some Philosophical Problems from the stand point of Artificial Intelligence. *Machine Intelligence*, **4**, 463-502.

Nilsson, N. J. (1982). *Principles of Artificial Intelligence*. New York, NY: Springer-Verlag.

Learning control of an inverted pendulum using a Genetic Algorithm.

Malachy Eaton
Dept. of Computer Science and Information Systems, University of Limerick
Limerick, Ireland

Abstract

This paper presents an approach to the general problem of the control of processes whose dynamic characteristics are not known or little known. It demonstrates how a system consisting of a relatively small number of neuronlike adaptive elements can solve a difficult learning control problem. The task posed involves learning to balance a pole which is hinged to a movable cart by applying forces of varying magnitude to the base of the cart. We assume that the equations of motion of the cart-pole system are not known and that the only feedback signal available is the angle that the pole makes with the cart at a point in time and the previous two sampling instants, together with an overall performance measure for each controller. It is very difficult to tell what is the proper output of the controller to cancel the error based on instantaneous measurements, it is better to adopt a more global approach and this is the method used in this paper. The procedure adopted is a type of simulated evolution process in which a group of controllers, each of which is represented by a small neural network, gradually improves over time. They do this by combining their connection weights, by small mutational changes to their weights, and by selective reproduction based on fitness values assigned to each network based on a global evaluation of their performance obtained using the Integral of Time by Absolute Error (ITAE) criterion.

1 Control of an inverted pendulum

To illustrate the control abilities of the system we choose the classic and fascinating problem of the control of an inverted pendulum. The cart must be moved so that the mass m is always in an upright position. This system is inherently unstable and is related to the control of a rocket about to take off.

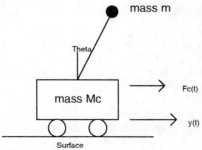

Fig 1. *Simulated system*

The system under simulation is shown in Fig. 1 where M_c = mass of cart, m = mass of pole, l = length of pole, and Fc = Force applied to centre of cart at time t.

In order to fully simulate this system a set of non-linear equations is required which take into account the coefficient of friction of the cart on the track and the pole on the cart, however by ignoring friction and assuming that the angle of rotation, θ, is small so that the equations are linearised, and assuming the mass of the pole is insignificant compared with the mass of the cart we obtain the following simplified equations.

$$M_c \frac{d^2x}{dt^2} + ml\frac{d^2\theta}{dt^2} - F_c = 0 \tag{1}$$

$$ml\frac{d^2x}{dt^2} + ml^2\frac{d^2\theta}{dt^2} - mgl\theta = 0 \tag{2}$$

where g = acceleration due to gravity.

This is the system modelled in this experiment. A general purpose simulation system was also developed in order to simulate this system and others. The specific parameters used are $Mc = 4$kg, $m = 0.2$kg, $l = 1.25$ metres.

The neural network structure chosen to attempt control of this system consists of a three layer network of neurons using linear activation functions with feed forward connections between the layers. It takes as input the current angle the pole makes with the cart, θ, and the angle the pole made at the previous 2 time instants, and it outputs the control signal Fc to push the cart left or right. The sampling/control period is chosen by the genetic algorithm. A novel encoding scheme is used combining elements of thermometer coding and straight binary coding. The overall structure of the 30 bit chromosome representing the neural controller is as follows

Sampling Period	unused	W0	W1	W2	W3	W4	W5
3 bits	3 bits	4 bits	4 bits	4 bits	4 bits	4 bits	4 bits

Fig 2. *Chromosome structure*

The architecture of the neural structure is as follows:

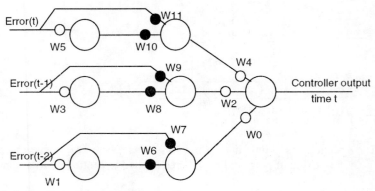

Fig 3. *Detailed structure of the neural controller*

Here the filled in small circles represent weights which are fixed, the remainder are chosen by the genetic algorithm. The fitness function used throughout these simulations is based on the ITAE criterion, mentioned earlier. W6, W8 and W10 have the value 1/16, W7, W9 and W11 are fixed at unity. Bits 0-2 are used to represent the sample/control period of the digital controller, it is allowed to vary between 4 seconds and 1/32 of a second. The sample/control period chosen by the best controller in the final generation is 1/16 of a second. The variable weights are coded in four bits each, taking values between minus 16 and 16. General results of this simulation are given in Fig. 4 showing both best and average controllers per generation. Excellent control is achieved after 25 generations of controllers demonstrating the utility of this approach.

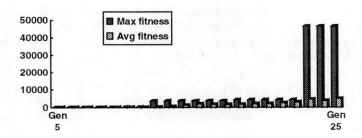

Fig 4. *Maximum and average fitness for inverted pendulum.*

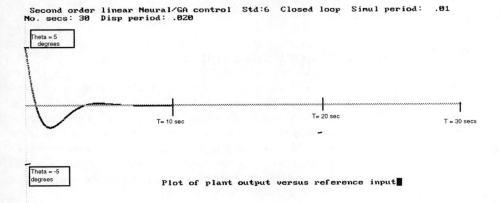

Fig 5. *Plant output for best controller after 25 generations*

2 Acknowledgment:

Portions of this work were done as part of work leading to a Ph.D degree in the College of Engineering and Science in the University of Limerick, under the supervision of Professor Eamonn McQuade, whose help and support is gratefully acknowledged.

Aspectual Prototypes

Antony Galton
Department of Computer Science, University of Exeter,
Exeter, U.K.

1 Aspect and aspectual categories

Aspect has been much studied in linguistics, both conventional and computational, and a loose consensus has emerged roughly as follows. A verb phrase describes a particular type of 'situation', which its aspect modifies by converting it into a different type. This yields two aspectual notions, for which I propose the terms *primary* and *secondary* aspect, corresponding to the traditional *Aktionsart* and *Aspekt*.

The distinction is primarily semantic, but it tends to be reflected syntactically too. Sometimes, indeed, syntax can provide the best guide to drawing the distinction satisfactorily. When aspect is expressed lexically (e.g., durative 'travel' *vs* punctual 'arrive') or contextually (e.g., telic 'eat a boiled egg' *vs* atelic 'eat scrambled egg'), we call it primary, but when it is expressed morphologically (e.g., 'eat' vs 'be eating' vs 'have eaten') we call it secondary. These syntactic distinctions are themselves not always clear: e.g., is the perfective/imperfective distinction in Slavonic lexical or morphological? Semantically, this amounts to asking whether it expresses primary or secondary aspect.

The detailed articulation of this picture requires a classification of situation-types and aspect-types. Verkuyl (1989) gives a full survey of the situation-type classifications. What makes such classification hard is that whatever system is proposed, it always seems possible to discover anomalies which don't fit it. This is where prototype theory can help.

2 Prototype theory and aspect

Prototype theory has been championed by many researchers in linguistics, cognitive science, and psychology, most conspicuously Lakoff (1987). It is an attempt to break the mould of traditional assumptions about categorization. The traditional view, it is claimed, is that categories are defined in terms of necessary and sufficient conditions for membership, and correspond to objective discontinuities in reality. Prototype theory claims that most categories cannot be defined in this way, and they correspond to the organization of our mental conception of the world rather than to the objective nature of the world itself.

A category is held to consist of, first, a nucleus of central or prototypical members characterized by the possession of certain salient attributes; and second, various more or less non-central or peripheral members which owe their membership of the category to their sharing some, but not all, of the salient attributes of the prototypical members. It may also be possible for something to belong to a category by virtue of attributes shared with non-central members of the category only.

Aspect theory is concerned with properties like *durativity, telicity, unitarity*, and *agentivity*. A prototypical situation has all of these: it is a single end-directed action which takes time, for example *John eats a pizza*. Situations close to the prototype but not themselves prototypical are obtained by removing one of the salient attributes, as *John eats* (not telic), *John starts eating* (not durative), *John eats pizzas* (not unitary), or *John falls down* (not agentive).

Prototypical meanings of secondary aspects are defined in terms of the attributes of prototypical situations which they apply to; these prototypical meanings will themselves be characterized in terms of lists of salient attributes. We can then discover non-central meanings of the aspects either by applying prototypical aspects to non-prototypical situations or by considering possible deviations from the prototypical aspects.

3 The Progressive

Let the uninflected verb-phrase VP refer to a telic, durative, unitary action A. Then the progressive form of VP prototypically refers to a state in which the agent is engaged in activity that is intended to form part of a to-be-completed action of type A. Thus the prototypical interpretation of *John is eating a pizza* is that John is engaged in the activity of eating a pizza (i.e., putting portions of the pizza into his mouth one after another, chewing and swallowing them), and he intends to continue doing this until the pizza is all eaten. The salient attributes are that (1) the activity is actually happening, and (2) there is an intention that it should continue to the end. By dropping one or other of these attributes we get variations on the prototypical meaning of the progressive, as follows.

First, the activity is not actually happening. Either (a) it has been interrupted (the 'broad' sense of the progressive) or (b) it hasn't started yet (the 'futurate' progressive): here the progressive may merely record a present intention for the future (*I am driving to London tomorrow*), or the activity may contribute to the completion of the event without forming part of it (e.g., *I am making a cake*, said while shopping for ingredients or getting them out of the pantry).

Second, the intention of an agent is not invoked. Instead, the progressive is justified by the fact that the activity was actually completed. This only applies to past tense sentences, because only then is the fact of completion available: *When Beethoven was writing the Eroica, ...* (the 'closed' sense of the progressive).

When the situation involved is a telic event, as here, the non-progressive contrasts strongly with the progressive: the former entails the completion of the event, the latter merely entails an activity directed towards that completion. So there are situations that *John is eating a pizza* applies to but *John eats a pizza* does not: namely those in which John starts eating a pizza but doesn't finish it. This is the 'imperfective paradox', though it is, surely, hardly paradoxical.

We now consider applying the progressive to non-prototypical situations.

First, the action is not telic. Here there cannot be an intention that it should continue to the end, only that it should continue. The contrast between stopping short of the end and continuing to the end is neutralised so that, e.g., *John ate* and *John was eating* apply to the same situations, differing only in 'narrative focus' (the former treats the situation as an event, the latter as a kind of state). So the 'imperfective paradox' does not arise for atelic situations.

Second, the action is not durative. Here there is no activity that can form part of the action, nor can the action be interrupted; the natural interpretation of the progressive is that an agent intends that the event *will* occur, or is actively engaged in preparing for it (as in the futurate progressive), e.g., *I'm leaving* said by someone putting his coat on in readiness to go.

Third, the action is not unitary. The situation is inherently 'interrupted', so a 'broad' reading applies (*frequentative* progressive): *John is eating pizzas.* Here non-unitarity is explicitly marked by the plural inflection on the object; but it can be implicit, as in *Someone is knocking at the door.* When a situation is neither telic, durative, nor agentive, nothing remains to support prototypical or futurate readings, so the frequentative reading is forced: *John is coughing, The light is flashing.*

Fourth, the event is non-agentive. The process must derive its end-directedness from a source other than an agent's intention. Possibilities are (a) the 'closed' progressive, discussed above (*When Vesuvius was erupting, ...*); (b) bare physical causality: *The tree is falling down*; (c) a mechanism, as in *Your money is being counted*, displayed at an autobank terminal.

4 The Perfect

Let the uninflected verb-phrase VP refer to a telic, durative, unitary action A. Then the perfect form of VP prototypically means that some state of affairs consequent upon the actual completion of an action of type A still obtains. For example *Mary has written a letter* refers to a state of affairs in which a letter by Mary exists. The salient attributes of this reading of the perfect are that (1) the action is complete, and (2) the result is a state which is specifiable independently of the action but which is directly caused by it. Non-prototypical senses of the perfect can be derived by dropping or modifying these attributes.

First, the action is not complete. For a telic action, this requires a verb in the progressive: *Mary has been writing a letter.* Here the result state is the existence not of a letter, but of part of one. While the progressive does not imply completion, it need not imply non-completion, so the 'part' may in fact be the whole (*Mary has been writing a letter; now she's going to post it*).

Second, the result is not specifiable independently of the action. (a) *Mary has run a mile* need imply no more than that she has a certain achievement to her credit ('experiential perfect'). (b) Or the 'result' may amount to no more than a *continuing relevance* of the fact that the action has been completed—as when *I've had lunch* alludes to the state of my not being ready for another meal. (c) Another kind of 'result' that is not independently specifiable is *recency*, as in *I've just seen Mary.*

Third, the result is only *indirectly* caused by the action, If a man sets in motion a chain of events which results in a woman dying, we might say *He's killed her*; but the state referred to (i.e., the woman's being dead) is only indirectly the result of the action which we describe as his killing her. Normally *He's killed her* suggests something more direct like an immediately fatal stabbing or shooting.

Turning to non-prototypical situations, we note that if the situation is atelic, the distinction between stopping and completing is neutralised, so *completion* is no longer salient. As Aristotle observed, with a telic situation the perfect is inconsistent with the progressive (if I have eaten a pizza, I can't still be eating it), but with an atelic situation they are compatible (I can have laughed and still be laughing).

5 Towards an implementation

For both the progressive and the perfect there is a family of senses that can be regarded as variations on a prototypical modification to a prototypical situation. They are derived from the prototype by dropping or altering one or more of its salient attributes, as regards either primary or secondary aspect. Our purpose was not to unearth new facts about English aspect, but to develop a perspective, within a particular theoretical framework, that accounts for the known facts as concisely and elegantly as possible.

This perspective has implications for the treatment of aspect in computational linguistics. We shall confine our remarks to language understanding rather than language production. A prototype-based language-understanding mechanism will associate with each lexical and morphological element of the language an enumeration of the salient attributes in the description of that element's prototypical meaning, together with an indication of relative saliences. When two elements co-occur in a sentence, their meanings will be determined by finding maximally salient, mutually compatible subsets of their respective prototypical attribute lists.

In particular, each syntactic aspectual construction will be associated with its prototypical attributes (e.g., the progressive will have the attributes 'activity now in progress' and 'intention to continue to the end' in an appropriate representation) and likewise for each bare verb phrase (e.g., a phrase of the form 'X eats Y' will have the attributes 'durative' and 'agentive', and the conditional attributes 'if Y is a singular count noun then telic' and 'if X and Y are singular nouns then unitary'). The meaning of the aspectualized verb-phrase will then be computed by attempting to match the salient attributes required by the secondary aspect against those of the bare verb phrase; the set of salient attributes that can be successfully matched then determines a suitable meaning for the whole.

The procedures sketched here resemble the phenomena of 'coercion' (Moens and Steedman, 1988). We advocate prototype theory as providing both a concise theoretical explanation of why coercion phenomena occur, and the basis for a set of practical procedures for handling these phenomena in a computational context.

References

Lakoff, G. (1987). *Women, Fire, and Dangerous Things: what categories reveal about the mind.* Chicago, IL: University of Chicago Press.

Moens, M. and Steedman, M. (1988). Temporal ontology and temporal reference. *Computational Linguistics*, **14**, 15–28.

Verkuyl, H. J. (1989). Aspectual classes and aspectual composition. *Linguistics and Philosophy*, **12**, 39–94.

Formal Logic For Expert Systems

Hugh McCabe
Department of Computer Science, University College Dublin
Dublin, Ireland [1]

1 Expert Systems and Inconsistency

The machinery of an expert system essentially consists of a *representation language* and an *inference method* (Luger and Stubblefield 1989). The representation language commonly used is that of *production rules*, simple *if-then* constructs which specify the conditions under which a conclusion can be said to follow. The inference method links sequences of these rules together by means of *backward* or *forward chaining* to extrapolate conclusions from the knowledge base.

In this paper we are concerned with the relationship between the above system and the predicate calculus. We see that there are two crucial differences.

1. Any formula of predicate logic can be expressed as an implication (see for example Manthey and Bry (1988)). Production rules can be thought of as implications with the restriction that disjunctions are not allowed on the right-hand side. So, syntactically, production rules are equivalent to the Horn clause subset of logic. This is fortunate for expert system designers since backward chaining will only work over this subset.

2. The production rule formalism has no semantics. Unlike logic there is no conception of what the syntactic statements mean in terms of truth values and in particular there is no conception of how one statement can contradict another.

The second limitation gives rise to the following problem. Production rules clearly *mean* the same thing as implications in logic, i.e. that the truth of the antecedent entails the truth of the consequent, but since these semantic notions are not explicitly defined there is no way of ensuring that the results of syntactic derivations accord with the implicit semantics. In particular, since there is no formal explanation of how one statement can contradict another, derivations can be constructed which include formulas and their negations within their chains of inference. A conclusion which depends on the truth of contradictory propositions is not a desirable one.

We cannot simply augment the production rule formalism with the standard semantics of predicate logic. Classical logic contains the principle that *everything follows from a contradiction*. So, if the premises of our proof, in our case the knowledge base of the expert system, contain contradictions, then any conclusion can be inferred. So an expert system based on predicate logic would be useless unless we could be sure that no such contradictions were present.

---

[1]Department of Computer Science, University College Dublin, Belfield , Dublin 4. Email: hmccabe@ccvax.ucd.ie

However it is unreasonable to expect this to be the case. The knowledge base of a typical expert system contains information gleaned from a number of different experts and we cannot expect these to concur. Furthermore an expert system essentially attempts to represent the beliefs of experts and the literature on knowledge and belief (Konolige, 1986) assures us that it is unreasonable to even expect an individuals beliefs to be consistent.

So, there is clearly a need for a logical system which can reason in the presence of inconsistency. This system must not fall prey to the property of classical logic that *everything follows from a contradiction* but, unlike production rules, it must be capable of identifying such contradictions so as to avoid producing derivations which are logically suspect.

2 Using Logic as a Representation Language

There are many non-classical logics which do not collapse in the presence of inconsistency. These logics are known as *paraconsistent* but are not suitable for our purposes because their semantics stray too far from that of classical logic and employing them would change the nature of our proposed expert system drastically. We wish therefore to use standard logic. In order to do this we must alter the conventional notion of how an inference is carried out. Typically, an expert system inference will use only a small fraction of the knowledge base. It therefore seems inappropriate to say that from a knowledge base S, we can infer a conclusion p, usually written $S \vdash p$, since most of S plays no part in the derivation of p. Bearing this in mind it also seems unreasonable that inconsistencies in sections of S that did not contribute to the derivation of p should affect it. So, instead of saying that from a set S we can derive p, we prefer to say that S *supports a proof of p*. We define this as follows:

> A set of clauses S, supports a proof of p, written $S \approx p$, iff there exists a consistent set s such that $s \subseteq S$ and $s \vdash p$, where $\vdash$ is the standard classical consequence relation.

We have changed the notion of what constitutes the premises of a proof. There is a large inconsistent knowledge base from which proofs can be carried out but the premises of these proofs are any *consistent subsets* of S rather than S itself. This notion of inference ensures that inconsistencies do not lead to logical chaos and, since the premises of proofs must be *consistent* subsets, no proofs can be constructed which depend on the truth of contradictory propositions.

3 Theorem Proving

This will be of no practical use unless we can design a theorem proving strategy which can carry out this notion of inference in an efficient manner. We immediately encounter a problem when we try to do this. Efficient automated theorem provers all use some variation on *proof by contradiction*. To prove a goal they merely add the negation of the goal to the premises (which are of course assumed to be consistent) and then show that a contradiction results. The theorem prover does not concern itself with whether the negated goal caused the contradiction itself, merely that one is present. This means that if the premises are already inconsistent then any proof will succeed. So a direct application of this to our situation would be futile. There are methods of proving theorems which do not rely on proof by contradiction but these methods, if they can be automated at all, are invariably highly inefficient. So we are left with no choice but to use some variation on proof by contradiction.

We have chosen to base our implementation on Manthey and Bry's model generation theorem prover, SATCHMO (Manthey and Bry, 1988). This takes a set of clauses S, and a formula p, and attempts to show that $S \models p$ by proving that $S \cup \neg p$ has no models. It expresses all formulas as sequents, with positive literals like A expressed as $true \Rightarrow A$ and negative literals such as $\neg A$ as $A \Rightarrow false$. It is impoprtant to note that negation is *not* represented explicitly. So, all clauses are of the form $a1, a2, .., an \Rightarrow b1; b2; ..; bm$.[2] If m=1 then the sequent is a Horn sequent, otherwise it is non-Horn and hence is disjunctive. SATCHMO looks for clauses whose left-hand sides must be part of the model (initially ones whose left-hand side is *true*) and then asserts that the right-hand side must be also. It forward chains in this way until hopefully it reaches a clause with *false* as a right-hand side. This means a contradiction has been reached and the clause set has no models. For example the set $[true \Rightarrow A, A \Rightarrow B, B \Rightarrow false]$ has no models. For sets of Horn sequents this is a simple and extremely efficient procedure. However to deal with full predicate calculus there must be an additional mechanism for non-Horn sequents. A non-Horn sequent has a disjunctive right-hand side so it presents the model generation process with alternatives. So, SATCHMO provides a mechanism for splitting this process to show that if we proceed to build a model with *any* of these alternatives, a contradiction will result.

We wish to adapt the model generation process so that it does not hunt for arbitrary contradictions in the knowledge base but rather hunts for consistent sets of clauses which entail the goal. The solution to this problem is to make the proof procedure *goal-directed*. We do not add the negation of the goal to the premises and instead of forward chaining over the sequents in search of *false*, we forward chain in search of the goal. In practice we use the equivalent procedure of backward chaining from the goal to *true*. In the case of Horn sequents this is entirely sufficient. The result of a proof will be a set of sequents of the form $true \Rightarrow a1, a1 \Rightarrow a2, ..., an \Rightarrow goal$. This *must* be consistent because there is no representation of negation except sequents with *false* on the right-hand side and these cannot form part of the chain. This procedure is essentially what the inference method of an expert system does except that has no mechanism for preventing the inclusion of inconsistencies in the proof sequence.

However it is our stated aim to cater for the full predicate calculus so we must deal with non-Horn sequents also. Consider the simple clause set $[true \Rightarrow A, A \Rightarrow B; C, C \Rightarrow false]$. This set clearly entails B but to prove this we must show that

---

[2] here we are using the Prolog convention of representing conjunction with a comma and disjunction with a semicolon

the remaining disjunct ,i.e. C, leads to a contradiction. So, we are forced to add some limited form of *proof by contradiction* into our theorem prover in order to close off the remaining branches of the disjunctions. This can be done but we must take care to ensure that the theorem prover does not simply pluck an irrelevant inconsistency from another part of the database and then declare the branch closed. In other words the clause set above is valid proof of B according to our definition of section 2 but the set $[true \Rightarrow A, A \Rightarrow B; C, true \Rightarrow D, D \Rightarrow false]$ is not. To avoid this happening we again take advantage of the fact that SATCHMO does not represent negation explicitly and try to prove a contradiction by forward chaining from the unclosed disjunct to *false*. We can thus be confident that the result is not an irrelevant inconsistency from the database but is dependent on the data already encountered in the proof sequence. This method of theorem proving can be made efficient if we also incorporate the improvements to Manthey and Bry's algorithm which are described in Ramsay (1991).

The preceding discussion was a synopsis of the main features of our theorem prover and many details have been omitted. Some additional checks are necessary to ensure a valid proof but the implementation of the above scheme successfully carries out our notion of inference efficiently.

References

Konolige, K. (1986). *A Deduction Model of Belief*. London: Pitman.

Luger, G.F., & Stubblefield, W.A. (1989). *Artificial Intelligence and the Design of Expert Systems*. Redwood City, CA: Benjamin/Cummings.

Manthey, R., & Bry, F. (1988). SATCHMO: a Theorem Prover in Prolog. *Conference on Automated Deduction 1988*.

Ramsay, A.M., (1991). Generating Relevant Models *Journal of Automated Reasoning* **7**.

Exceptions in Multiple Inheritance Systems

Chabane Oussalah
Laboratoire d'Etudes et de Recherche en Informatique
Nîmes, France

Martine Magnan
Laboratoire d'Etudes et de Recherche en Informatique
Nîmes, France

Lucile Torrès
Laboratoire d'Etudes et de Recherche en Informatique
Nîmes, France

Abstract

In this paper, we are interested in the representation and management of multiple inheritance systems with exceptions in both semantic networks and object-oriented languages. Exception management raises different problems, particularly in the presence of contradictions. The different methods used to solve these contradictions are presented.

1 Introduction

In this paper, we are interested in inheritance systems with exceptions whether it be in semantic networks, object programming languages or knowledge representation languages. Exceptions in inheritance systems are useful because they are in agreement with reusability and incremental design.

In the second section, we outline the problems due to the presence of exceptions : three types of contradictions may be identified.

In the third section, the main algorithms used in semantic networks to resolve the three types of exceptions are presented.

Finally, in the fourth section, we are concerned with exception representation and management in object languages. In these languages, two types of exceptions may be identified: exceptions by cancellation of an inheritance link and exceptions by cancellation of a property.

2 Exceptions in inheritance systems

Exceptions do not allow inheritance from an object to take place. The main effect of exception is to cause the inheritance relation to lose its transitivity (Ducournau & Habib, 1991). When one object does not inherit from another, it does not inherit from this object nor from any of the ancestors of this object. The main problem when confronted by the presence of exceptions is to determine the inheritability set, i.e. all the inheritable objets.

Contradictions appear when two inheritability paths exist simultaneously which may or may not allow inheritance from an object to take place. Three types of contradictions may be identified (cf. Figure 1).

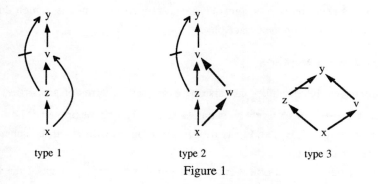

Figure 1

<u>type 1</u> : The first type concerns a contradictory redundancy because one of the two paths passes over a transitivity link of the inheritance relation.

<u>type 2</u> : This second type of contradiction is a generalization of the previous case (Sandewall, 1986). Here, the transitivity link is interrupted by the node w.

<u>type 3</u> : The third type is an ambiguity. For this type of contradiction, all objects of the contradictory inheritability paths (except the origin and the extremity) are not comparable.

3 Management of exceptions in semantic networks

We present here how the different methods used in semantic networks allow the different contradictions to be dealt with (Rychlik, 1989).

The shortest-path algorithm was developed in NETL (Fahlman, 1979). Although this algorithm is appropriate for multiple inheritance systems that do not admit exceptions, it often gives bad results in the presence of contradictions, whether it be a contradiction of type 1, 2 or 3.

Inferential distance or on-path preemption proposed by Touretzky (1986) allows contradictions of type 1 to be dealt with. In the presence of contradictions of types 2 and 3, all possible extensions are generated which correspond to a credulous attitude.

As an improvement, Sandewall (1986) proposes off-path preemption which allows both contradictory redundancy and contradictions of type 2 to be solved.

As concerns the third type of contradiction, either a credulous attitude or a skeptical attitude may be adopted. In the first case, all conclusions are generated (Touretzky, Horty & Thomason, 1987), whereas, in the second case, no conclusion is drawn (Horty, Thomason & Touretzky, 1987).

4 Exceptions in object-oriented languages

Until now, we have only considered exceptions by cancellation of an inheritance link which we call object-object exceptions. We now introduce a second type of exception: exceptions by cancellation of a property or object-property exceptions, which do not allow inheritance from a property to take place.

4.1 Object-object exceptions

In object languages, inheritability can also be expressed in terms of properties. The inheritance link $x \longrightarrow y$ (resp. $x \not\longrightarrow y$) is translated by x having the property "I am a y" (resp. "I am not a y"). Inheritability is therefore a property which can be inherited, masked or present conflicts.

The problematics of object-object exceptions remains the management of the three types of contradictions.

The first two types of contradiction may be solved by masking. In both cases, the property " I am not a y" in z overrides the property "I am a y" in v. The third type of contradiction remains the crucial problem. In this case, masking cannot be applied. For this type of contradiction, either the skeptical attitude or the credulous attitude may be adopted, and, in this last case, a conclusion may be chosen arbitrarily either in giving priority to the affirmation or in giving priority to the negation.

4.2 Object-property exceptions

Object-property exceptions indicate that an object does not have a property defined in its ancestors. The object is then said to be atypical. An object-property exception in x on the property P is noted as except(P) and means that x does not have property P. In the presence of object-property exceptions, the problem consists in determining the set of properties which are inherited by an object.

The problematics of object-property exceptions is linked to multiple inheritance. Two classes of contradictions may be distinguished. The first is the class of contradictions which can be solved by masking, i.e. the objects which possess and except the property may be compared by the inheritance relation. In the example in Figure 2, the exception on property P defined in z is more specific than the declaration of P in y; x does not have property P. The second class of contradictions cannot be solved by masking; objects which possess and except the property may not indeed be compared by the inheritance relation (cf. Figure 3). The management of these contradictions constitutes the main problem of the management of object-property exceptions. As for object-object exceptions, some give priority to the affirmation or, on the contrary, to the negation.

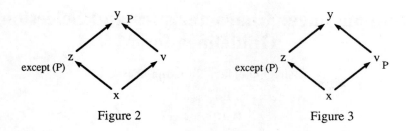

Figure 2   Figure 3

5 Conclusion

In this paper, we have seen how exception management may be achieved both in semantic networks and object-oriented languages according to a graph-oriented approach. Another possible approach is the logic approach (Rychlik, 1989) but it was not developed here. As object-oriented languages have been widely developed during the last few years, the methods for handling exceptions in these languages seem to be the most appropriate.

References

Ducournau, R. , & Habib, M. (1991). Masking and Conflicts or to Inherit is not to Own!. In M. Lenzerini, D. Nardi & M. Simi (Eds.), *Inheritance Hierarchies in Knowledge Representation and Programming Languages* (pp. 223-244). John Wiley & Sons.

Fahlman, S. E. (1979). *NETL : a system for representing and using real-world knowledge.* Cambridge, MA : MIT Press.

Horty, J. F., Thomason, R. H. , & Touretzky, D. S. (1987). A Skeptical Theory of Inheritance in Nonmonotonic Semantic Networks. *Proceedings AAAI'87, Seattle, Washington* (pp. 358-363) : Morgan Kaufmann.

Rychlik, P. (1989). Multiple Inheritance Systems with Exceptions. *Artificial Intelligence Review*, **3** (2,3), 159-176.

Sandewall, E. (1986). Nonmonotonic Inference Rules for Multiple Inheritance with Exceptions. *Proceedings of the IEEE* , **74** (10), 1345-1353.

Touretzky, D. S. (1986). *The Mathematics of Inheritance Systems.* Los Altos, CA : Morgan Kaufman.

Touretzky, D. S. , Horty, J. F. , & Thomason, R. H. (1987). A Clash of Intuitions: The Current State of Nonmonotonic Multiple Inheritance Systems. *Proceedings of the 10th IJCAI, Milano, Italy* (pp. 476-482).

Sublanguage: Characteristics and Selection Guidelines for M.T.

Elaine Quinlan and Sharon O'Brien

National Centre for Language Technology, D.C.U.

Dublin, Ireland

1. Introduction

The fact that the TAUM-METEO MT system is arguably the most successful MT system currently developed, makes a claim for further investigation into sufficiently constrained subject fields or *sublanguages* for MT. Sublanguage has been taken to mean the restricted language of a particular domain, which is used by a specialist community. However, a complete and comprehensive definition of the term "sublanguage" has yet to be formalised. Also, the notions of text and communicative function and how they effect sublanguage have yet to be fully addressed. Many definitions have been proposed, yet few offer a general description of sublanguage or guidelines by which sublanguages may be chosen for NLP and in particular MT. The latter is what we aim to achieve in this paper.

2. Analysis of a sublanguage corpus

The sublanguage we analysed was that of sewing instructions, which exhibits an extremly limited range of text types. A subcorpus was indexed which had a total of 1999 words and 255 unique words. More patterns were then added to create a larger corpus with a total of 4932 words and 413 unique words. From this we can deduce that the sublanguage is significantly closed because an addition of 2933 words only resulted in an increase of 158 in the number of unique words.

2.1 Restrictions

The corpus was analysed for lexical, syntactic and semantic restrictions. Restriction of the lexical items and lexical categories contributes to the closure of this sublanguage. The following lexical categories occur frequently: nouns, verbs, prepositions, coordinating conjunctions, and adjectives, while adverbs, indefinite articles, and subordinating conjunctions occur infrequently. These categories never occur: relative pronouns, personal pronouns, the impersonal pronoun *it*, and the definite article *the*. If other pattern types were to be added to the corpus, the initial result would be an increase in the number of lexical items, due to the addition of new garment types.

However, the sewing instructions will remain unchanged. Therefore, the growth rate of the lexicon should finally decrease and eventually stop. This is due to the fact that the domain of sewing is very limited and covers a finite set of possibilities.

One of the most outstanding features of this sublanguage text is the high use of the imperative form which accounts for 68% of the total verb forms. The predominance of the imperative verb form is significant in reducing processing problems, in that in a verb-initial sublanguage such as this one, where the verb is the governor of the sentence, parsing problems may be reduced. The present progressive is the next most frequently occurring verb form (20%). There are no occurrences of past tense forms. Voice is essentially active, subordinating conjunctions are rare, while *and* is the only occurring coordinating conjunction.

Semantic ambiguity is avoided when there is categorial restriction, i.e. when a word occurs in only one category in the sublanguage whereas it may occur in several categories in language as a whole. In this subanguage there are many examples of categorial restrictions. For example, *circle* and *view* occur only as nouns and never as verbs. Similarly, ambiguity is avoided when the semantic range of a lexical item is restricted, for example, *fuse* does not refer to a concept in the domain of electricity, and *tack* does not refer to the concept in the domain of sailing. Another problematic area for MT is the occurrence of *empilages*. They present major problems when it comes to parsing the nominal group, because proper bracketing of an empilage requires an understanding of the semantic/syntactic relations between the components. The sublanguage of sewing instructions has relatively few empilages. The longest in the corpus consisted of five lexical items, and there were only two cases of these, i.e. *front skirt side seam allowance* and *facing centre back seam allowance*.

3. Conclusions

It can be concluded from the analysis, that this sublanguage text is highly restricted, lexically, syntactically, and semantically. Until now restrictions on the lexicon, syntax and semantics of various sublanguages have been examined without due regard being given to the communicative function which creates the text type. If the corpus was extended by addition of more patterns for other types of garment, the initial result would be an increase in the number of lexical items. The increase in the number of unique lexical items will eventually stop. As the function of the texts in the corpus remains the same, i.e. to instruct, the syntactic structures should also remain the same.

This indicates that the lexicon and the semantics are governed by the sublanguage, while the syntactic structure is governed by the communicative function. Outlined below is a set of characteristics by which a sublanguage may be identified, taking communicative function and text type into account.

(i) A sublanguage is a proper subset of natural language, where sublanguage is taken to mean a complete set of linguistic phenomena i.e. the grammar, lexicon, communicative function etc. of a particular text type or text types.

(ii) A sublanguage will display coherence, cohesion and closure where cohesion and coherence are text-type related and closure is sublanguage related. Cohesion and coherence are referred to here in the text linguistic sense, as outlined in De Beaugrande and Dressler (1981). Sublanguage theorists mistakenly attribute cohesion and coherence to properties of the sublanguage when, in fact, both are governed by the text and not by the sublanguage.

(iv) A sublanguage must contain restrictions, where the restrictions on syntax are text type related and restrictions on semantics and vocabulary are sublanguage related.

(v) Sublanguage creativity is limited to the possible neologisms and loan words which may enter the lexicon.

(vi) The degree of closure of a sublanguage or the extent to which general language exists in the sublanguage, is dependent on the purpose, topic, communicative function and users of the sublanguage text.

4. Guidelines for identifying suitable sublanguages for MT

Outlined below are proposed guidelines by which suitable sublanguages may be identified for MT:

(i) A sublanguage should exhibit all the characteristics outlined above.

(ii) A sublanguage should be identified by a sublanguage expert. The sublanguage expert should also be a native speaker of the natural language in question. As it is possible for the educated lay-person to recognise and understand highly specialised domains, only a sublanguage expert is qualified to estimate the closure level of the sublanguage, indicate the possible range of text types that occur in the sublanguage, and be familiar with the development of the subject area on a global scale. This last criteria is particularly important if the sublanguage requires comparative analysis across languages for MT.

(iii) If machine translation of the sublanguage is the intention, it should be ensured that the sublanguage has developed at the same rate and to the same extent in each language community, so as to guarantee that the concepts exist in each language. Similarly, it should be ensured that the sublanguage is unlikely to develop rapidly.

(iv) The aim of the sublanguage analysis will influence the sublanguage texts chosen for investigation. Sublanguage texts for machine translation should belong to a text type driven by a restricted number of non-complex, communicative functions. It is preferable that the sublanguage texts do not contain sub-texts, but display uniform text structure throughout. It is recommended that a comprehensive corpus of texts be created, which would be representative of the possible text types that occur in the sublanguage. A number of factors will influence the content, size, and format of the corpus. The corpus should be large enough to measure closure and to contain the complete range of lexical items and syntactic constructions. The corpus should be available in machine-readable form. A similar corpus should be created in the target language of the MT system. In this case, it is important that *parallel* sublanguage texts are included in the foreign language corpus and not *translations* of the source language texts.

(v) It is essential that suitable text retrieval, indexing, analysis and statistical software is available for analysis and parsing of the sublanguage texts.

Bibliography

de Baugrande, R., & Dressler, W. (1981). *Introduction to Text Linguistics*. New York: Longman Group.

Grishman, R., & Kittredge, R. (1986). *Analyzing Language in Restricted Domains: Sublanguage Description and Processing*. Hillsdale, N.J.: Lawrence Erlbaum Associates.

Harris, Z. (1968). *Mathematical Structures of Language*. New York: Wiley Interscience.

On the Interface between Depth-From-Motion and Block-Worlds Systems: Some Psychophysical Data

A. H. Reinhardt-Rutland
Department of Psychology, University of Ulster at Jordanstown
Newtownabbey, Northern Ireland

1 Introduction

Observer motion may play a role in depth perception by way of cues such as motion parallax, the relative visual motion resulting from points at different distances from the observer. Even minimal observer motion permits full detection of VDU-simulated relative depth (Rogers & Graham, 1979), which argues for autonomous motion systems (Lawton, Rieger, & Steenstrup, 1987). However, evidence from *real* stimuli - in which many other cues may be available - shows that pictorial cues severely limit the effectiveness of motion (Reinhardt-Rutland, 1990, in press). For example, monocular observers viewed rectangular and trapezoidal surfaces, the latter simulating the rectangular surfaces (Fig 1). Even with 200 mm lateral head motion, perceived orientation was not wholly veridical. Although the specificity of pictorial information - as in the "block-world" systems popular some years ago - is at odds with the search for general-purpose systems (Garnham, 1988), similar conclusions apply to perceived structure of rotating objects viewed by static observers (Cowie, 1987).

Pertinent to the present paper, Reinhardt-Rutland (1992) investigated whether background pattern behind experimental surfaces might increase veridical responding: motion information might be enhanced because surface edges visually move relative to the pattern, or - more holistically - because the surface's perspective transformation is more detectable. Surprisingly, pattern had no effect. Experiments 1 and 2 replicate this finding. In addition, Experiment 2 reveals that background pattern can elicit overestimation of non-frontal orientation of a real surface.

Rectangularity orientation = 60°   Rectangularity orientation = 90°

FIG 1: *Postulated rectangularity: the left stimulus appears as a non-frontal rectangular surface, while the right stimulus would be a frontal rectangular surface. The pictorial information is powerful - but specific to a limited range of quadrilateral stimuli.*

2 Experiment 1

Observers, 30 undergraduates, viewed white, free-standing card stimuli along a table top in ultra-violet light under monocular-static, monocular-moving (150 mm lateral head-motion) and binocular-static conditions. Of the six stimuli, three were rectangular (180 mm X 100 mm) and horizontally slanted $30°$, $45°$ or $60°$ to the line-of-sight; the other three were frontal trapezoids simulating the rectangular stimuli. Viewing distance was 750 mm. One of two backgrounds was matt-black with no discernible pattern. The other included a grid of seven 100 mm by 3 mm vertical stripes in a frontal plane 160 mm behind the stimulus' left edge. Stripes were of orange retroreflective paper equally-spaced across a rectangular transparent perspex sheet 250 mm X 100 mm. Luminance of stimuli and grid stripes was about 0.2 cd/m^2. To each stimulus observers made a forced response of "frontal" or "non-frontal".

Results (Fig 2) show that viewing condition greatly affected responding, but background did not - conclusions confirmed by ANOVA. Motion cues were less effective than binocular cues, but neither was fully effective. However, the dependent variable in Experiment 1 - although readily implemented by observers - gives a rudimentary idea of how surface-orientation is perceived. In Experiment 2, viewing was moving-monocular alone and observers attempted to quantify perceived orientation.

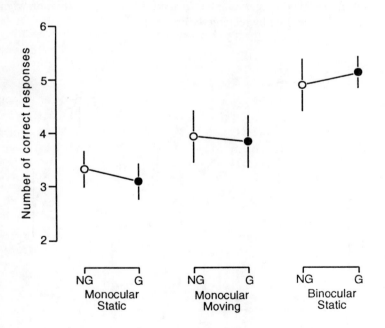

FIG 2: *Means and SEs from Experiment 1. Unfilled circles: no grid present (NG). Filled circles: grid present (G). If perception is veridical, 6 correct responses would be predicted; if perception is dominated by rectangularity, 3 correct responses would be predicted.*

3 Experiment 2

12 undergraduate observers viewed four stimuli. Two were rectangular, one frontal at $90°$ to the observer's line of sight and labelled R90T90 - R is rectangularity orientation and T is true orientation - and the other, R60T60, at $60°$. The other stimuli, R90T60 and R60T90, were trapezoidal, simulating R90T90 and R60T60 respectively. Orientation was judged by way of analog time. Observers considered the table top as a clock face. Each stimulus was the minute hand; the left edge corresponded to the clock centre. Responses were in minutes past the hour - which can be converted to degrees of orientation by multiplying responses by six.

Results (Fig 3) show again that motion in itself has an incomplete effect on veridical responding. However, the grid had a strongly significant influence: for all stimuli except R90T90 judged orientation was altered away from frontal when the grid was present. As in Experiment 1, the grid did not overall improve veridicality: R90T60 responses indicate greater veridicality, *but* R60T90 and R60T60 responses indicate decreased veridicality. The grid's effect is comparable to that in Graham and Rogers' (1982) VDU displays. A simulated surface appeared frontal during observer motion; however, surrounded by simulated surfaces receding, say, to the left, it appeared to recede to the right. Experiment 2 shows that similar effects can occur for real non-frontal surfaces - in this context, the grid is presumably equivalent to a continuous frontal surface. In addition, results for R60T90 suggest that such *orientation contrast* can occur when non-frontal orientation is specified by rectangularity.

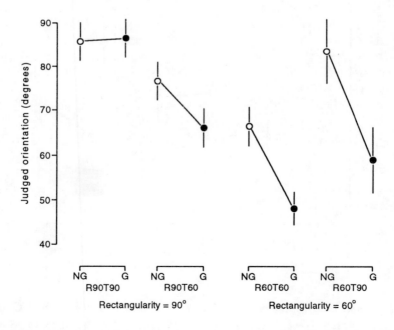

FIG 3: *Means and SEs from Experiment 2. Unfilled circles: no grid present (NG). Filled circles: grid present (G).*

4 Conclusions

(a) As previously, observer motion has an incomplete effect on veridical depth perception. This is not confined to an isolated surface: it also affects the more ecologically-plausible case of a patterned background. Pictorial cues explain these results, indicating that modular depth-from-motion systems may address a restricted aspect of human depth perception. Block-world systems may hold more clues to human perception than has recently been believed.

(b) A novel finding is that perceived orientation of real surfaces is affected by orientation contrast. Contrast is recognised as important in aspects of low-level processing such as brightness; it must also be given more consideration in modelling later perceptual processes. Its purpose is probably to extract relative - as opposed to absolute - information. Perception of absolute depth may be often unimportant in human perception: unless very precise manipulations are required, it may be sufficient to know, say, that one surface is relatively more frontal than another.

References

Cowie, R. I. D. (1987). Rubber rhomboids: Nonrigid interpretation of a rigid structure moving. *Perception and Psychophysics, 42,* 407-408.

Garnham, A. (1988). *Artificial intelligence.* London: Routledge & Kegan Paul.

Graham, M. E. & Rogers, B. J. (1982). Simultaneous and successive contrast effects in the perception of depth from motion and stereoscopic information. *Perception, 11,* 247-262.

Lawton, D., Rieger, J., & Steenstrup, M. (1987). Computational techniques in motion processing. In M. A. Arbib & A. R. Hanson (Eds.) *Vision, brain and cooperative computation.* (Pp. 419-488.) Cambridge MA: Bradford.

Reinhardt-Rutland, A. H. (1990). Detecting orientation of a surface: The rectangularity postulate and primary depth cues. *Journal of General Psychology, 117,* 391-402.

Reinhardt-Rutland, A. H. (1992). Primary depth cues and background pattern in the portrayal of slant. *Journal of General Psychology, 119,* 29-35.

Reinhardt-Rutland, A. H. (in press). Perceiving surface orientation: Information based on inferred rectangularity can be overriden during observer motion. *Perception.*

Rogers, B. & Graham, M. (1979). Motion parallax as an independent cue for depth perception. *Perception, 8,* 125-134.

A Connectionist Approach to the Inference of Regular Grammars from Positive Presentations Only.

Donie O Sullivan
CSIS Dept., University of Limerick
Limerick, Ireland

Kevin Ryan
CSIS Dept., University of Limerick
Limerick, Ireland

Abstract

In this paper a new method for inferring regular grammars from example strings is outlined. This inference mechanism is itself trained by example and is partly based upon connectionist technology. As it is trained by example, neither negative presentations nor indications of desired complexity are required in order to infer new grammars as these criteria are induced from the examples upon which the inference mechanism was trained.

1 Grammatical Inference

The principle of grammatical inference is simple. The user, or teacher, provides the computer with a series of strings of a language. The computer, after consideration will return a suitable grammar describing these strings. The solution grammar should display the results of some inductive inference. It is not a difficult task to return a grammar describing exactly the set of strings submitted however it is more useful to return the grammar describing the language that the algorithm perceives (induces) the person had in mind. Consider the following example.

Example 1.
The teacher submits the following strings;
   abc
   aabc
   aaabc
The computer returns the following result,
   a*bc

a grammar describing not only the original strings but also

>aaaabc
>aaaaabc
>aa..aaabc

The computer has inferred that the teacher had in mind the language composed of any number of "a"s followed by a "bc" from the strings submitted. Normally the degree of induction involved in the inference process is controlled by the teacher.

2 Grammatical Inference and Presentations

There are three main ways in which information about the language to be inferred can be presented to a grammatical inference algorithm.

- **Positive Presentation.** - Strings which are part of the language are presented to the inference algorithm as above[Example 1].
- **Complete Presentation.** - As well as a positive presentation, a negative presentation is also given. This consists of a set of strings which do not belong to the language.
- **Presentation by Informant.** - The inference algorithm can query a teacher as to whether certain strings belong to the language or not. This is, effectively, equivalent to a complete presentation.

3 Identification of the Solution Language

Gold (1967) showed that no language containing an infinite component can be recognised from positive presentation only. This applies to most useful languages, regular languages included. Biermann and Feldman (1972) introduced the notion that for any given set of strings there will be an infinite number of grammars capable of generating those strings. One can imagine these grammars, they say, as being ordered on a "fit" spectrum with the grammars which generate the largest languages(poorest fit) on the left and those which give a tighter representation(better fit) on the right. Grammatical inference therefore is not a process for selecting a solution grammar but rather for selecting the most suitable solution grammar from an infinite set of candidates.

4 Negative Presentations and Complexity

The negative presentation together with a possible indication of complexity can determine how the elements of the language which could have ambiguous interpretations are to be handled. Consider the following example.

Example 2.

positive presentation	negative presentation	complexity
a	aaaaa	4 (states in FSA)
aa	aaaaaa	
aaa		
aaaa		

Working from only the positive presentation there are an infinite number of possible solution grammars. However taking the negative presentation or the desired complexity indicator into account would lead us to select the language containing only the strings in the positive presentation.

5 The Connectionist Approach

Instead of using a negative presentation and/or an indication of desired complexity to decide how to handle ambiguous elements of the sample strings, we can consider training a neural network to make these decisions.

The proposed method is to train a neural network by back propagation to interpret the possibly ambiguous elements and to handle them according to defined criteria. Once the user has trained the network to give the required interpretation, it can subsequently be used for the inference of as many grammars as are desired. When a new interpretation or level of complexity is required, a new network can be trained for this purpose. The main advantage of the process is that in the areas where, in general, ambiguous elements are handled in a consistent manner, after training a network once, the user will not be required to include negative presentations or an indication of complexity for the subsequent inference of any grammar, and the solutions provided should be satisfactory.

6 Results

A method has been designed whereby the correct interpretations of the possibly ambiguous elements can be represented in the outputs of a neural network. Such a neural network, when presented with the strings of the positive presentation, will produce a list of the main patterns present in the input strings together with a list of the terminal symbols which comprise these patterns. Given the original positive presentation and the outputs from this network it is not a difficult task to create the most suitable grammar which describes the language.

The program to do so operates as follows. For each pattern type a set of productions is created which will generate that pattern type. Wherever one pattern is followed by another, a production which will link both is created. Where a pattern type is the last in a string, a production which will link it to the terminating state is created. The grammars produced are non-deterministic as these are the easiest to generate. However, it is straightforward to convert from a non-deterministic regular grammar to a deterministic one and there are numerous algorithms to do so e.g. Aho and Ullman (1977).

The preliminary task of training a neural network to produce the pattern definitions needed, has proven much more difficult and has not been solved so far.

References

Aho, A.V., & Ullman, J.D. (1977). Principles of Compiler Design. Reading, MA : Addison-Wesley.

Biermann, A.W., & Feldman, J.A. (1972). A survey of the results in Grammatical Inference. In S. Watanabe (Ed.), *Frontiers of Pattern Recognition* (pp. 31-54). London: Academic Press.

Gold, E.M., (1967). Language Identification in the Limit. *Information Control*, **10**, 447-474.

Author Index

Abbruzzese, F.	44
Al-Asady, R.	291
Armando, A.	295
Bahler, D.	307
Baird, B.W.	157
Baker, K.L.	299
Beardon, C.	263
Bell, D.A.	16
Blöchl, B.	303
Bluff, K.	71
Bowen, J.	307
Bringsjord, S.	277
Byrne, R.M.J.	59
Catanzariti, E.	251
Cesta, A.	84
Cimatti, A.	311
Conte, R.	84
Costello, F.	7
Cowie, R.	237, 315
Creaney, N.	197
Cunningham, P.	179
D'Aloisi, D.	319
Dai, H.	16
Dragoni, A.F.	323
Eaton, M.	327
Farley, A.M.	31
Freeman, K.	31
Galton, A.	330
Gilmore, J.F.	141
Giunchiglia, E.	295
Green, P.	299
Handley, S.J.	59
Hickey, R.J.	157
Hughes, J.G.	16
Keane, M.T.	7
Lewis, J.D.	166
MacDonald, B.A.	166

Maga, G.	319
Magnan, M.	338
Matthews, P.	111
McCabe, H.	334
McFetridge, P.	101
Miceli, M.	84
Minicozzi, E.	44
Nardiello, G.	44
Nicolson, R.I.	299
O Sullivan, D.	350
O'Brien, S.	342
Oussalah, C.	338
Pearson, J.	121
Piazzullo, G.	44
Probert, G.	315
Quinlan, E.	342
Reinhardt-Rutland, A.H.	346
Ryan, K.	350
Sandewall, E.	3
Smyth, B.	179
Spalazzi, L.	311
Sutcliffe, R.F.E.	210
Torrès, L.	338
Traverso, P.	295, 311
Tsinas, L.	303
Wallace, J.G.	71
Walsh, F.	237
Way, A.	224
Williams, S.M.	299

Published in 1990–92

AI and Cognitive Science '89, Dublin City University, Eire, 14–15 September 1989
A. F. Smeaton and G. McDermott (Eds.)

Specification and Verification of Concurrent Systems, University of Stirling, Scotland, 6–8 July 1988
C. Rattray (Ed.)

Semantics for Concurrency, Proceedings of the International BCS-FACS Workshop, Sponsored by Logic for IT (S.E.R.C.), University of Leicester, UK, 23–25 July 1990
M. Z. Kwiatkowska, M. W. Shields and R. M. Thomas (Eds.)

Functional Programming, Glasgow 1989
Proceedings of the 1989 Glasgow Workshop, Fraserburgh, Scotland, 21–23 August 1989
K. Davis and J. Hughes (Eds.)

Persistent Object Systems, Proceedings of the Third International Workshop, Newcastle, Australia, 10–13 January 1989
J. Rosenberg and D. Koch (Eds.)

Z User Workshop, Oxford 1989, Proceedings of the Fourth Annual Z User Meeting, Oxford, 15 December 1989
J. E. Nicholls (Ed.)

Formal Methods for Trustworthy Computer Systems (FM89), Halifax, Canada, 23–27 July 1989
Dan Craigen (Editor) and Karen Summerskill (Assistant Editor)

Security and Persistence, Proceedings of the International Workshop on Computer Architecture to Support Security and Persistence of Information, Bremen, West Germany, 8–11 May 1990
John Rosenberg and J. Leslie Keedy (Eds.)

Women into Computing: Selected Papers 1988–1990
Gillian Lovegrove and Barbara Segal (Eds.)

3rd Refinement Workshop (organised by BCS-FACS, and sponsored by IBM UK Laboratories, Hursley Park and the Programming Research Group, University of Oxford), Hursley Park, 9–11 January 1990
Carroll Morgan and J. C. P. Woodcock (Eds.)

Designing Correct Circuits, Workshop jointly organised by the Universities of Oxford and Glasgow, Oxford, 26–28 September 1990
Geraint Jones and Mary Sheeran (Eds.)

Functional Programming, Glasgow 1990
Proceedings of the 1990 Glasgow Workshop on Functional Programming, Ullapool, Scotland, 13–15 August 1990
Simon L. Peyton Jones, Graham Hutton and Carsten Kehler Holst (Eds.)

4th Refinement Workshop, Proceedings of the 4th Refinement Workshop, organised by BCS-FACS, Cambridge, 9–11 January 1991
Joseph M. Morris and Roger C. Shaw (Eds.)

AI and Cognitive Science '90, University of Ulster at Jordanstown, 20–21 September 1990
Michael F. McTear and Norman Creaney (Eds.)

Software Re-use, Utrecht 1989, Proceedings of the Software Re-use Workshop, Utrecht, The Netherlands, 23–24 November 1989
Liesbeth Dusink and Patrick Hall (Eds.)

Z User Workshop, 1990, Proceedings of the Fifth Annual Z User Meeting, Oxford, 17–18 December 1990
J.E. Nicholls (Ed.)

IV Higher Order Workshop, Banff 1990
Proceedings of the IV Higher Order Workshop, Banff, Alberta, Canada, 10–14 September 1990
Graham Birtwistle (Ed.)

ALPUK91, Proceedings of the 3rd UK Annual Conference on Logic Programming, Edinburgh, 10–12 April 1991
Geraint A.Wiggins, Chris Mellish and Tim Duncan (Eds.)

Specifications of Database Systems
International Workshop on Specifications of Database Systems, Glasgow, 3–5 July 1991
David J. Harper and Moira C. Norrie (Eds.)

7th UK Computer and Telecommunications Performance Engineering Workshop
Edinburgh, 22–23 July 1991
J. Hillston, P.J.B. King and R.J. Pooley (Eds.)

Logic Program Synthesis and Transformation
Proceedings of LOPSTR 91, International Workshop on Logic Program Synthesis and Transformation, University of Manchester, 4–5 July 1991
T.P. Clement and K.-K. Lau (Eds.)

Declarative Programming, Sasbachwalden 1991
PHOENIX Seminar and Workshop on Declarative Programming, Sasbachwalden, Black Forest, Germany, 18–22 November 1991
John Darlington and Roland Dietrich (Eds.)